Structures under Shock and Impact XVII

WIT*PRESS*

WIT Press publishes leading books in Science and Technology.
Visit our website for the current list of titles.
www.witpress.com

WIT*eLibrary*

Home of the Transactions of the Wessex Institute.
Papers contained in this volume are archived in the WIT eLibrary in volume 215 of WIT Transactions on the Built Environment (ISSN 1743-3509).
The WIT eLibrary provides the international scientific community with immediate and permanent access to individual papers presented at WIT conferences.
Visit the WIT eLibrary at www.witpress.com.

SEVENTEENTH INTERNATIONAL CONFERENCE ON STRUCTURES UNDER SHOCK AND IMPACT

SUSI 2025

Conference Chairmen

S. Syngellakis
Wessex Institute, UK

F. Teixeira-Dias
University of Edinburgh, UK

International Scientific Advisory Committee

Organised by
Wessex Institute, UK
University of Edinburgh, UK

Sponsored by
WIT Transactions on the Built Environment

WIT Transactions

Wessex Institute
Ashurst Lodge, Ashurst
Southampton SO40 7AA, UK

We would like to express our thanks to the conference Chairs and members of the International Scientific Advisory Committees for their efforts during the 2025 conference season.

Conference Chairs

International Scientific Advisory Committee Members 2025

Structures under Shock and Impact XVII

Editors

S. Syngellakis
Wessex Institute, UK

F. Teixeira-Dias
University of Edinburgh, UK

WIT*PRESS* Southampton, Boston

Editors:

S. Syngellakis
Wessex Institute, UK

F. Teixeira-Dias
University of Edinburgh, UK

Published by

WIT Press
Ashurst Lodge, Ashurst, Southampton, SO40 7AA, UK
Tel: 44 (0) 238 029 3223; Fax: 44 (0) 238 029 2853
E-Mail: witpress@witpress.com
http://www.witpress.com

For USA, Canada and Mexico

Computational Mechanics International Inc
25 Bridge Street, Billerica, MA 01821, USA
Tel: 978 667 5841; Fax: 978 667 7582
E-Mail: infousa@witpress.com
http://www.witpress.com

British Library Cataloguing-in-Publication Data

A Catalogue record for this book is available
from the British Library

ISBN: 978-1-78466-495-4
eISBN: 978-1-78466-496-1
ISSN: 1746-4498 (print)
ISSN: 1743-3509 (on-line)

The texts of the papers in this volume were set individually by the authors or under their supervision. Only minor corrections to the text may have been carried out by the publisher.

Preface

This volume comprises a selection of papers presented at the 17th International Conference on Structures under Shock and Impact (SUSI), which took place in Edinburgh, UK from 9 to 11 June 2025. The meeting was organised by the Wessex Institute and The University of Edinburgh.

Current global events, on-going conflict and the pace of growth in certain regions demonstrate the vulnerability and fragility of the built environment and the need to mitigate and protect it against high dynamic loads. Safety of personnel and society remains a high priority with engineers developing new designs and assessing the capability of existing structures.

The papers cover a variety of topics relevant to anyone interested in impact and blast effects on structural integrity. These are areas of active research among academics as well as practitioners in civil, mechanical, nuclear, transportation, petrochemical and process industries, focused on the survivability of physical facilities, structures and the protection of people.

Particular attention is given to the assessment and modelling of propagating shock waves due to blast loading of various origins. The capacity of structures to absorb energy, protection against explosions in the built environment, as well as the dynamic behaviour of materials under shock and impact are among the topics explored in the book. Such investigations rely on small- and large-scale experiments or advanced computer codes. Special cases explored in the book include: the deformation of buried pipelines due to subsoil detonation; the mitigation of naval equipment vibrations through resilient mounting systems; the rapid prediction of interaction between blast waves and the built environment; and the damage of steel columns due to vehicle explosions.

Advanced composite material systems, as well as concrete, are numerically assessed for their capacity to resist ballistic impact. There is particular focus on the response of bi-layer ceramic/metal composite and wrought maraging steel targets.

The complexity of impact phenomena requires special rig designs for their experimental investigation. Two such examples are included in this volume: the first combines impact with axial load and heat, the second uses a compression-to-tension jig set for high strain assessment in a split Hopkinson pressure bar. Experimentation is combined with numerical modelling for bridge columns and polymeric gyroid triply periodic minimal surface structures.

Glass and gradient hexachiral auxetics are materials whose response is investigated under blast and impact load, respectively. Artificial neural networks are applied to yield design parameters from inverse analysis of blast-loaded structures.

Finally, several papers are dealing with the mitigation of impact and blast effects; the effectiveness of composite shields and barriers are, respectively, investigated. Observation and shock wave tracking are proposed as means of enhancing critical infrastructure and transport security.

In summary, the volume contains contributions essential to extending and deepening the knowledge of how to assess structural and material behaviour under various types of dynamic loads. This volume brings together the work of researchers and other professionals actively involved in developing more efficient and reliable analysis methods to simulate impact and blast events and their effects, as well as validating computer models using scaled and full-size experimental techniques.

The papers contained in this volume, as well as others presented at the Wessex Institute conferences are archived in the Institute's elibrary (witpress.com/elibrary) where they are permanently available in open access format to the international community.

The editors are grateful to all the authors for the quality of their contributions as well as their colleagues of the International Scientific Advisory Committee who helped to review the papers and hence ensure the quality of this volume.

The Editors, 2025

Contents

Section 3: Terminal ballistics

Section 4: Experiments and numerical modelling

Section 5: Dynamic behaviour of materials and structures

Section 6: Hazard mitigation and security

SECTION 1
IMPACT AND
BLAST LOADING

NECESSITY OF CONSIDERING CHEMICAL REACTION EQUILIBRIUM IN QUASI-STATIC PRESSURE THERMODYNAMIC MODEL FOR CONFINED TNT EXPLOSIONS

YANG HUANG[1] & SUWEN CHEN[1,2]
[1]College of Civil Engineering, Tongji University, China
[2]State Key Laboratory of Disaster Reduction in Civil Engineering, Tongji University, China

ABSTRACT
The thermodynamic model is often used to describe the quasi-static pressure with the mass-to-volume ratio (*m/V*) in confined explosions and further derive physical quantities such as adiabatic index based on chemical components and temperature. However, there is a limited study to investigate the necessity of considering reaction equilibrium in the thermodynamic model of TNT confined explosions. This study first conducted a comparative study of thermodynamic model results with the reaction equilibrium considered or not. The results reveal that introducing reaction equilibrium causes a maximum relative difference of less than 20% in quasi-static pressure. However, the *m/V* for solid carbon generation shifts from 0.371 kg/m^3 to 3.850 kg/m^3, the *m/V* corresponding to peak temperature transitions from 0.371 kg/m^3 to 0.680 kg/m^3, and discrepancies in material components and temperature results become increasingly significant when $m/V > 0.1$ kg/m^3. Therefore, models considering reaction equilibrium are essential for calculating physical quantities related to product components and temperature. Then, a simplified calculation method for determining mole numbers of components, temperature and pressure in the quasi-static stage of TNT confined explosion was proposed, which showed good consistency with the thermodynamic model results considering reaction equilibrium. Finally, a practical solution for calculating afterburning energy in numerical models for quasi-static pressure simulation was proposed, which reached great accuracy and achieved a broader applicability range than the two existing methods. The research contributes to a theoretical understanding of the necessity for considering reaction equilibrium in the thermodynamic model, as well as achieving the rapid calculation of thermodynamic parameters in the quasi-static stage and accurate simulation of numerical models.
Keywords: chemical reaction equilibrium, confined explosion, quasi-static pressure, thermodynamic model.

1 INTRODUCTION

The explosive detonation in confined space typically exhibits shockwave convergence at space corners and significant thermal effects. Structural damage arises from both the reflected pressure during wave propagation and the quasi-static pressure after the flow field stabilizes. For negative oxygen balance explosives like TNT, detonation products can further react with oxygen in the confined space, resulting in the afterburning effect, which increases the temperature within space and the load on internal walls [1], [2]. The quasi-static stage of confined explosions is characterized by a long duration, with the quasi-static pressure being notably affected by the afterburning effect and strongly influencing the structural response [3], [4]. Therefore, accurately calculating the thermodynamic state in the quasi-static stage of confined explosions is crucial for designing structures to resist internal explosions and assessing structural damage.

Currently, TNT is widely used in structural anti-explosion research and blast-resistant engineering design. Researchers have conducted a series of experimental and numerical studies for TNT confined explosions. Existing experimental research primarily involves

WIT Transactions on The Built Environment, Vol 215, © 2025 WIT Press
www.witpress.com, ISSN 1743-3509 (on-line)
doi:10.2495/SUSI250011

conducting explosion experiments in spherical or cylindrical fully confined containers with varying mass-to-volume ratios m/V. The explosives were mainly placed at the space centre, and piezoelectric, piezoresistive, and thermocouple sensors on internal walls were used to measure pressure and temperature during the quasi-static stage, respectively. Fig. 1 presents quasi-static pressure obtained from experiments [2], [5]–[12] in comparison with the UFC curve [13] derived from experimental fitting. The experimental results demonstrate good agreement with the UFC curve. Existing numerical studies primarily adopt combustion models [14] or incorporating afterburn energy into the equation of state [1], [15] to simulate afterburning effects and have demonstrated that numerical methods such as combustion models can effectively replicate both wall loadings and structural damage observed in experiments.

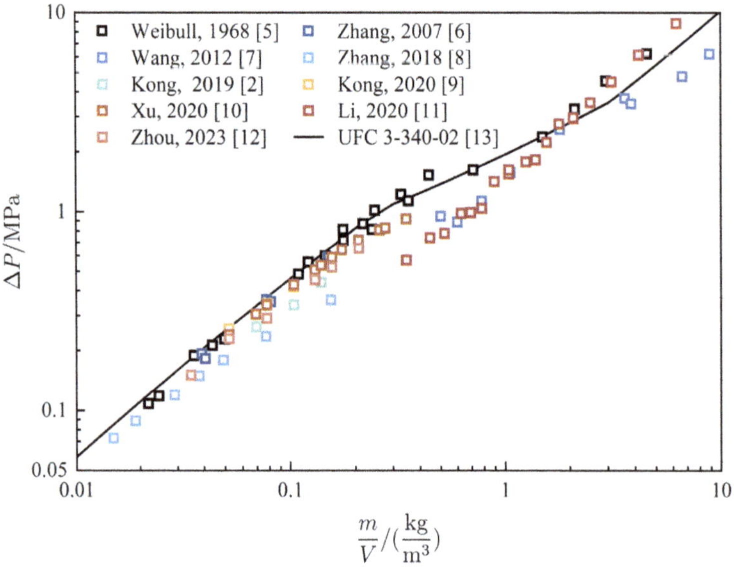

Figure 1: Quasi-static pressure from TNT confined explosion experiments and the corresponding UFC curve.

To accurately calculate the thermodynamic parameters in the quasi-static stage of confined explosions, thermodynamic models based on the assumptions of constant-volume processes and complete reactions have been proposed and widely adopted. The thermodynamic models can effectively capture actual chemical processes and calculate parameters such as mole numbers of components, temperature, pressure, and gas adiabatic index in the quasi-static stage. Kinney et al. [16] proposed a thermodynamic model that incorporates chemical reaction equilibrium, which demonstrated good agreement between the calculated pressure and experimental results. However, the model requires iteration of both temperature and mole numbers of components to ensure reaction equilibrium, and the computational stability depends on the selection of initial values, making the calculation relatively complex. Feldgun et al. [17] proposed a simplified thermodynamic model without chemical reaction equilibrium. The model also shows good agreement with experimental pressure results and only requires temperature iteration, making the solution process simpler, leading to its widespread application in determining gas adiabatic indices and energy release parameters for theoretical and numerical simulations [15], [17]. Nevertheless, the model cannot consider the influence of temperature on reaction equilibrium.

In summary, pressure results obtained from thermodynamic models with and without the reaction equilibrium for confined TNT explosions exhibit good agreement with experiment pressure results and enable the derivation of physical parameters including chemical compositions, temperature, and adiabatic index. However, there is a limited study to conduct a systematic investigation to consider the necessity of introducing reaction equilibrium in the thermodynamic model for different physical parameters. Considering that thermal effects in confined explosions may alter reaction equilibrium, thermodynamic parameters including temperature and adiabatic coefficients from models without chemical equilibrium could deviate significantly from those obtained through reaction equilibrium analysis, thereby impeding accurate representation of real physical processes in theoretical or numerical models. This study first conducted a comparative analysis of thermodynamic model results with the reaction equilibrium considered or not to demonstrate the necessity of considering reaction equilibrium for different physical quantities. Subsequently, a simplified calculation method was proposed to determine mole numbers of product compositions, temperature, and pressure in the quasi-static stage of TNT confined explosions when reaction equilibrium is accounted for. Finally, a practical solution was proposed to determine the value of additional afterburn energy for numerical modelling, thereby achieving effective quasi-static pressure simulation.

2 THE NECESSITY OF CONSIDERING REACTION EQUILIBRIUM IN THE THERMODYNAMIC MODEL

To investigate the effect of reaction equilibrium on thermodynamic model results, the results of thermodynamic models with and without reaction equilibrium were compared based on a unified model solution framework, and then the necessity of considering reaction equilibrium was demonstrated.

2.1 The framework for solving thermodynamic models

Fig. 2 shows the framework for solving thermodynamic models in this study. The solution framework was utilized to ensure that the differences between two model results are solely attributed to the consideration of reaction equilibrium or not. The initial step for solving models involves inputting the *m/V* and setting the initial value of the adiabatic combustion temperature T_{ad}. For the thermodynamic model without considering reaction equilibrium, the number of moles of the product can be calculated directly based on the chemical reaction equation, where elemental mass conservation is inherently satisfied. The mass conservation equation for element *e* can be expressed as:

$$\sum_k x_k l_{e,k} = \sum_j x_j l_{e,j}. \tag{1}$$

$l_{e,k}$ and $l_{e,j}$ represent the numbers of element *e* atoms in the *k*th reactant and *j*th product component, respectively, while x_k and x_j denote the molar coefficients of the *k*th reactant and *j*th product in the equation. For the thermodynamic model considering chemical equilibrium, initial mole numbers n_i for each product *i* were first estimated. Then, the equilibrium constant K_p at temperature T_{ad} was calculated. The residuals of the mass conservation equations were computed, and n_i were iteratively adjusted until the mass conservation equations converge. After the mole numbers of products are determined, T_{ad} was iterated based on energy conservation until energy convergence was achieved:

$$H_{\text{reac}} - H_{\text{prod}} - R(n_{\text{reac}}T_{\text{init}} - n_{\text{prod}}T_{\text{ad}}) = 0. \tag{2}$$

H_{reac} and H_{prod} are the standard formation enthalpies of the reactants and products, respectively, while n_{reac} and n_{prod} denote the total molar amounts of gaseous reactants and products. After T_{ad} was determined, the confined-space overpressure was calculated using the ideal gas law:

$$\Delta p = \frac{n_{\text{prod}}RT_{\text{ad}}}{1\,\text{mol} \cdot V_{\text{prod,gas}}} - p_0. \tag{3}$$

In the above equation, p_0 denotes the initial atmospheric pressure, and $V_{\text{prod,gas}}$ represents the spatial volume occupied by the gaseous products after the reaction is completed.

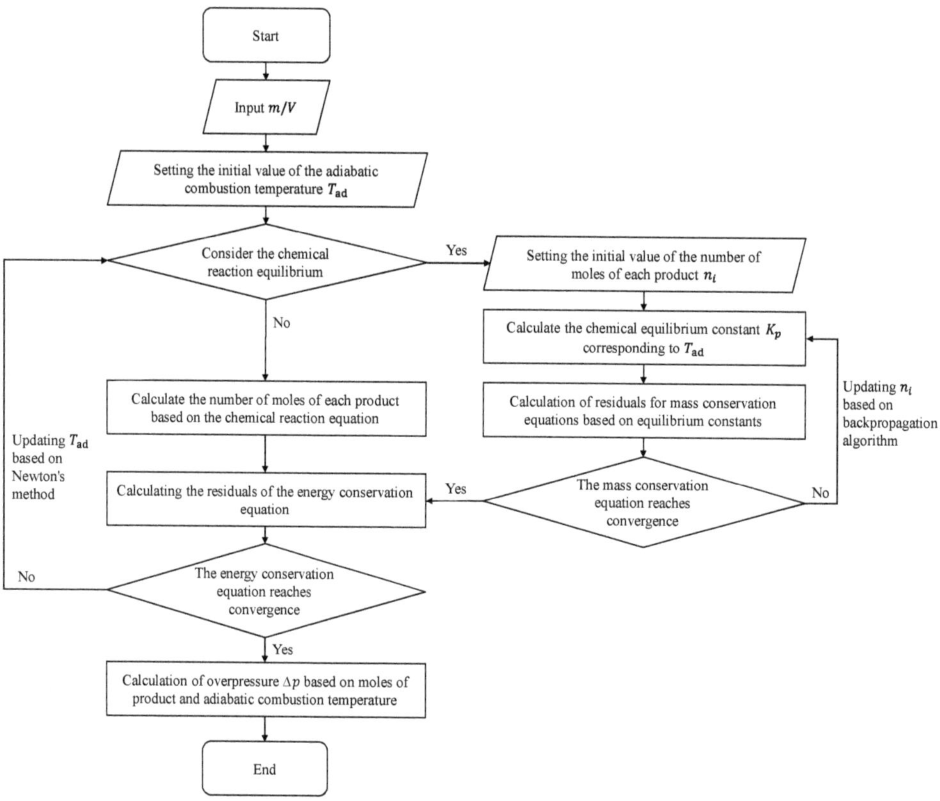

Figure 2: The solution framework of thermodynamic models.

During the solving process, Newton's method was used to solve T_{ad} until the residual remained below 0.1 K, while the backpropagation algorithm was employed to iterate n_i until its residual remained below 10^{-6} mol. By initializing the temperature at 500 K and using the n_i values from the non-equilibrium model as initial values for the model with chemical equilibrium, the stability of both algorithms can be effectively ensured.

2.2 The thermodynamic model without reaction equilibrium

The thermodynamic model without reaction equilibrium employs a simplified model proposed by Feldgun et al. [17] with several modifications. Define Φ as the molar number of air per unit mole of TNT:

$$\Phi\left(\frac{m}{V}\right) = \frac{n_{\text{Air}}}{n_{\text{TNT}}} = \frac{p_0 M_{\text{TNT}}}{RT_{\text{init}}}\left(\frac{1}{m/V} - \frac{1}{\rho_{\text{TNT}}}\right) = 9.284\left(\frac{1}{m/V} - \frac{1}{1630}\right). \tag{4}$$

In the above equation, n_{Air} and n_{TNT} represent the molar quantities of air and TNT, respectively. M_{TNT}= 0.227 kg/mol denotes the molar mass of TNT, and ρ_{TNT} = 1630 kg/m^3 is the density of TNT. Since the complete combustion of one mole of TNT requires 5.25 moles of oxygen, the condition for achieving stoichiometric combustion using ambient air (where oxygen constitutes 21% by volume) is expressed as $0.21\Phi = 5.25$. Solving this equation yields the transitional mass-to-volume ratio $(m/V)_{\text{trans},1}$ = 0.371 kg/m^3. In this way, the reaction equation for confined explosion can be derived. When $m/V < 0.371$ kg/m^3, it is assumed that the explosive undergoes complete reaction, leading to the following expression:

$$\begin{gathered} C_7H_5N_3O_6 + \Phi(0.21O_2 + 0.79N_2) \\ (0.79\Phi + 1.5)N_2 + (0.21\Phi - 5.25)O_2 + 2.5H_2O + 7CO_2. \end{gathered} \tag{5}$$

When $m/V \geq 0.371$ kg/m^3, under the assumption that all oxygen participates in the combustion process, the proportion of explosive undergoing complete combustion is given by $\mu = 0.21\Phi/5.25 = \Phi/25$, while the remaining fraction undergoes no-afterburning detonation. The reaction equations can be expressed as:

$$\begin{gathered} C_7H_5N_3O_6 + \Phi(0.21O_2 + 0.79N_2) \\ (0.79\Phi + 1.5)N_2 + 2.5H_2O + 7\mu CO_2 + 3.5(1-\mu)CO + 3.5(1-\mu)C. \end{gathered} \tag{6}$$

By incorporating eqns (5) and (6) into the solution framework introduced in Section 2.1, the model without reaction equilibrium can be solved.

2.3 The thermodynamic model with reaction equilibrium

The thermodynamic model with reaction equilibrium mainly employs the model proposed by Kinney et al. [16]. Assuming that the reaction products from confined TNT detonation comprise carbon (C), seven gaseous molecular species (CO, CO_2, O_2, H_2, H_2O, N_2, NO), and four free radical species (OH, H, O, N). The reaction equations can be expressed as

$$\begin{gathered} C_7H_5N_3O_6 + \Phi(0.21O_2 + 0.79N_2) \\ n_C C + n_{CO}CO + n_{CO_2}CO_2 + n_{O_2}O_2 + n_{H_2}H_2 + n_{H_2O}H_2O + n_{N_2}N_2 + n_{NO}NO \\ + n_{OH}OH + n_H H + n_O O + n_N N. \end{gathered} \tag{7}$$

The equation system contains 12 variables, thus requiring 12 independent equations for solution. Based on the mass conservation, four governing equations can be established:

$$\begin{cases} n_{\mathrm{C}} + n_{\mathrm{CO}} + n_{\mathrm{CO_2}} = 7 \\ 2n_{\mathrm{H_2}} + 2n_{\mathrm{H_2O}} + n_{\mathrm{OH}} + n_{\mathrm{H}} = 5 \\ 2n_{\mathrm{N_2}} + n_{\mathrm{NO}} + n_{\mathrm{N}} = 3 + 2 \cdot 0.79\Phi \\ n_{\mathrm{CO}} + 2n_{\mathrm{CO_2}} + 2n_{\mathrm{O_2}} + n_{\mathrm{H_2O}} + n_{\mathrm{NO}} + n_{\mathrm{OH}} + n_{\mathrm{O}} = 6 + 2 \cdot 0.21\Phi. \end{cases} \tag{8}$$

The remaining eight equations are established based on chemical equilibrium principles. Assuming that the following eight chemical reactions occur during the combustion process, and all components reach equilibrium upon completion of the reactions:

$$\mathrm{H_2} \rightleftharpoons 2\mathrm{H}, \qquad K_1 = \frac{p_{\mathrm{H}}^2}{p_{\mathrm{H_2}}}. \tag{9}$$

$$\mathrm{O_2} \rightleftharpoons 2\mathrm{O}, \qquad K_2 = \frac{p_{\mathrm{O}}^2}{p_{\mathrm{O_2}}}. \tag{10}$$

$$\mathrm{N_2} \rightleftharpoons 2\mathrm{N}, \qquad K_3 = \frac{p_{\mathrm{N}}^2}{p_{\mathrm{N_2}}}. \tag{11}$$

$$\mathrm{H_2O} \rightleftharpoons \mathrm{H_2} + 0.5\mathrm{O_2}, \qquad K_4 = \frac{p_{\mathrm{H_2}}^2 p_{\mathrm{O_2}}^{0.5}}{p_{\mathrm{H_2O}}}. \tag{12}$$

$$\mathrm{H_2O} \rightleftharpoons \mathrm{OH} + 0.5\mathrm{H_2}, \qquad K_5 = \frac{p_{\mathrm{OH}} p_{\mathrm{H_2}}^{0.5}}{p_{\mathrm{H_2O}}}. \tag{13}$$

$$\mathrm{H_2O} + 0.5\mathrm{N_2} \rightleftharpoons \mathrm{H_2} + \mathrm{NO}, \qquad K_6 = \frac{p_{\mathrm{H_2}} p_{\mathrm{NO}}}{p_{\mathrm{H_2O}} p_{\mathrm{N_2}}^{0.5}}. \tag{14}$$

$$\mathrm{CO_2} \rightleftharpoons \mathrm{CO} + 0.5\mathrm{O_2}, \qquad K_7 = \frac{p_{\mathrm{CO}} p_{\mathrm{O_2}}^{0.5}}{p_{\mathrm{CO_2}}}. \tag{15}$$

$$\mathrm{CO} \rightleftharpoons \mathrm{C} + 0.5\mathrm{O_2}, \qquad K_8 = \frac{p_{\mathrm{O_2}}^{0.5}}{p_{\mathrm{CO}}}. \tag{16}$$

In the above equations, p_i represents the dimensionless partial pressure of product i, and K_1–K_8 represent the equilibrium constants for the respective reactions, which can be calculated based on the chemical kinetics database [18]. By solving eqns (8)–(16) and incorporating eqn (7) into the solution framework introduced in Section 2.1, the model with reaction equilibrium can be solved. The carbon generation point of the model can be determined to be 3.850 kg/m^3 using the bisection method.

2.4 Result comparison of the models

Fig. 3 compares the results of product components, temperature, and pressure derived from the two models within 0.01 kg/m^3 $\leq m/V \leq$ 10 kg/m^3. Figs 3(a) and 3(b) indicate that discrepancies in components and temperature emerge when $m/V >$ 0.1 kg/m^3 and become pronounced with increasing m/V. The introduction of reaction equilibrium shifts the m/V for carbon generation from 0.371 kg/m^3 to 3.850 kg/m^3, reduces the peak adiabatic flame temperature from 3578 K to 3060 K, and alters the m/V corresponding to maximum temperature from 0.371 kg/m^3 to 0.680 kg/m^3. Fig. 3(c) shows that pressure differences between models remain insignificant when $m/V <$ 0.1 kg/m^3 or $m/V >$ 1 kg/m^3. At

m/V = 0.371 kg/m^3, the model results without reaction equilibrium (1.394 MPa) exhibit a maximum relative error of 19.3% compared to the results from the reaction equilibrium model (1.168 MPa).

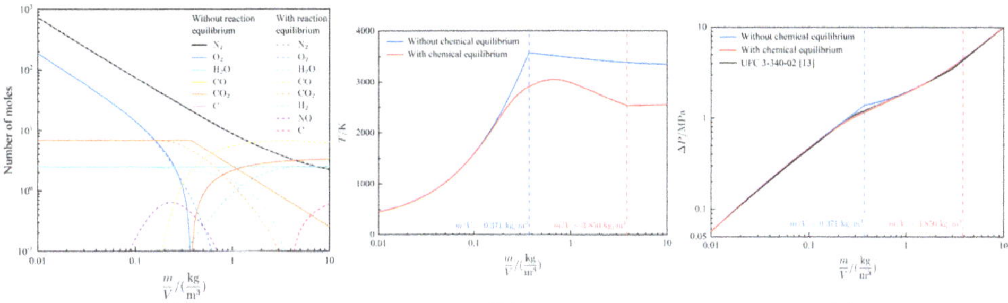

(a) (b) (c)

Figure 3: Model results comparison. (a) Product components; (b) Temperature; (c) Pressure.

The main reason for differences between two model results in product components and temperature is that as the m/V ratio increases, the elevated T_{ad} drives the dissociation reactions of CO_2 and H_2O toward the forward direction. Consequently, the m/V corresponding to CO generation occurs earlier than the thermodynamic model without reaction equilibrium, while the m/V corresponding to O_2 depletion is delayed. Concurrently, the dissociation reactions also increase the proportion of partially combusted products such as CO and H_2 in space. However, since the overpressure is proportional to both n_{prod} and T_{ad}, the influence of product components and temperature is not significantly reflected in the pressure result. Given that the components and temperature in model results exhibit significant differences when $m/V > 0.1$ kg/m^3, the model considering reaction equilibrium is a more rational choice for calculating physical quantities directly related to components and temperature.

3 A SIMPLIFIED CALCULATION METHOD FOR DETERMINING COMPONENTS, TEMPERATURE, AND PRESSURE

To facilitate engineering application of the thermodynamical model with reaction equilibrium, this section presents a simplified method for determining the mole numbers of product composition, temperature, and pressure in the quasi-static stage of TNT confined explosions based on the thermodynamic model result and symbolic regression technique.

3.1 Components

Neglecting the free radical and NO terms in chemical reaction products, list the mass conservation equations of CHON elements for the remaining seven products (C, CO, CO_2, O_2, H_2, H_2O, N_2):

$$\begin{cases} 2n_{N_2} = 3 + 2 \cdot 0.79\Phi \\ n_C + n_{CO} + n_{CO_2} = 7 \\ 2n_{H_2} + 2n_{H_2O} = 5 \\ n_{CO} + 2n_{CO_2} + 2n_{O_2} + n_{H_2O} = 6 + 2 \cdot 0.21\Phi. \end{cases} \quad (17)$$

From eqn (17), the expression for n_{N_2} can be derived:

$$n_{N_2}/\text{mol} = 1.5 + 0.79\Phi \approx \frac{7.34}{m/V} + 1.50. \tag{18}$$

The mass conservation equations for elements C, H, and O involve six unknowns. To make the equations solvable, the molar quantities of at least three products must be known. By applying symbolic regression algorithm to fit the thermodynamic model results and incorporating eqns (5) and (6), the following formula can be derived:

$$n_C/\text{mol} = \begin{cases} 0, & 0.01\ \text{kg/m}^3 \le m/V < 3.85\ \text{kg/m}^3 \\ 1 - \frac{3.85}{m/V}, & 3.85\ \text{kg/m}^3 \le m/V \le 10\ \text{kg/m}^3 \end{cases}. \tag{19}$$

$$n_{H_2}/\text{mol} = \begin{cases} 3.41 \cdot \exp\left(-\frac{1.28}{m/V}\right), & 0.01\ \text{kg/m}^3 \le m/V < 3.85\ \text{kg/m}^3 \\ 2.5, & 3.85\ \text{kg/m}^3 \le m/V \le 10\ \text{kg/m}^3 \end{cases}. \tag{20}$$

$$n_{O_2}/\text{mol} = \begin{cases} \frac{1.95}{m/V} - 5.25, & 0.01\ \text{kg/m}^3 \le m/V < 0.25\ \text{kg/m}^3 \\ 2.97 \cdot \frac{0.0018^{m/V}}{m/V}, & 0.25\ \text{kg/m}^3 \le m/V < 0.64\ \text{kg/m}^3 \\ 0, & 0.64\ \text{kg/m}^3 \le m/V \le 10\ \text{kg/m}^3 \end{cases}. \tag{21}$$

After n_C, n_{H_2}, n_{O_2} are determined, the expressions for the mole numbers of H_2O, O, O_2 can be derived using eqn (17):

$$n_{H_2O}/\text{mol} = 2.5 - n_{H_2}. \tag{22}$$

$$n_{CO_2}/\text{mol} = \frac{3.90}{m/V} - 1 + n_C - n_{H_2O} - 2n_{O_2}. \tag{23}$$

$$n_{CO}/\text{mol} = -\frac{3.90}{m/V} + 8 - 2n_C + n_{H_2O} + 2n_{O_2}. \tag{24}$$

Eqn (24) shows a relatively significant cumulative error than eqns (22) and (23) compared with the thermodynamic mode result. To address this problem, n_{CO} was refitted with piecewise functions and symbolic regression:

$$n_{CO}/\text{mol} = \begin{cases} 0, & 0.01\ \text{kg/m}^3 \le m/V < 0.18\ \text{kg/m}^3 \\ \exp[1.97 - (2.39 \cdot m/V)^{-1.7}], & 0.18\ \text{kg/m}^3 \le m/V < 3.85\ \text{kg/m}^3 \\ 7, & 3.85\ \text{kg/m}^3 \le m/V \le 10\ \text{kg/m}^3 \end{cases}. \tag{25}$$

Comparison with thermodynamic model results shows that the relative error of the proposed eqns (18)–(23) and (25) for calculating the number of moles of each component is less than 15% when the molar quantities exceed 0.7 mol.

3.2 Temperature

T_{ad} was fitted using symbolic regression:

$$T_{ad}/\text{K} = 2541 + \frac{1045.5}{2.22^{m/V}} - \frac{3304.3}{292.3^{m/V}}. \tag{26}$$

The relative error of eqn (26) to the thermodynamic model result is less than 3% in $0.01\ \text{kg/m}^3 \leq m/V \leq 10\ \text{kg/m}^3$.

3.3 Pressure

To calculate the quasi-static pressure, assuming the volume occupied by solid carbon in the products is negligible, the expression for $V_{\text{prod,gas}}$ can be simplified to:

$$V_{\text{prod,gas}}/(\text{m}^3/\text{mol}) \approx \frac{M_{\text{TNT}}}{m/V} = \frac{0.227}{m/V}. \tag{27}$$

n_{prod} can be expressed as

$$n_{\text{prod,gas}}/\text{mol} \approx n_{\text{CO}} + n_{\text{CO}_2} + n_{\text{H}_2} + n_{\text{H}_2\text{O}} + n_{\text{N}_2} + n_{\text{O}_2} \approx \frac{7.34}{m/V} + 11 - n_{\text{C}} + n_{\text{O}_2}. \tag{28}$$

Based on the expressions for n_C and n_{O_2}, n_{prod} can be determined. After obtaining $V_{\text{prod,gas}}$, n_{prod}, and T_{ad}, the quasi-static pressure can be calculated using eqn (3). The calculated values exhibit relative errors of less than 4% compared to thermodynamic model results and less than 8% relative to UFC specification curves within $0.01\ \text{kg/m}^3 \leq m/V \leq 10\ \text{kg/m}^3$.

Finally, the limitations of the simplified method should also be noted. The algorithm accounts for only a limited number of chemical species, so the results should be treated as approximate for actual product compositions within the studied component range. Additionally, while the framework is theoretically applicable to other explosives, the current formulations are validated only for TNT.

4 A PRACTICAL SOLUTION FOR QUASI-STATIC PRESSURE SIMULATION CONSIDERING AFTERBURNING ENERGY

By incorporating afterburn energy into the equation of state for detonation products, the afterburning process in confined explosions can be effectively simulated without directly resolving chemical reactions, thereby significantly reducing the computational costs of numerical simulations. The magnitude of the afterburn energy directly determines the quasi-static pressure in the numerical model. Consequently, this section proposes a theoretical and practical method for determining the afterburn energy in confined explosion simulations.

4.1 Determination of the afterburning energy

In current numerical simulations of confined explosions, the air is typically modelled with the ideal gas equation of state:

$$p = (\gamma - 1)\rho_{\text{Air}} e_{\text{Air}}. \tag{29}$$

In eqn (29), γ denotes the adiabatic index, ρ_{Air} denotes the density of the air, and e_{Air} is the internal energy of the air per unit mass. The explosive materials are generally simulated with the JWL equation:

$$p = A\left(1 - \frac{\omega}{R_1\theta}\right)\exp(-R_1\theta) + B\left(1 - \frac{\omega}{R_2\theta}\right)\exp(-R_2\theta) + \omega\rho e_{\text{explosive}}. \tag{30}$$

In eqn (30), $\theta = \rho_{\text{explosive}}/\rho$, $\rho_{\text{explosive}}$ is the initial density of the explosive, ρ is the density of the detonation products, A, B, R_1, R_2, ω are constants related to the explosive material, and $e_{\text{explosive}}$ is the internal energy of the explosive per unit mass. To simulate the development of afterburning energy, $e_{explosive}$ can be decomposed into the internal energy induced by detonation e_{det} and augmented by afterburning e_{ab}. The afterburning-induced internal energy increases gradually over time and eventually reaches a steady-state value E_{ab} that represents the afterburning energy per unit mass of the explosive [1], [15], [20]. The JWL equation considering the afterburning energy can be written as:

$$p = A\left(1-\frac{\omega}{R_1\theta}\right)\exp(-R_1\theta) + B\left(1-\frac{\omega}{R_2\theta}\right)\exp(-R_2\theta) + \omega\rho(e_{det}+e_{ab}). \quad (31)$$

To facilitate the derivation of E_{ab} under a given quasi-static pressure, assuming that the afterburn energy is instantaneously released in the numerical model. Since the assumption does not alter the total energy of the confined flow field, it does not influence the theoretical quasi-static pressure corresponding to the steady-state flow field [17], [18]. Based on this assumption, e_{ab} in the initial state equals E_{ab}, and $e_{det} = E_{det}$ which represents the detonation energy per unit mass of the explosive. The energy conservation equation of the flow field under this assumption can be expressed as:

$$\frac{p_0}{\gamma-1}\left(1-\frac{m/V}{\rho_{TNT}}\right) + (E_{det}+E_{ab})\frac{m}{V} = \frac{p_{qs}}{\gamma-1}\cdot\frac{V_{Air}}{V} + \left[p_{qs} - A\left(1-\frac{\omega}{R_1\theta}\right)\exp(-R_1\theta) - B\left(1-\frac{\omega}{R_2\theta}\right)\exp(-R_2\theta)\right]\cdot\frac{m/V}{\omega\rho}. \quad (32)$$

In eqn (32), p_0 is the initial ambient pressure, p_{qs} is the quasi-static pressure from the thermodynamic model or the UFC standard, and V_{air} is the volume occupied by air. Since there is no mass exchange between the detonation products and the air, ρ can be expressed as

$$\rho = \frac{m}{V-V_{Air}} = \frac{m/V}{1-V_{Air}/V}. \quad (33)$$

Assuming that after the afterburning reaction, the air and detonation products possess identical densities, which is consistent with the quasi-static flow characteristic of the numerical model result. Under this condition, the following equation can be formulated:

$$\frac{\rho_{Air}\left(1-\frac{m/V}{\rho_{explosive}}\right)}{V_{Air}/V} = \frac{m/V}{1-V_{Air}/V}. \quad (34)$$

By solving eqns (31)–(34), the value of E_{ab} under a specified p_{qs} can be determined.

4.2 Validation of the proposed solution

The proposed method was validated based on numerical simulation. The numerical model is a 1/8 symmetric model constructed based on LS-Dyna with the ALE (arbitrary Lagrange–Euler) algorithm in which a spherical TNT explosive was detonated at the centre of a 1 m × 1 m × 1 m confined space, as shown in Fig. 4. The m/V ratio was set as 0.01, 0.05, 0.1, 0.5, 1, 5, and 10 kg/m^3 in different cases, respectively. The equations of state EOS for air and the explosive were eqn (29) and (31), respectively, with material parameters detailed in Table 1. The average pressure at the wall's normal reflection point during the 20–25 ms after the explosion was taken as the quasi-static pressure. The initial internal energy of TNT was set as $E_{det} + E_{ab}$, and E_{ab} was determined based on eqns (31)–(34). The mesh size employed is 0.30 cm, which is sufficient for model result convergence in all cases.

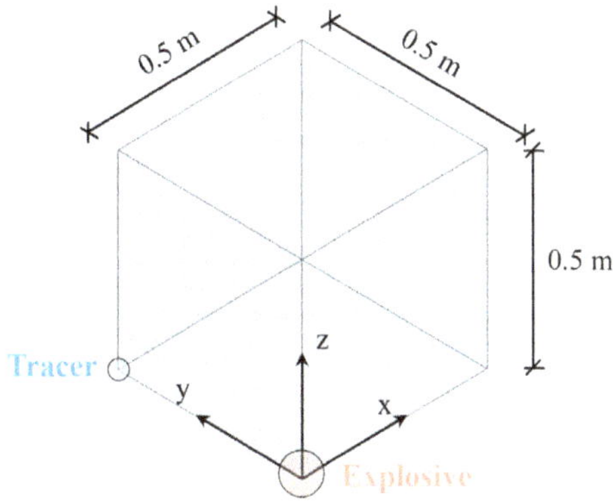

Figure 4: Quasi-static pressure from TNT confined explosion experiments and the corresponding UFC curve.

Table 1: Material parameters of the air and TNT for numerical simulation.

Material	Parameter	Value	Parameter	Value
Air	ρ_{Air}	1.2929 kg/m^3	γ	1.4
	p_0	0.101 MPa		
TNT	ρ_{TNT}	1630 kg/m^3	A	371.25 GPa
	B	3.231 GPa	R_1	4.15
	R_2	0.95	ω	0.30
	E_{det}	4294.5 KJ/kg		

Table 2 compares the accuracy of the proposed method and the simplified method proposed by Hernandez et al. [21] and the method proposed by Zhou et al. [15]. Hernandez provided an empirical method for calculating E_{ab}, while Zhou's method was proposed based on the ideal gas equation. As shown in Table 2, the proposed method achieves an absolute relative error within 3% in all cases. The relative error of Hernandez's method measured in this study does not fully align with the results reported in Hernandez et al. [21]. This discrepancy arises primarily because Hernandez's method cannot account for the influence of equation-of-state (EOS) parameter selection on the results, and the EOS parameters used

Table 2: Accuracy comparison of quasi-static pressure derived from different methods.

m/V	Relative error of the proposed method	Relative error of Hernandez's method	Relative error of Hernandez et al.'s method from [21]	Relative error of Zhou's method
0.01	+0.01%	−6.22%	/	+34.89%
0.05	−1.18%	+0.76%	+1.41%	+16.38%
0.1	−1.24%	+4.63%	+7.65%	+11.20%
0.5	+0.15%	+8.82%	−3.42%	+4.15%
1	+1.03%	+13.42%	+5.23%	+0.71%
5	+2.84%	+35.93%	+5.03%	−6.46%
10	+2.59%	/	/	−8.50%

for simulating air and the explosive in the two studies differ. Although the methods from Hernandez or Zhou achieve lower relative errors in some cases, they fail to maintain satisfactory accuracy across the entire *m/V* range investigated, demonstrating that the proposed method exhibits broader applicability compared to the two existing methods.

5 CONCLUSIONS

This study first conducted a comparative analysis to investigate the necessity of considering reaction equilibrium in the quasi-static pressure thermodynamic model for TNT confined explosion. Then, a simplified calculation method was proposed to determine mole numbers of product compositions, temperature, and pressure in the quasi-static stage. Finally, a practical solution was proposed for determining additional afterburning energy in numerical models for accurate quasi-static pressure simulation. Here are the main conclusions:

1. The thermodynamic model considering reaction equilibrium is a more rational choice for calculating physical quantities related to components and temperature. The introduction of reaction equilibrium in the thermodynamic model shifts the *m/V* for carbon generation from 0.371 kg/m^3 to 3.850 kg/m^3 and reduces the peak temperature from 3578 K to 3060 K. The discrepancies in product components and temperature results become increasingly significant when $m/V > 0.1$ kg/m^3.
2. The proposed simplified calculation method shows good consistency with the thermodynamic model results considering reaction equilibrium. Within the range of 0.01 kg/m$^3 \leq m/V \leq 10$ kg/m^3, the simplified method yields relative errors of less than 15% in the mole number of each component when the mole number exceeds 0.7 mol. The relative errors for temperature and pressure remain below 3% and 4%, respectively.
3. The proposed practical solution for calculating the afterburning energy in numerical models achieves an absolute relative error of quasi-static pressure within 3% and shows a broader applicability range than the two existing methods in the range of 0.01 kg/m$^3 \leq m/V \leq 10$ kg/m^3.

Research in future should conduct additional experiments to measure parameters like quasi-static temperature and heat of detonation in confined explosions across varying *m/V*, which will help validate the applicability scope of the thermodynamic model based on the complete reaction assumption from multiple perspectives.

ACKNOWLEDGEMENTS

This work was supported by the National Natural Science Foundation of China (grant number 51678448), the Fundamental Research Funds for the Central Universities through Tongji University (grant number 2023-2-ZD-05), and the Natural Science Foundation of Shanghai (grant number 22dz1202900).

REFERENCES

[1] Edri, I., Feldgun, V.R., Karinski, Y.S. & Yankelevsky, D.Z., On blast pressure analysis due to a partially confined explosion: III. afterburning effect. *International Journal of Protective Structures*, **3**(3), pp. 311–331, 2012. https://doi.org/10.1260/2041-4196.3.3.31.

[2] Kong, X. et al., An experimental study on the mitigation effects of fine water mist on confined-blast loading and dynamic response of steel plates. *International Journal of Impact Engineering*, **134**, 103370, 2019. https://doi.org/10.1016/j.ijimpeng.2019.103370.

[3] Kong, X., Wang, Z., Kuang, Z., Zhou, H., Zheng, C. & Wu, W., Experimental study on the mitigation effects of confined-blast loading. *Explosion And Shock Waves*, **41**(6), 062901-1-062901-14, 2021. (In Chinese.) https://doi.org/10.11883/bzycj-2020-0193.
[4] Grisaro, H.Y., Pressure-impulse diagrams for assessment of structural response due to a fully confined explosion. *Engineering Failure Analysis*, **146**, 107103, 2023. https://doi.org/10.1016/j.engfailanal.2023.107103.
[5] Weibull, H.R., Pressures recorded in partially closed chambers at explosion of TNT charges. *Annals of the New York Academy of Sciences*, **152**(1), pp. 357–361, 1968. https://doi.org/10.1111/j.1749-6632.1968.tb11987.x.
[6] Zhang, F., Anderson, J. & Yoshinaka, A., Post-detonation energy release from TNT-aluminium explosives. *AIP Conference Proceedings*, **955**(1), pp. 885–888, 2007. https://doi.org/10.1063/1.2833268.
[7] Wang, D., Zhang, D., Li, Y., Wang, C., Liu, W. & Wang, H., Experiment investigation on quasi-static pressure in explosion containment vessels. *Acta Armamentarii*, **33**(12), pp. 1493–1497, 2012. (In Chinese.) https://doi.org/10.3969/j.issn.1000-1093.2012.12.014.
[8] Zhang, Y., Su, J., Li, Z., Jiang, H., Zhong, K. & Wang, S., Quasi-static pressure characteristic of TNT's internal explosion. *Explosion And Shock Waves*, **38**(6), pp. 1429–1434. 2018. (In Chinese.) https://doi.org/10.11883/bzycj-2017-0170.
[9] Kong, X., Kuang, Z., Zheng, C. & Wu, W., Experimental study of afterburning enhancement effect for blast load in confined compartment space. *Acta Armamentarii*, **41**(1), pp. 75–85, 2020. (In Chinese.) https://doi.org/10.3969/j.issn.1000-1093.2020.01.009.
[10] Xu, J., Study on the similarity laws of structural dynamic response under internal explosion load in compartments. Master's thesis, Wuhan, pp. 32–69, 2019. (In Chinese.) https://doi.org/10.27381/d.cnki.gwlgu.2019.001805.
[11] Li, Y., Zhang, L., Du, Z., Ren, X., Zhang, X. & Fang, D., Theoretical and experimental study on formation of quasi-static pressure in internal blast. *Shipbuilding of China*, **61**(2), pp. 28–34, 2020. (In Chinese.)
[12] Zhou, H. et al., TNT equivalency method in confined space based on steel plate deformation. *International Journal of Impact Engineering*, **178**, 104587, 2023. https://doi.org/10.1016/j.ijimpeng.2023.104587.
[13] United States Department of Defence, Structures to resist the effects of accidental explosions. UFC 3-340-02. https://stg.wbdg.org/dod/ufc/ufc-3-340-02. Accessed on: 1 Sep. 2014.
[14] Fedina, E. & Fureby, C., Investigating ground effects on mixing and afterburning during a TNT explosion. *Shock Waves*, **23**, pp. 251–261, 2013. https://doi.org/10.1007/s00193-012-0420-9.
[15] Zhou, H. et al., Dynamic behaviour of steel plates subjected to confined blast loading considering afterburning effect. *International Journal of Impact Engineering*, **188**, pp. 104934, 2024. https://doi.org/10.1016/j.ijimpeng.2024.104934.
[16] Kinney, G.F., Sewell, R.G.S. & Graham, K.J., Peak overpressures for internal blast. NWC TP 6089. U.S. Naval Weapons Center. https://apps.dtic.mil/sti/pdfs/ADA071312.pdf. Accessed on: 11 Jul. 2024.
[17] Feldgun, V., Karinski, Y., Edri, I. & Yankelevsky, D., Prediction of the quasi-static pressure in confined and partially confined explosions and its application to blast response simulation of flexible structures. *International Journal of Impact Engineering*, **90**, pp. 46–60, 2016. https://doi.org/10.1016/j.ijimpeng.2015.12.001.
[18] NIST Standard Reference Database. https://webbook.nist.gov/chemistry/.

[19] Yue, X., Zhou, H., Kong, X., Zheng, C. & Wu, W., Experimental and simulation study of afterburning effect for blast load in confined cabin. *Chinese Journal of Ship Research*, **18**(4), pp. 223–232, 2023. (In Chinese.) https://doi.org/10.19693/j.issn.1673-3185.02708.
[20] Cao, W., He, Z. & Chen, W., Experimental study and numerical simulation of the afterburning of TNT by underwater explosion method. *Shock Waves*, **24**, pp. 619–624, 2014. https://doi.org/10.1007/s00193-014-0527-2.
[21] Hernandez, F., Hao, H. & Abdel-Jawed, M., Additional afterburning energy value to simulate fully confined trinitrotoluene explosions. *International Journal of Protective Structures*, 7(2), pp. 232–264, 2016. https://doi.org/10.1177/2041419616640113.

SIMPLIFIED EULERIAN SIMULATION OF BLAST WAVE PROPAGATION

ASSAL HUSSEIN
Civil Engineering Department, University of Diyala, Iraq

ABSTRACT

The computational analysis of blast waves involves various techniques and models to simulate the complex phenomena of blast wave generation, propagation and interaction with obstacles. In this study, a numerical simulation using the Eulerian model in ABAQUS is applied to analyse the blast wave generated from an unconfined explosion of 2.3 kg TNT charge. The study offers an efficient approach to simulate blast wave propagation in free air which can provide an accurate prediction to blast wave parameters and reduce time of analysis. Three-dimensional non-linear finite element analysis is performed to predict blast peak-overpressure, time of arrival and blast shock wave velocity. The results of Eulerian model are compared and validated with empirical model. The findings of this study clarified an excellent match with empirical model by Kingery–Bulmash. The advanced numerical simulation models can precisely capture complex phenomena of explosions in different burst events without necessity to conduct costly and unsafe blast field tests.

Keywords: Euler's model, blast, shockwave, ABAQUS, Kingery–Bulmash, empirical model.

1 INTRODUCTION

Simulating blast wave propagation in different media such as air, water, and soil is a challenge and complicated task. This complexity involves non-linear nature of blast shock wave which mean the characteristics changing rapidly at a very short period [1]. Hence, a small change in initial conditions can make a tremendous difference in blast wave intensity and make capturing shock wave propagation difficult [1], [2]. Rapid energy release from the source of the explosion generating shock waves in air travels away from the explosion location. Engineers have considered empirical equations to estimate blast shock wave intensity and blast response due to accuracy and ease of computations [3]–[5]. Blast wave parameters have been published in blast standards and graphically simulated with respect to scaled distance (Z) [6]–[8]. The empirical blast models had not taken in account flow environment conditions like temperature and assumed idealized tests environment [9]. Moreover, empirical blast models' accuracy is questionable for the near-field explosion [9]. The limitations of empirical models have been considered in the advanced hydrocodes models to accurately simulate blast and predict blast wave characteristics in different ambient conditions [10]–[12]. Various hydrocodes software blast models have been developed to overcome this issue with specified capability and range of applicability of presenting materials flow at high pressure and temperature [2]. Conventional Eulerian models can give a precise prediction of blast wave characteristics for the near-field burst but require small-mesh size and processing time is costly [1], [2]. Moreover, three-dimensional Euler model equations are complex to solve, and simulation of non-linear blast shock wave propagate requires high computation cost. The current study offers an efficient approach to simulate blast wave propagation in free air which can provide an accurate prediction to blast wave parameters and reduce time of analysis. In this study, a proposed explosion scenario resulted from an unconfined explosion of in free air is considered. Conventional coupled simulation methods of blast shock wave propagation have been considering 3D models for complex multi-phase flow problems. These approaches measure blast wave propagation environment precisely with high computation cost. For far-field burst, blast shock wave propagation and

WIT Transactions on The Built Environment, Vol 215, © 2025 WIT Press
www.witpress.com, ISSN 1743-3509 (on-line)
doi:10.2495/SUSI250021

pressure within the Euler domain is uniform [13]. This study suggests a simple approach to model explosive as built-in instance within the Eulerian domain. The proposed Eulerian analysis is primarily focused on capturing the in-plane dynamic blast shock wave characteristics without adding complexity of modelling the out-of-plane parameters. This method can significantly reduce the computational time and maintain the accuracy of results. The blast peak-overpressure (P_{so}), time of arrival (t_a) and blast shock wave velocity (U) are predicted. The results of Eulerian model in ABAQUS are compared and validated with empirical model by Kingery–Bulmash [6].

2 EULER MODEL EQUATIONS OF BLAST SHOCK WAVE

This study adopted Euler's model equations to simulate blast wave phenomena in free air. The governing equations of this model include conservation form of mass, momentum, and energy [9]. The general mathematical expressions of these partial differential equations (PDEs) are:

Mass conservation:

$$\frac{\partial \rho}{\partial t} + \nabla.(\rho v) = 0. \tag{1}$$

Here, ρ is the air density, v is velocity vector of air, and t is the time.

Momentum equation:

$$\frac{\partial (pv)}{\partial t} + \nabla.(\rho v \otimes v) + \nabla p = 0. \tag{2}$$

Here, p is the air pressure, v is velocity vector of air, and t is the time.

Conservation of energy:

$$\frac{\partial (pe)}{\partial t} + \nabla.(\rho v e + pv) = 0. \tag{3}$$

Here, p is the air pressure, v is velocity vector of air, e is internal energy per unit mass of air, and t is the time.

Moreover, the ideal gas law correlates the relation among pressure, density, and temperature od air as in the following mathematical expression.

$$p = \rho RT. \tag{4}$$

Here, p is the air pressure, R is the specific gas constant, and T is the temperature.

3 KINGERY–BULAMSH EMPERCAL BLAST MODEL

Kingery and Bulmash introduced the most admitted empirical blast model equations to estimate blast wave parameters of spherical and hemispherical bursts [6]. The developed equations provide data of incident peak-overpressure (P_{so}), reflected peak-overpressure (P_r), incident impulse (i_s), reflected impulse (i_r), duration of positive pressure phase (t^+_o), time of arrival of the shock wave (t_a) and shock front velocity (U) [5], [7]. The data were presented graphically in design graphs in terms of scaled distance (Z) [7], [8]. Thereafter, the negative phase blast wave parameters were included [7]. The only limitation of Kingery–Bulmash model is the accuracy of the data in close-field burst due to high generated temperature at the early time of explosion [14]. The estimated blast wave parameters and blast response are accurate and valid for freely blast wave propagation without interrupt by obstacles [15]. In this case, blast load-structure interaction is negligible.

The mathematical expression of Kingery–Bulmash model equations applicable ranges of scaled distance, 0.053 m/kg$^{1/3}$ ≤ Z ≤ 40 m/kg$^{1/3}$, and from 0.067 m/kg$^{1/3}$ ≤ Z ≤ 40 m/kg$^{1/3}$ for spherical and hemispherical TNT, respectively [14], [16], [17]. Eqn (5) shows the Kingery–Bulmash equation

$$Y = C_o + C_1U + C_2^2 + \cdots + C_N^N, \tag{5}$$

where $U = K_0 + K_1T$, Y, is the measured blast parameter, T is the base 10 logarithms of (Z), and N is the order of fit. These equation parameters are only open for the U.S. military and contractors [18].

4 NUMERICAL ANALYSIS RESULTS

4.1 Finite element model: Summary

Three-dimensional (3D) non-linear dynamic finite element model using Eulerian model in ABAQUS/Explicit [19]. The dimensions of the Eulerian domain are 10 m × 5 m × 1 cm. The Eulerian domain boundaries were considered based on studied standoff distances to ensure that shock wave does not leave the domain and becomes undetectable to the solver. Moreover, the Eulerian element in ABAQUS can only be applied for 3D models, therefore, the third dimension is given only small value to reduce computations cost. The TNT equivalent mass of the explosive charge is 2.3 kg, the height of the explosive is 2.5 m. The explosive charge is modeled within Eulerian domain and Eulerian–Eulerian contact is performed without need to define contact interaction [19]. Fig. 1 shows the model geometry description and Finite element meshed parts and explosive charge setup of the Eulerian model considered in this study. The air part is modeled inside Eulerian domain which is meshed as a volume element using 8-node linear Eulerian brick, reduced integration, hourglass control (EC3DR) element. This approach has a perfect capability to support multi-materials flow and good for large deformation. The Eulerian domain size has been selected based on considered scaled distances and guarantee blasts wave is not extended out of the domain boundaries.

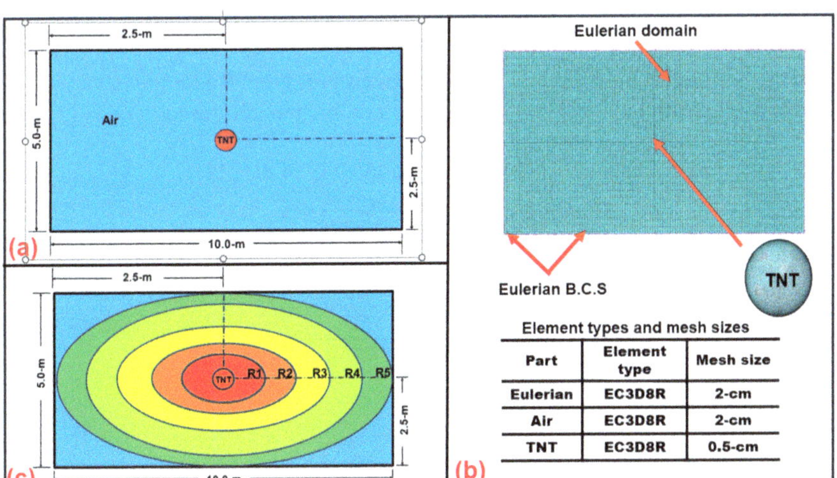

Part	Element type	Mesh size
Eulerian	EC3D8R	2-cm
Air	EC3D8R	2-cm
TNT	EC3D8R	0.5-cm

Figure 1: (a) Eulerian model geometry details; (b) Meshed parts and mesh sizes; and (c) Standoff distances where blast wave parameters measured.

4.2 Material models

The TNT explosive charge is modeled using the Jones–Wilkins–Lee (JWL) equation of state (EOS) [20]. The JWL model is a reliable method to simulate blast pressure generated from explosives burst [20], [21]. The generated blast pressure can be calculated using the equation-of-state-JWL as shown in eqn (6)

$$P = A\left(1 - \frac{\omega}{R_1 V}\right)e^{-R_1 V} + \left(1 - \frac{\omega}{R_2 V}\right)e^{-R_2 V} + \frac{\omega E}{V}. \tag{6}$$

Here, P is the pressure, A and B, are pressure parameters, ω, R_1 and R_2 are dimensionless nonlinear pressure parameters, V is the relative volume, E is the internal energy/unit volume [22]. The JWL TNT material parameters are listed in Table 1 [23].

Table 1: JWL model parameters of TNT [23].

Parameters	Value
P (kg/m^3)	1630
velocity (m/sec)	6930
A (GPa)	3.738
B (GPa)	3.747
R_1	4.15
ω	0.35
V_o	1.0
E (MPa)	6.0

The air domain is simulated suing ideal gas equation of state. This model can capture the air behaviour at low pressure level utilizing the proportional of air volume to temperature [19]. The ideal gas model equation of state cab be expressed as in eqn (7). Table 2 lists the material properties and EOS model parameters of air [24]

$$p + p_A = \rho R(\theta - \theta^Z). \tag{7}$$

Here p is the generated pressure, p_A, is the ambient pressure, R is a gas constant, θ is the current initial temperature, and θ^Z *is* absolute zero on the temperature scale [19].

Table 2: EOS model parameters of air [24].

Parameters	Value
P (kg/m^3)	1.297
p_A (kPa)	101.325
R (J/kg-k)	287
C_p	717.6
θ^Z (k)	0

4.3 Blast wave parameters

The Eulerian model analysis results of blast wave parameters are validated with empirical model of Kingery–Bulmash [7]. The Eulerian volume fraction (EVF) is an efficient method often used in fluid dynamics, specifically for multi-phase flow simulations. It describes the portion of the element volume filled with material. The value of EVF is equal to zero when the domain is empty and equal to one when it is filled with material [25]. The volume fraction of each particle, expressed as a Eulerian value, and associated with the mesh cell [26]. For elements partially filled with material, the EVF value is interpolating between adjacent elements [19]. The simulation is performed using grid-based technique which is often apply for classical multi-phase flow. While the simulation provides highly accurate results of blast shock wave propagation in open air, the accuracy for other blast scenarios such as confined explosions, non-ideal explosives, material behaviours, and multi-phase interactions requires further validation and examination [19]. Fig. 2(a) and (b) shows two-dimensional simulation of the EVF contour plot of air and TNT propagation within Eulerian domain, respectively.

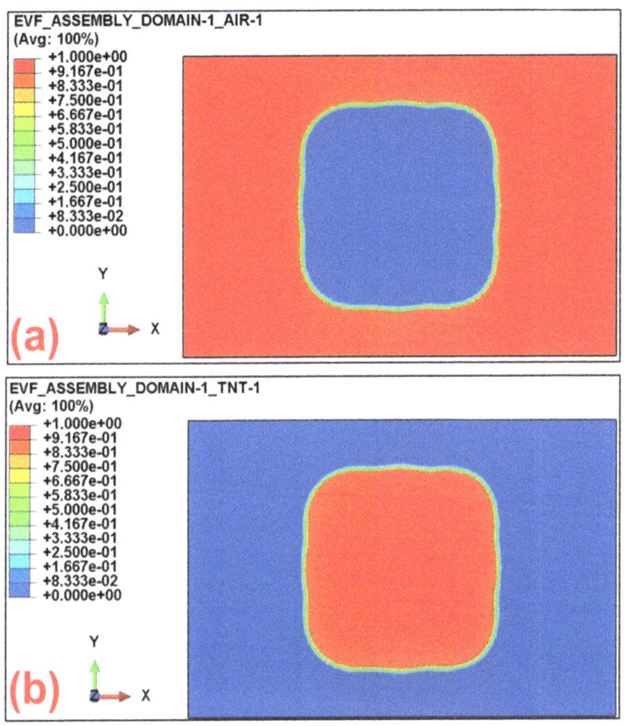

Figure 2: 2D Eulerian volume fraction contour. (a) Air; and (b) TNT.

4.3.1 Incident blast peak-overpressure

The considered Eulerian analysis model is applied to simulate blast wave extending in free air. Table 3 lists the measurement variables, standoff distance, R (m), explosive charge mass, W (kg), and scaled distance, Z (m/kg$^{1/3}$) of the current study (see Fig. 1(c)). Peak incident peak-overpressure is computed and validated with Kingery–Bulmash model data [7]. The results of the Eulerian model of incident peak-overpressure showed a good convergence with empirical data with maximum difference of 27.78 as shown in Fig. 3.

4.3.2 Time of arrival

The Empirical models of blast wave have measured time of arrival of blast shock wave considering optical-photo techniques [5], [7]. Fig. 4 shows the prediction of the time of

Table 3: Measurement variables.

P_t #	R (m)	W (kg)	Z (m/kg$^{1/3}$)
R_1	1.0		0.76
R_2	2.0		1.52
R_3	3.0	2.3	2.3
R_4	4.0		3.0
R_5	5.0		3.8

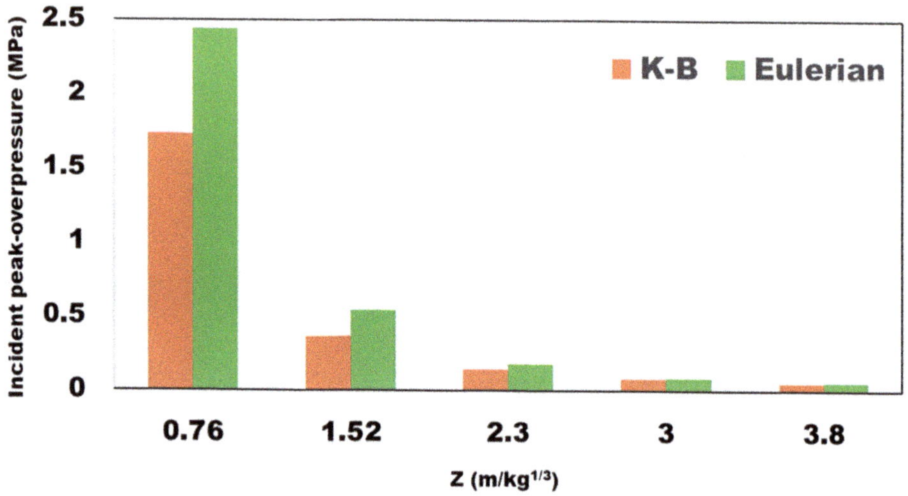

Figure 3: Incident peak-overpressure of Eulerian model and Kingery–Bulmash data.

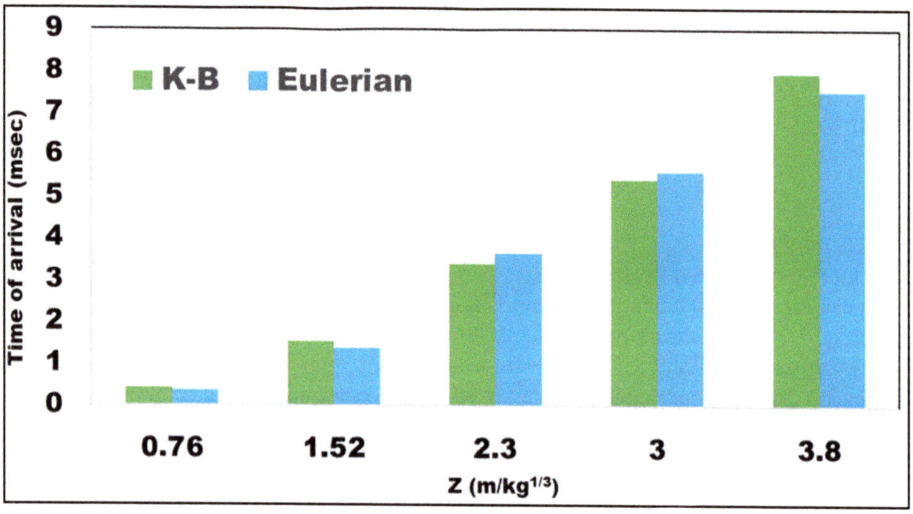

Figure 4: Time of arrival of Eulerian model compared to Kingery–Bulmash data.

arrival of blast wave (t_a) of the proposed Eulerian model with respect to Kingery–Bulmash data. The results of Eulerian demonstrated a good match with empirical model with maximum difference of 7.52%.

4.3.3 Blast shock wave velocity

The blast shock wave travels through the domain at a higher speed of sound. The amplitude of the shock wave velocity depends on different factors like density of the medium, ambient temperature, released energy and distance from source of explosion [27]. Fig. 5 shows the comparison of the current Eulerian model prediction and Prediction of the empirical model by Kingery–Bulmash. The results of the proposed model showed a good agreement with empirical model data with maximum difference of 20.13%.

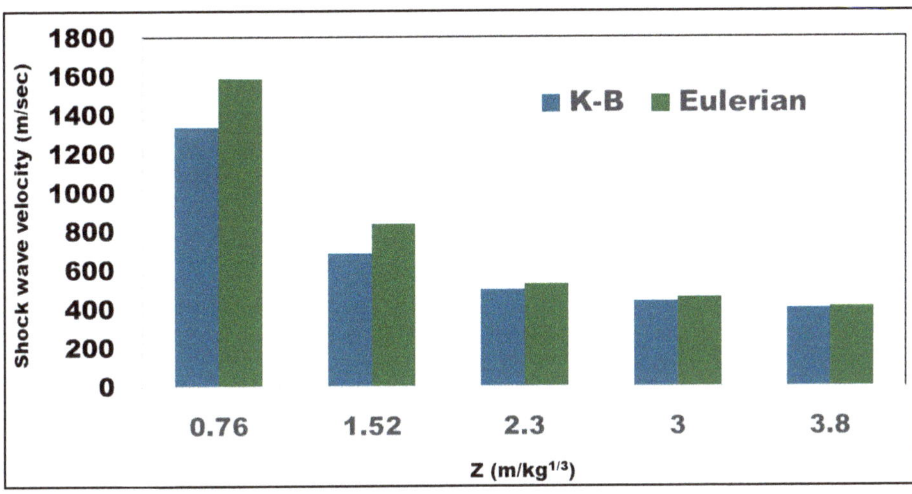

Figure 5: Comparison of Eulerian model Peak blast shock wave velocity with Kingery–Bulmash data.

5 CONCLUSIONS

In summary, the proposed Eulerian model indicates that analyzing blast shock wave propagation in open air using a Eulerian approach can be effectively represented by a simplified model that accurately predicts the characteristics of the blast wave with minimal computational effort. The results of the study highlighted the effectiveness of this approach in capturing the blast wave phenomena, eliminating the need for unsafe blast field tests or costly computational models. However, the accuracy of other blast scenarios such as confined explosions, non-ideal explosives, material behaviours, and multi-phase interactions requires further validation and examination. Future research should focus on developing simplified approaches that can be used by engineers to predict blast load parameters and blast response of structural systems.

REFERENCES

[1] Kim, J., Yoo, H.S., Jo, Y.B. & Kim, E.S., Simulation of shockwave propagation characteristics in nuclear reactor cavity during external steam explosion using unified SPH. *Nuclear Engineering and Technology*, 103274, 2024. https://doi.org/10.1016/j.net.2024.10.036.

[2] Peton, N. & Lardjane, N., A Eulerian version of geometrical blast dynamics for 3D simulations. *Shock Waves*, **32**(3), 2022. https://doi.org/10.1007/s00193-022-01070-w.
[3] Goel, M.D., Blast: Characteristics, loading and computation-an overview. *Advances in Structural Engineering: Mechanics, Volume One*, pp. 417–434, 2015. https://doi.org/10.1007/978-81-322-2190-6_36.
[4] Vannucci, P., Masi, F. & Stefanou, I., A comparative study on the effects of blast actions on a monumental structure. Research Report, 2018. https://hal.science/hal-01720557v2.
[5] Hussein, A. & Heyliger, P., On the accuracy of CEL blast simulations: Validation and application. *Asian Journal of Civil Engineering*, **26**(2), pp. 843–866, 2024. https://doi.org/10.1007/S42107-024-01226-2/METRICS.
[6] Kingery, C.B.G. & Bulmash, G., *Airblast Parameters from TNT Spherical Air Burst and Hemispherical Surface Burst*. Ballistic Research Laboratories, Aberdeen Proving Ground, MD, 1984.
[7] UFC 3-340-02, Structures to Resist the Effects of Accidental Explosions. Unified Facilities Criteria, U.S. Army Corps of Engineers, December, 2008.
[8] U.S. Army, Structures to resist the effects of accidental explosions, TM 5-1300. U.S. Department of the Army Technical Manual, Washington, DC, 1990.
[9] Nguyen-Dinh, M., Lardjane, N., Duchenne, C. & Gainville, O., Direct simulations of outdoor blast wave propagation from source to receiver. *Shock Waves*, **27**(4), 2017. https://doi.org/10.1007/s00193-017-0711-2.
[10] Shin, J., Whittaker, A.S. & Cormie, D., Incident and normally reflected overpressure and impulse for detonations of spherical high explosives in free air. *Journal of Structural Engineering*, **141**(12), 2015. https://doi.org/10.1061/(asce)st.1943-541x.0001305.
[11] Taylor, A.G., Numerical analysis of blast effects and mitigation in the far-field from small explosions. *Applied Sciences (Switzerland)*, 12(17), 2022. https://doi.org/10.3390/app12178824.
[12] Abd-El-Nabi, E., El-Helloty, A. & Summra, A., Numerical analysis of reinforced concrete buildings subjected to blast load. *Structural Concrete*, **24**(3), 2023. https://doi.org/10.1002/suco.202200726.
[13] Trasborg, P., Olmati, P. & Naito, C., Increasing the ductility of reinforced concrete panels to improve blast response. *Proceedings of the 2012 Critical Infrastructure Symposium*, 2012.
[14] Guzas, E.L. & Earls, C.J., Air blast load generation for simulating structural response. *Steel and Composite Structures*, 10(5), 2010. https://doi.org/10.12989/scs.2010.10.5.429.
[15] Wadley, H.N.G. et al., An active concept for limiting injuries caused by air blasts. *Int. J. Impact. Eng.*, **37**(3), 2010. https://doi.org/10.1016/j.ijimpeng.2009.06.006.
[16] Karlos, V. & Solomos, G., *Calculation of Blast Loads for Application to Structural Components*. Publications Office of the European Union: Luxembourg, 2013.
[17] Karlos, V., Solomos, G. & Larcher, M., Analysis of blast parameters in the near-field for spherical free-air explosions. *JRC Technical Reports*, 2016.
[18] Hetherington, J. & Smith, P., *Blast and Ballistic Loading of Structures*, 2014. https://doi.org/10.1201/9781482269277.
[19] ABAQUS, Abaqus User's Manual, version 2022, 2022.
[20] Lee, E.L., Hornig, H.C. & Kury, J.W., Adiabatic expansion of high explosive detonation products. University of California Radiation Laboratory at Livermore, Livermore, CA, USA, 1968.

[21] Mandal, J., Agarwal, A.K. & Goel, M.D., Numerical modeling of shallow buried tunnel subject to surface blast loading. *Journal of Performance of Constructed Facilities*, **34**(6), 2020. https://doi.org/10.1061/(asce)cf.1943-5509.0001518.
[22] Lai, J., Guo, X. & Zhu, Y., Repeated penetration and different depth explosion of ultra-high performance concrete. *Int. J. Impact. Eng.*, **84**, 2015. https://doi.org/10.1016/j.ijimpeng.2015.05.006.
[23] Mussa, M.H., Mutalib, A.A., Hamid, R., Naidu, S.R., Radzi, N.A.M. & Abedini, M., Assessment of damage to an underground box tunnel by a surface explosion. *Tunnelling and Underground Space Technology*, **66**, 2017. https://doi.org/10.1016/j.tust.2017.04.001.
[24] Sielicki, P.W. & Stachowski, M., Implementation of sapper-blast-module, a rapid prediction software for blast wave properties. *Central European Journal of Energetic Materials*, **12**(3), 2015.
[25] Smojver, I. & Ivančević, D., Bird strike damage analysis in aircraft structures using ABAQUS/Explicit and coupled Eulerian Lagrangian approach. *Compos. Sci. Technol.*, **71**(4), 2011. https://doi.org/10.1016/j.compscitech.2010.12.024.
[26] Diggs, A. & Balachandar, S., Evaluation of methods for calculating volume fraction in Eulerian-Lagrangian multiphase flow simulations. *J. Comput. Phys.*, **313**, 2016. https://doi.org/10.1016/j.jcp.2016.02.066.
[27] Hussein, A.T., Blast protection wall systems: Literature review. *WIT Transactions on the Built Environment*, WIT Press: Southampton and Boston, 2020. https://doi.org/10.2495/SUSI200081.

INVESTIGATION INTO THE EFFECTS OF WATER-BASED CONTACT CHARGES

SOPHIE TRÉLAT[1] & MICHEL-OLIVIER STURTZER[2]
[1]DEND, French Institute for Protection and Nuclear Security,
Department of Explosion and Weapons Effects, Fontenay-aux-Roses, France.
[2]ISL, French German Research Institute of Saint-Louis,
Department of Explosive Protection, Saint-Louis, France.

ABSTRACT

Law enforcement agencies commonly use combinations of explosive charges and water containers. Commercial companies already propose off-the-shelf products based on water-tamped explosive devices using the principle of impulse profile modification. The objective of this concept is to characterize the efficiency of bare explosive charges while minimizing surrounding blast effects in order to mitigate collateral damage, as a significant part of the energy of the explosion is transferred to the water in the form of kinetic energy. Based on existing and already-in-use geometries and volumes, we designed reference explosive charges comprising C4 parallelepipedal charges and water containers. In this study, the loading of the system was evaluated using a ballistic pendulum able to record the total mechanical load transferred. By recording the trajectory of the pendulum, it is possible to determine the transferred impulse. Influence of the ratio between explosive and water was investigated to analyse the system breaching efficiency and potential collateral damages. Collateral damage includes not only water ram effect, but also air blast effects. A side-on pressure sensor was consequently installed near to the pendulum to evaluate the influence of the water on peak overpressure propagating in the breach environment. Initial propagation of the blast wave occurs in aqueous medium. Comparison with high-speed images recorded during previous underwater detonation tests conducted close to water surface or reflective surfaces was thus possible. This analysis provided guidelines for future work on the optimization of water-based contact charges.

Keywords: blast, explosion effects, combined effects, water-based experiments, protection of infrastructures, pressure sensors, high-speed videos.

1 INTRODUCTION

The use of water in close range to an explosive event to mitigate blast and fragment effect damage radius is well documented in literature. It is especially efficient for enclosed ammunition storage facility and flexible in the case of overseas military operations [1]. Water is cheap, readily available in most places, easy to scale up even for very large amounts of explosives and does not present any transport or storage challenge.

Demolition devices are often composed of a breaching charge responsible of the severance of metal elements and a kicker charge consequently able to break out of balance the damaged structure to achieve the destruction. Bauer et al. [2] investigated the use of water-cased kicker charges. The goal is to reduce the fragmentation of steel members, and the quantity of explosive needed due to the increased density, incompressibility, and impedance mismatch water provides, using numerical simulations and experimental tests. In Baciu et al. [3], the authors present a comparative study between breaching charges effects on reference doors, experimental results and numerical simulation. Different types of doors are analysed, showing the complexity of the interaction between a multiphase water-based charge and a realistic door. In order to unify the charge choice for the different types of door resistances classes defined in well-established international standards, Sedláček and Palasiewicz [4] proposed a global approach suitable for operational forces. The objective is to secure the entry while minimizing the necessary amount of explosive used for Czech Army applications.

WIT Transactions on The Built Environment, Vol 215, © 2025 WIT Press
www.witpress.com, ISSN 1743-3509 (on-line)
doi:10.2495/SUSI250031

In Rus et al. [5], the authors focused on the determination of collateral effects (on personnel and environment) of metal breaching explosive charges. They consequently tend to reduce the charge mass and increase their efficiency. On this occasion, they demonstrated that there are very large differences between the effect obtained with similar devices on surface compared to the effect obtained in immersion. Most of the time a quantity of explosive that on land led to the desired effect, under water only deformed the same metal structure; in order to obtain similar results, a significantly larger quantity had to be used. Moldovan et al. also investigated human vulnerability of breaching charges in Moldovan et al. [6]. To create breaches, two different kinds of special explosive devices were used, according to the obstacle: cutting and pushing charges. Cutting charges are based on metal projection, pushing charges on water projection, using the properties of a layer of water to alter the pressure load on the target.

Finally, water-based charges are also commonly used for blasting rocks. In Li et al. [7], the authors present a numerical study of the water coupling coefficient on the blasting effect of red sandstone samples analysed with CT scanning and 3D reconstruction. Their goal is to predict the observed damage on the samples by defining an optimal coupling coefficient.

At DEND (the French Institute for Protection and Nuclear Security), we decided to go one step back, from realistic pieces of opening devices to reference target, however realistic in size, submitted to explosive charges combined with water. One of DEND missions is to carry out studies to support expertise in the security of nuclear and radioactive materials, their installations and their transport, against malicious acts. Concretely, our actions consist in assessing the performance and vulnerability of a physical protection system such as a wall, a steel container, a concrete infrastructure, windows or multi-material security doors for instance. To achieve that, we have to conduct relevant experiments to evaluate the load transferred by these explosive devices. Given that water is an incompressible medium, it is natural that the effects of similar explosive charges notably differ between air and underwater environments [8]. Consequently, the direct output of a hybrid water explosive charge should also be different when the objective is to examine the resistance of a material. We believe that future research focusing on measuring the pressure evolutions produced by the shock wave front resulting from underwater detonations of various quantities of explosives would be useful to better understand mechanical effects of these specific explosive devices. In this work, we therefore propose an approach linked with theoretical and applied studies based on experimental setups designed to characterize shock waves generated by either submarine or aerial explosions. We used a ballistic pendulum in addition to pressure gauges to evaluate the effect induced by a C4 reference explosive charge combined with water onto a metallic plate placed in contact as well as through the surrounding environment in open field.

2 EXPERIMENTAL SETUP – METROLOGY

2.1 Explosive charges

Fig. 1 presents the 50 g C4 parallelepipedal charge combined or not with a 1,000 g water pack and an exploding bridgewire (EBW) RP83 detonator used for ignition.

Charge is placed in contact with a S235 steel (structural steel for building and welding) plate fixed with adhesive to the ballistic pendulum system described hereafter.

Three consecutive tests (#2735, #2728 and #2731) were conducted with the same C4 explosive charge. Test #2735 is performed with bare C4 charge, whereas the two other ones include water pack: water envelope is on the plate side for #2728 (namely front water pack)

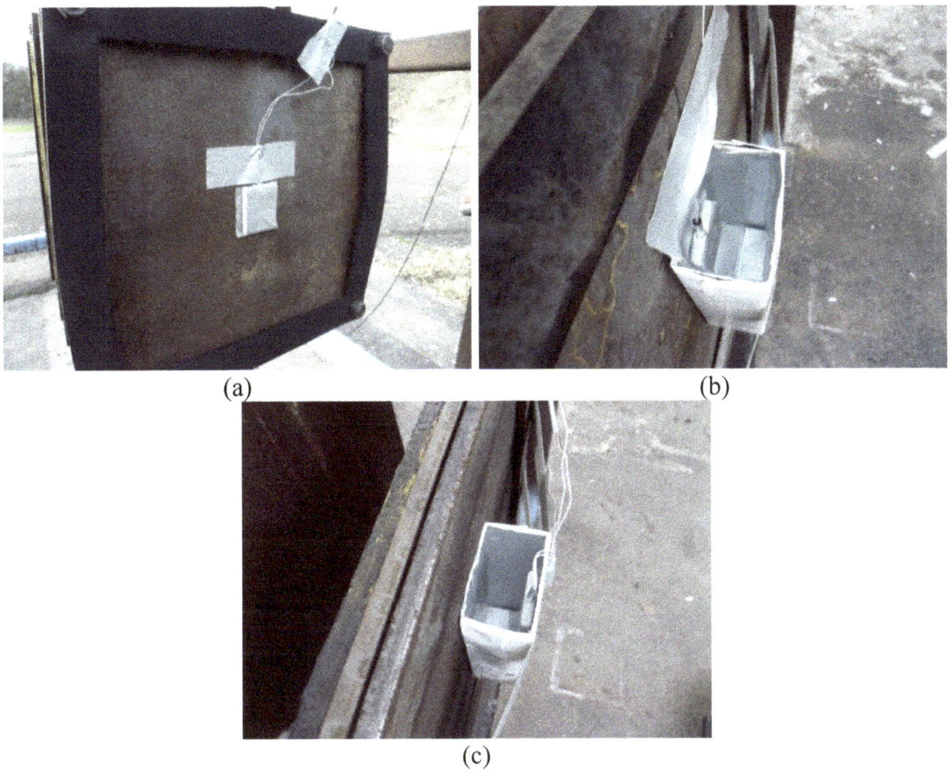

Figure 1: Explosive charges. (a) Bare charge; (b) 1,000 g water pack in the back; and (c) 1,000 g water pack in front.

Table 1: Charge specifications and configurations.

Charge reference	C4 mass (g)	Front water pack (g)	Back water pack (g)
2735	50	–	–
2728	50	1,000	–
2731	50	–	1,000

and on the free field side for #2731 (namely back water pack). The three charges specifications and configurations are summarized in Table 1.

2.2 Pressure sensors

Two PCB 137 B pressure gauges were installed at 3 m in front of the ballistic pendulum, facing the centre of the plate. One gauge was recording the side-on evolution (or incident) while the other was recording the face-on evolution of pressure (or reflected). PCB pencil probes are shown in Fig. 2(a). Blast pencil probes are flush mounted on 1.5 m posts. Fig. 2(b) and Fig. 3 present schematic representations of the experimental configurations with or without water pack.

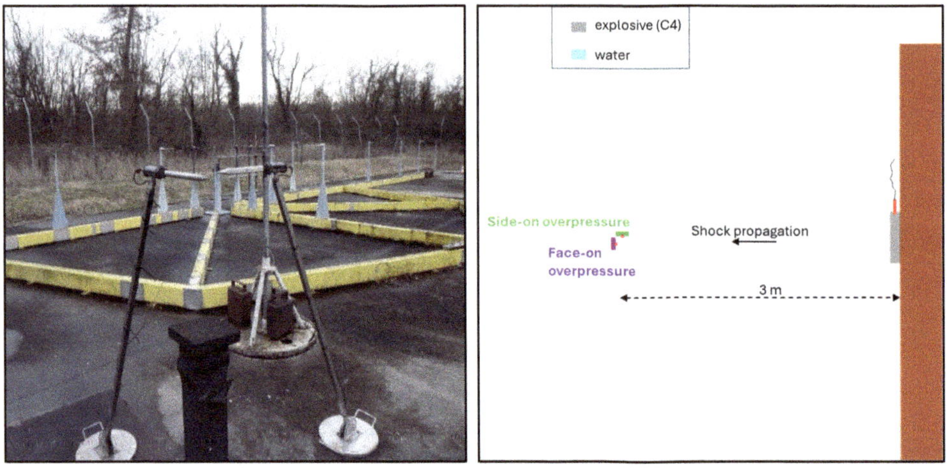

(a) (b)

Figure 2: (a) Side-on and face-on pressure gauges; and (b) Schematic of experimental setup with bare charge.

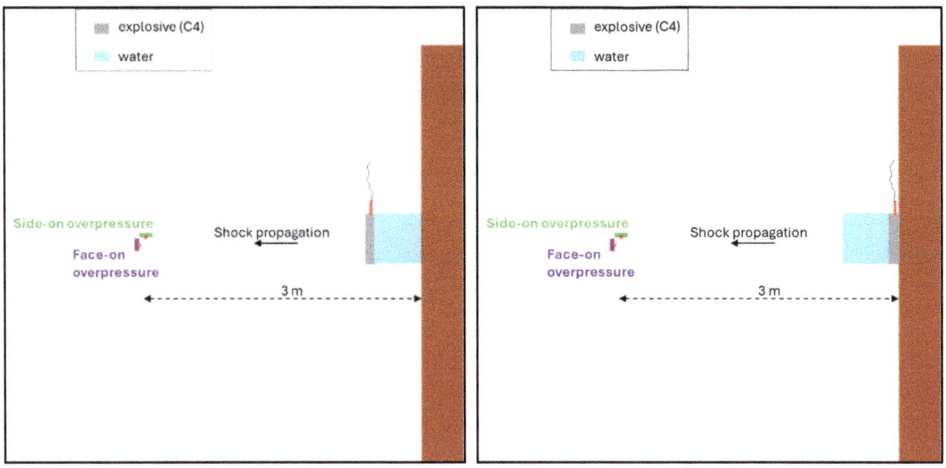

(a) (b)

Figure 3: Schematics of experimental setup with hybrid water charge. (a) Front water pack; and (b) Back water pack.

2.3 Ballistic pendulum

To evaluate the total loading generated by the various charges, we used ISL ballistic pendulum (Fig. 4), developed to assess heterogeneous blast effects [9]. The pendulum provides an evaluation of the total specific impulse (see eqn (1) below) transferred to its surface by measuring the maximum angle of deflection and using the principle of total energy conservation between the initial and the final position of the pendulum.

$$i_B = \frac{I_B}{A} = \frac{1}{A.b}\sqrt{2J_{Pz}\text{mgs}(1 - \cos\varphi_{\max})}. \tag{1}$$

Quantity	Value
Distance s	0.47 m
Distance b	1.25 m
Surface of the impact plate (load impact area) A	0.25 m²
Total pendulum mass m	155 kg
Moment of inertia at pivot point P J_{Pz}	88 kg m²
Gravitational acceleration g	9.81 m/s²

Figure 4: ISL Ballistic pendulum. *(Source: Sturtzer, 2008.)*

3 IMPULSE AND OVERPRESSURE

Table 2 presents ballistic pendulum results: maximum angle of deflection in degrees, corresponding impulse in bar.ms, and relative impulse in comparison to the bare charge (in %). Relative impulse is defined as the ratio between hybrid charge impulse and bare charge impulse.

Table 2: Ballistic pendulum results.

Charge reference	Water configuration	Angle (°)	Impulse (bar.ms)	Relative impulse (%)
2735	–/– (bare)	15	0.16	100
2728	1,000/– (front water)	23	0.24	153
2731	–/1,000 (back water)	37.5	0.39	246

We observe that the back water charge (C4 directly in contact with the pendulum plate) induces the highest increase of mechanical load on the steel plate with a 146% increase factor. This increase is almost three times higher than the one provided by the front water charge (seen in Fig. 5).

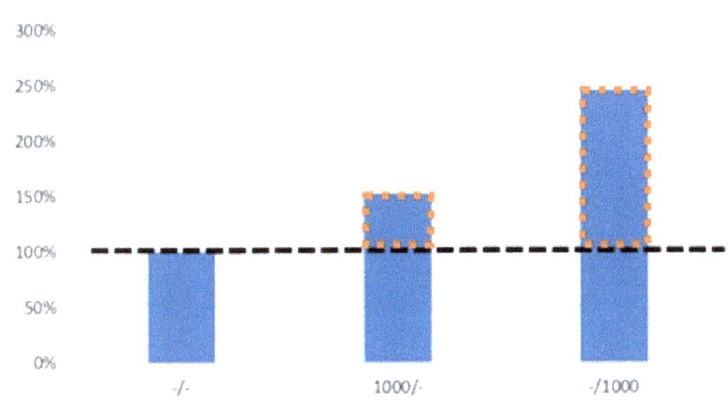

Figure 5: Relative impulses generated by explosive charges (the impulse increase provided by hybrid charges is represented in orange).

Table 3 reports pendulum results and experimental overpressures recorded at a 3 m distance by PCB sensors. P_i refers to incident overpressure measured by the side-on gauge, and P_r to reflected overpressure measured by the face-on gauge. Recorded times of arrival are respectively denoted ta_i and ta_r. For the sake of comparison, Kinney and Graham [10] (namely KG) theory values (incident overpressure, time of arrival and impulse) are presented for the bare charge in Table 4. Against all odds, pendulum impulse for the bare charge is equal to KG impulse obtained at a 3 m distance.

Table 3: Ballistic pendulum results in green, in comparison with pressure gauges measures at 3 m (in grey), and Kinney and Graham theory at 3 m for a bare TNT charge.

Charge reference	Water configuration	Impulse (bar.ms)	P_i (bar)	P_r (bar)	ta_i (ms)	ta_r (ms)
2735	–/– (bare)	0.16	0.2	0.5	5.15	5.16
2728	1,000/– (front water)	0.24	0.2	0.5	5.09	5.16
2731	–/1,000 (back water)	0.39	0.18	0.39	6.52	6.56

Table 4: Kinney and Graham (KG) theory blast parameters at 3 m for a bare TNT charge.

Charge reference	KG bare charge			
	Impulse (bar.ms)	P_i (bar)	P_r (bar)	ta_i (ms)
2735	0.16	0.22	0.49	5.7

Recorded incident and reflected overpressures as well as incident time of arrival are also in good correlation with KG theory with a difference of less than 10%. Hybrid charge with water in front provides same levels of overpressures at 3 m, meaning that influence of water is not noticeable at a reduced distance of 8 m/kg$^{1/3}$ for this charge. This tendency is slightly different when it comes to the hybrid charge with water in back, which leads to a 10% attenuation at 3 m for incident overpressure, a 22% attenuation for reflected overpressure (represented by grey dotted lines in Fig. 6), and a 27% attenuation for time of arrival.

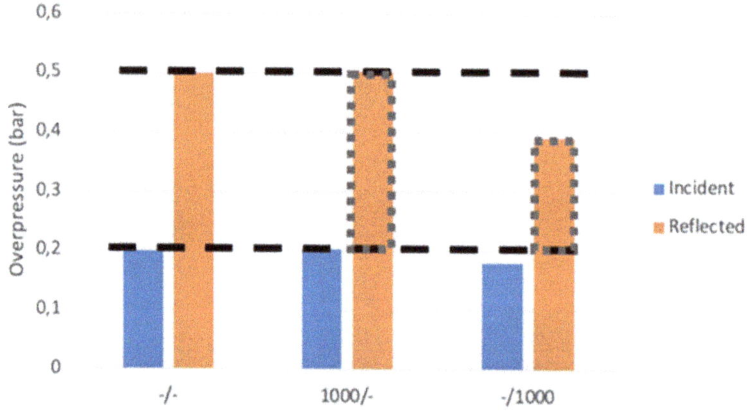

Figure 6: Peak overpressures (reflected in orange, incident in blue) at 3 m for 50 g C4 charges (the reflected overpressure attenuation provided by the hybrid back water charge is represented in grey).

4 HIGH-SPEED VISUALIZATION

Each test was filmed using a Phantom T3610 high-speed camera placed in the pendulum plate plane, at a 758 × 512 pixels resolution and 59,000 images per second.

Fig. 7 shows the detonation of the bare C4 charge. The pendulum plate is located on the right edge of the image. On the first image, the initial charge reaction is captured, with the intense light emission. On the second image, the fireball is developing, and the leading shock is progressively appearing on some part of the interface with fresh air. On the following images, the afterburning reactions intensity decreases, and the leading shocks separate from them, except for a visible central spike still propagating at a higher velocity, creating a conical shock wave. The remaining wave tends to become hemispherical.

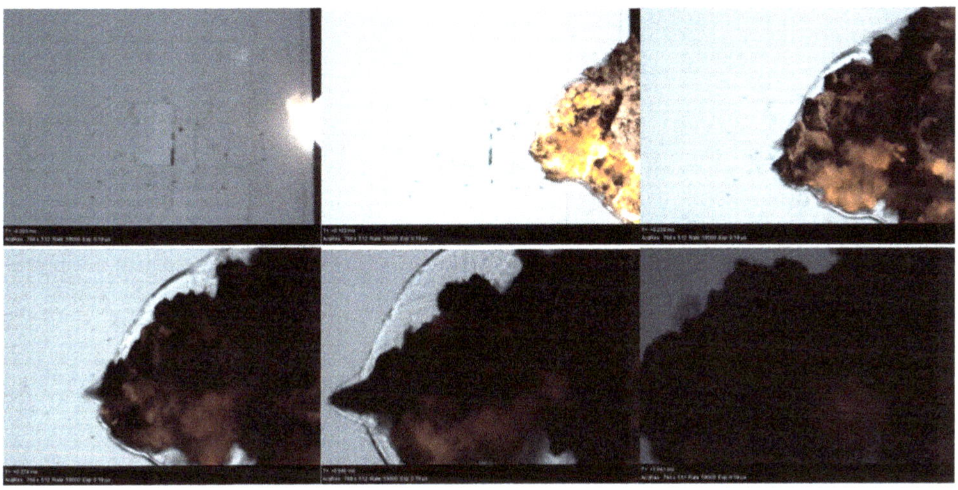

Figure 7: High-speed visualization of the 2735 bare charge (–/–).

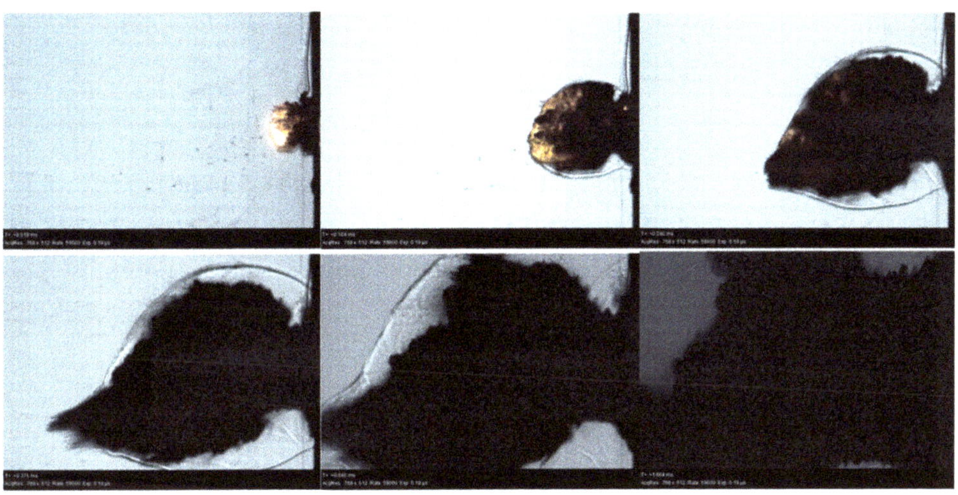

Figure 8: High-speed visualization of the 2728 charge with water in front (1,000/–).

In Fig. 8, the influence of front location of the water is visible: the fireball intensity is reduced in comparison with the bare charge as the afterburning reactions are quenched by the energy absorption of the dispersal and vaporization of water. From the third image, light emission is not visible anymore and the water volume jets out through the leading shock wave. The leading shock average propagation does not seem to be significantly affected by the presence of water, confirming the peak overpressure measurements at 3 m.

By setting the water volume behind the explosive charge (represented in Fig. 9), the propagation velocity of the leading shock is clearly slower as on the last image, it remains visible, unlike for the previous two tests. The water dispersion and vaporization decrease even further the intensity of afterburning reactions by blocking their access to fresh air and cooling them rapidly. The hemispherical shape of the leading is once again altered by a spike of projected water travelling at higher velocity.

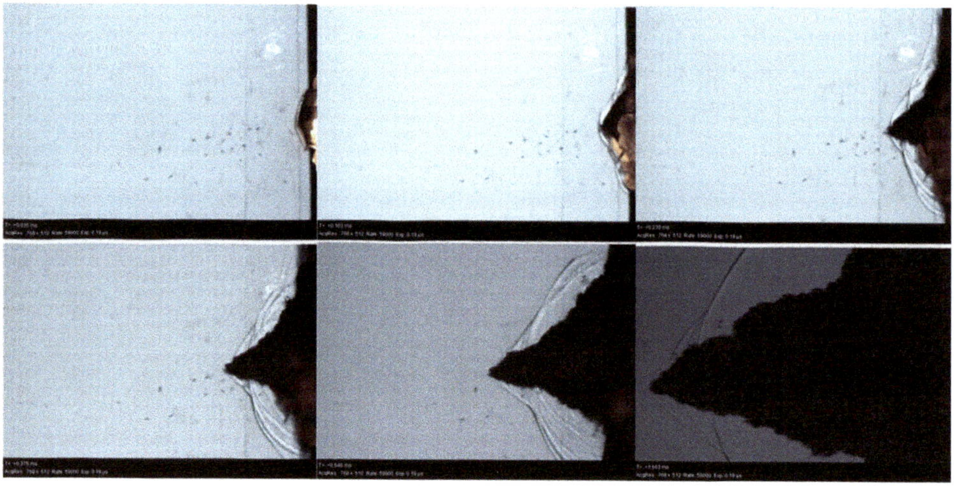

Figure 9: High-speed visualization of the 2731 charge with water in the back (–/1,000).

5 CONCLUSIONS

The influence of adding a volume of water to a reference explosive charge placed in contact to a metallic plate was assessed in this study. 1,000 g of water were placed in front and in the back of a 50 g C4 parallelepiped charge.

Total impulse and blast effects were assessed using a ballistic pendulum on which the explosive charge was fixed, free-field sensors facing the pendulum at 3 m and a lateral high-speed camera. Results showed a 53% and 146% increase of the total impulse transferred by the C4 charge by adding respectively the water in front of and behind it. Pressure measurements and high-speed images showed a significant decrease of the leading shock wave intensity for the water in the back and almost no alteration for the water in front.

Future work will include:

- the extension of the analysis of the peak overpressure field around the pendulum, using additional pressure sensor gauges with varying distances and azimuths, but also a larger field of view for the high-speed imaging, in order to record the shock velocity evolution with time and space, to compare both series of data;
- the further investigation of the ratio between the explosive and the water masses;

- the scalability assessment to larger explosive charges;
- a comparison with pressure profiles and high-speed images for underwater blast studies conducted in reduced-scale tanks (shown in Fig. 10 for instance).

Figure 10: High-speed visualization of an underwater shock wave propagating in a reduced-scale fishtank [11].

REFERENCES

[1] Absil, L. & Bryntse, A.N., Blast mitigation by water. FOI – Swedish Defense Research Agency Report FOI-R, 2049SE, 2006.
[2] Bauer, R.L., Pitzel, P., Johnson, E.M. & Johnson, C.E., Water-cased kicker charges for use in explosive demolition. *Buildings*, **13**(2), pp. 378–390, 2023.
[3] Baciu, C., Lupoae, M., Cotovanu, A., Constantin, D. & Craioveanu, M., Simulation and experimental tests of door breaching using multifunctional devices. *U.P.B. Sci. Bull., Series B*, **85**(2), pp. 199–214, 2023.
[4] Sedláček, M. & Palasiewicz, T., Unification of explosive charges in breaching for different door resistance classes. *Proceedings of the International Conference on Military Technologies*, pp. 1–5, 2023.
[5] Rus, S., Kaiter, E.H., Iftimie, B., Munteanu, C., Bursuc, A., Chiriac O.G. & Rus N., Creation of breaches, with the help of explosives, in metal structures under water and on the ground surface. *Technium Sustainability,* **2**(1), pp. 54–66, 2022.
[6] Moldovan, M., Lupoae, M., Baciu, C. & Constantin, D., Urbanism: Breaching in light obstacles. *Arhitectura. Constructii,* **6**(3), pp. 35–44, 2015.
[7] Li, Y., Yang, R., Wang, Y. & Fu, D., Influence of water coupling coefficient on the blasting effect of red sandstone specimens. *International Journal of Coal Science and Technology*, **11**(8), pp. 2–19, 2024.
[8] Cole, R.H., *Underwater Explosions*, 1st ed., Princeton University Press, 1948.
[9] Reck, B., Sturtzer, M.O., Allen, R. & Eckenfels, D., Ballistic pendulum for blast wave impulse measurement: Analysis and optimisation. Presented at *20th International Symposium on the Military Aspects of Blast and Shock*, Oslo, Norway, 2008.
[10] Kinney, G.F. & Graham, K.J., *Explosives Shocks in Air*, 2nd ed., Springer-Verlag: Berlin, 1985.

[11] Trélat, S. & Sturtzer, M.-O., Blast effects of an underwater explosion: A small-scale approach using high-speed imaging. Presented at *EUROPYRO International Conference*, Saint-Malo, France, 2023.

SECTION 2
SHOCK/BLAST WAVE DYNAMICS AND INTERACTION WITH STRUCTURES

IMPACT OF AN UNDERWATER EXPLOSION ON CRITICAL INFRASTRUCTURE: AN EXPERIMENTAL SMALL-SCALE INVESTIGATION

ORIANE DEGLETAGNE[1], SOPHIE TRÉLAT[1] & MICHEL-OLIVIER STURTZER[2]
[1]DEND, French Institute for Protection and Nuclear Security,
Department of Explosion and Weapons Effects, Fontenay-aux-Roses, France
[2]ISL, French German Research Institute of Saint-Louis,
Department of Explosive Protection, Saint-Louis, France

ABSTRACT

This study investigates the effects of underwater explosions on critical infrastructures, with a focus on structures commonly found in the nuclear sector. The research addresses the potential consequences of detonation (accidental or deliberate) on industrial pools and other containing facilities designed for watertightness, often through the use of thin stainless steel liners. The final goal is to characterize the mechanical load induced by consecutive pressure waves in close range to assess the damage to these immersed structures, focusing on both local deformations related to the overall structural response and the direct effects of the explosions, including the liner perforation. The investigation is conducted by the French Institute for Protection and Nuclear Security (DEND), also conducting experimental research into the aerial blast effects of explosions. The institute is developing scaled-down experiments to study shockwave propagation and interaction with reference shapes, such as parallelepiped and cylindrical. Each test is equipped with pressure sensors and high-speed cameras. The French German Research Institute of Saint-Louis (ISL) interest in underwater explosion lies in the study of the mechanisms ruling the transmission of shockwaves through water contained in stainless steel envelopes. Applications include protection of mobile platforms and infrastructures. Preliminary tests have been conducted in free field (without any obstacle near the explosion) inside a reduced-scale basin filled with water, subjected to shockwaves generated by small underwater high-explosive charges. These tests focused on the visualization of the primary shockwave and its spherical propagation in close range, as well as the behaviour of the gas bubble created by the explosion. This paper explores the dynamics of this gas bubble: shape, size evolution, oscillations and migration. Additionally, the study seeks to improve the knowledge of TNT equivalency in aquatic medium, providing insights into the potential hazards associated with underwater explosions and their impact on critical infrastructure.
Keywords: aquatic blast, high-speed imaging, explosive, underwater.

1 INTRODUCTION

The French Institute for Protection and Nuclear Security (DEND) is investigating the impact of aerial explosions, whether accidental or deliberate, on critical infrastructure and transportation systems, with the aim of safeguarding the nuclear industry. To support this objective, scaled experiments are being developed to study the propagation of air blasts and their interaction with target shapes, such as rectangular or cylindrical objects. Diagnostic tools, including pressure sensors and high-speed cameras, are used to collect data.

Since 2017, IRSN has been collaborating with the French German Research Institute of Saint-Louis (ISL) to establish guidelines for conducting reduced-scale explosive tests. In December 2022, the two institutes jointly filed a patent for a novel imaging system. This technology has also been applied to the study of underwater explosive events. Preliminary experiments were carried out at the ISL Explosive Test Range, where glass tanks filled with water were subjected to the pressure effects of a small high-explosive charge placed at the centre.

WIT Transactions on The Built Environment, Vol 215, © 2025 WIT Press
www.witpress.com, ISSN 1743-3509 (on-line)
doi:10.2495/SUSI250041

This paper focuses on capturing the primary shock wave produced by a small-scale underwater explosion and its interaction with a nearby cylindrical object. The goal of these experiments was to determine the feasibility of visualizing and measuring the detailed structure of underwater shock waves as they interact with a convex surface, particularly focusing on incident, reflected, and Mach waves. Objectives also include shockwave propagation investigation, development of small-scale testing protocols in a specifically designed basin, with future work planned in the form of a long-term research project. This involves characterizing wave propagation and the interaction of shock waves with the basin walls and understanding their impact on surrounding structures. Another key goal is to compare the effects of explosions at different depths and positions within the basin, evaluating the differences between seabed and water column detonations. Additionally, the study aims to characterize the TNT equivalency of these explosive charges, determining their explosive potential and evaluating their specific impact in terms of energy release.

Ultimately, the collected data will be used to model these risks and develop predicting tools, ensuring effective protection against potential explosive event. These objectives will be useful to improve the understanding of underwater explosion impacts in confined environments (in spent fuel pools for instance) and to enhance security measures for such scenarios.

2 LITERATURE REVIEW

As in the case of an aerial explosion, a condensed explosive underwater detonation is characterized by a large amount of energy release, a rapid emission of hot pressurized gas (burnt products bubble) and the subsequent transmission of a supersonic pressure wave into the surrounding environment [1].

2.1 Explosive shockwave characteristics in underwater explosions

A typical pressure signal recorded at a certain distance is presented in Fig. 1. When the wave arrives at the considered point, the initial ambient pressure PH (for hydrostatic pressure) undergoes a sharp quasi-instantaneous increase (crossing shockwave duration between 10 and 100 μs) until peak overpressure P_{max}. Pressure then exponentially decreases back to its ambient value. We must keep in mind that recorded pressure in water can reach levels significantly higher than in air (until a few tens of kilobars) for the same considered distances and charges. The underwater wavefront is then much more concentrated than in air, with shockwaves propagating approximately four times faster (about 1480 m/s in water compared to 340 m/s in air). Subaquatic shockwave models by Cole [1] and Swisdack [2] are commonly used to predict parameters such as peak pressure and impulse based on charge size and distance. These models are established for spherical TNT charges and imply peak pressure values ranging between 3.4 and 138 MPa (i.e. 34 and 1,380 bar).

Fig. 2 shows the Kinney and Graham [3] chart (in dotted lines), which predicts the maximum overpressures in air as a function of reduced distance for a spherical TNT charge, in addition to threshold overpressure values for pyrotechnic effects defined by French regulation in Legifrance [4]. The Cole chart, shown in solid lines for the same explosive on the same graph, indicates the maximum overpressures in water as a function of reduced distance. A significant difference in pressure levels is evident depending on the propagation medium. The same explosive charge generates overpressure over much greater distances in water compared to air. For instance, a 1 kg TNT charge produces an overpressure of only 10 bar in air at 1 m, while in water at the same distance, the overpressure remains at 524 bar.

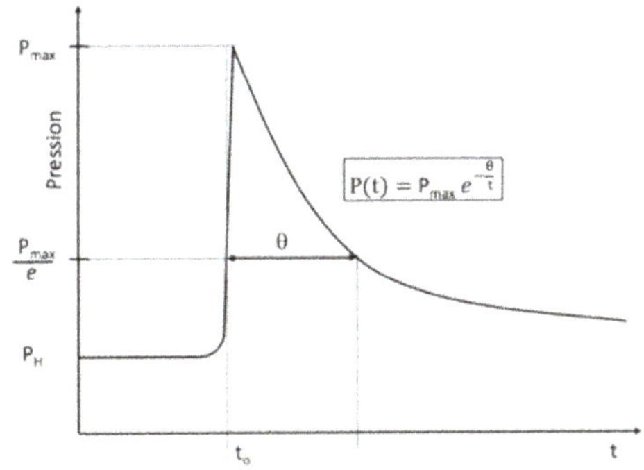

Figure 1: Pressure evolution as a function of time in water medium [1].

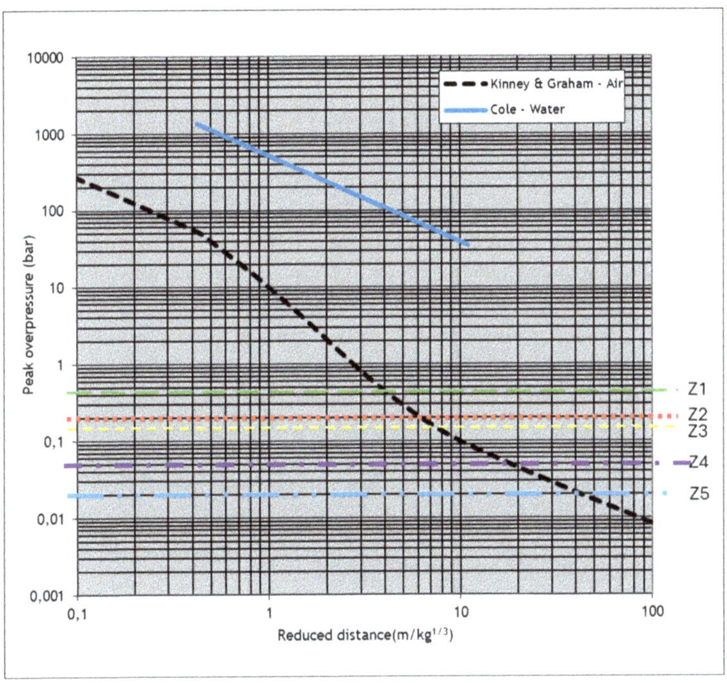

Figure 2: Peak overpressure evolution as a function of reduced distance in water and air medium [1]–[4].

As seen in Fig. 1, the characteristic time θ is defined as the time required for the pressure to fall to a value of $P_{max/e}$. Fig. 3 shows the Cole chart representing this reduced time constant $\theta/W^{1/3}$ where W is the TNT equivalent charge weight in kg, as a function of reduced distance for a spherical TNT charge. A critical parameter to the assessment of underwater blast effects is the distance separating the detonation point from the measuring point location (position of a potential target). Indeed, peak overpressure values rapidly decrease with the distance as shown in Fig. 2. The characteristic time θ also progressively increases with the distance as represented in Fig. 3.

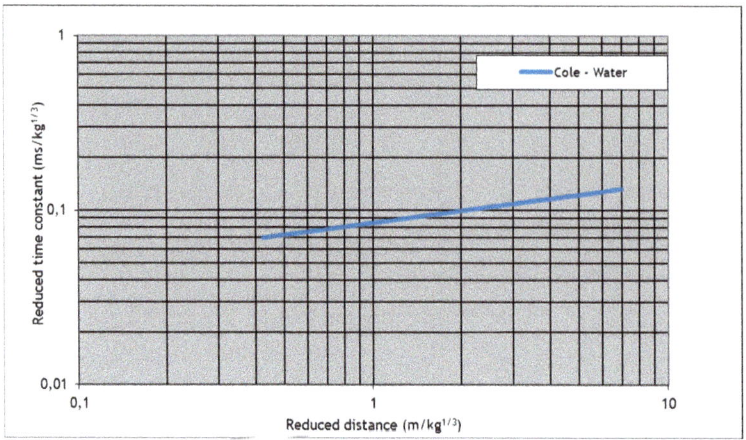

Figure 3: Reduced time constant evolution as a function of reduced distance in water medium [1].

The zones of effects defined in the decree of 2007 [4] setting the rules relating to risk assessment and accident prevention in pyrotechnic establishments and represented as horizontal lines Z_i ($i \in [1;5]$) in the graph of Fig. 2, are specifically designed to assess the risks linked to pyrotechnic activities in the aerial medium. These zones of effects are determined based on various criteria, such as the power and type of pyrotechnic material, to delimit the zones where dangerous effects can occur (for example, detonations, flying debris, bursts, flames, etc.). However, these areas of effect are not supposed to be identical in water. Indeed, the propagation parameters of pyrotechnic effects in an aquatic environment significantly differ from those in the air. Underwater phenomena such as sound propagation, shock waves, and flying debris are influenced by water density, depth, salinity, and other environmental factors (water volume, nature of seabed material, etc.). Therefore, risk assessment and definition of areas of effects underwater must consider these specific differences, such as, for instance, gas bubble dynamics which are qualitatively described in the section hereafter.

2.2 Gas bubble dynamics

After its initial expansion, the burnt products bubble starts to oscillate around the local hydrostatic pressure, while progressively travelling to the surface, as represented in Fig. 4. The end of each new bubble compression transmits a new pressure wave (secondary, then tertiary, etc.) to the ambient medium. This pulsating behaviour is due to the pressure difference between the burnt gases inside the bubble and the surrounding water. The size of the bubble and its initial expansion rate depend on the energy released by the explosion and the depth of detonation.

2.3 Shockwave interaction with target

The mechanical load, consecutive to an underwater explosion, applied to a specific point of an immerged target can be decomposed into three types of waves as illustrated in Fig. 5:

- the incident (or direct) wave that impacts the target with a similar profile as in Fig. 1;

- the compression waves reflected by high impedance surfaces (seabed, subaquatic obstacle, etc.)
- the expansion wave resulting from the reflection on the air interface (low impedance medium).

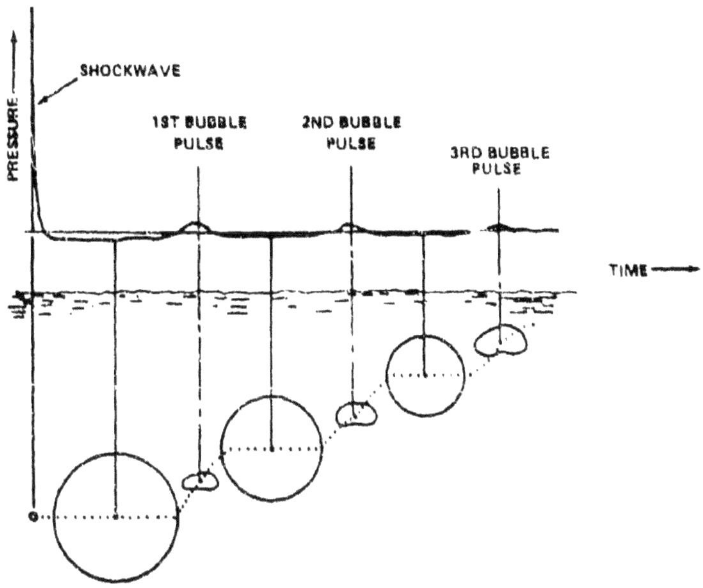

Figure 4: Pressure wave and bubble phenomena of underwater explosion. The upper part shows a pressure–time plot, the lower, the positions and size of the bubble for specific moments which correspond to the curve above as indicated by the vertical lines. *(Source: According to Swisdak [2].)*

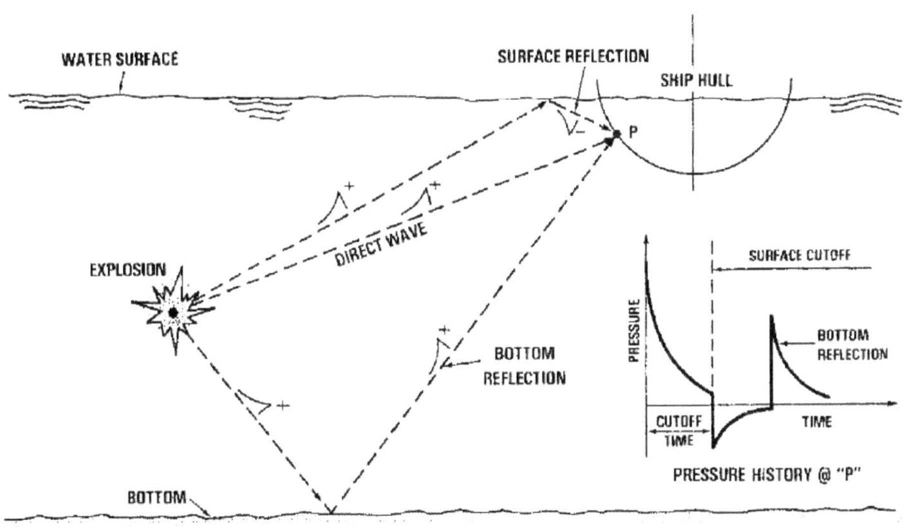

Figure 5: Interaction of underwater explosion waves with a ship hull [5].

In this work, we choose to evaluate underwater explosion effects in free field and close to the tank walls and bed by performing reduced scale experiments.

The experimental studies are conducted in small-scale facilities which allow enhanced monitoring by combining different diagnostic methods and providing the ability to better control and to vary the conditions for shock wave generation and the interaction with various geometrical configurations.

3 EXPERIMENTAL SETUPS

3.1 Underwater blast tanks

A 310.5 litre water steel blasting tank (100 × 60 × 57.5 cm) equipped with polycarbonate windows (presented in Fig. 6) was therefore developed. This set up filled with a mean water height of 54 cm is used to characterize free field overpressures by means of pressure gauge measurements (free field means without any obstacle placed under water, except from tank walls (Fig. 6).

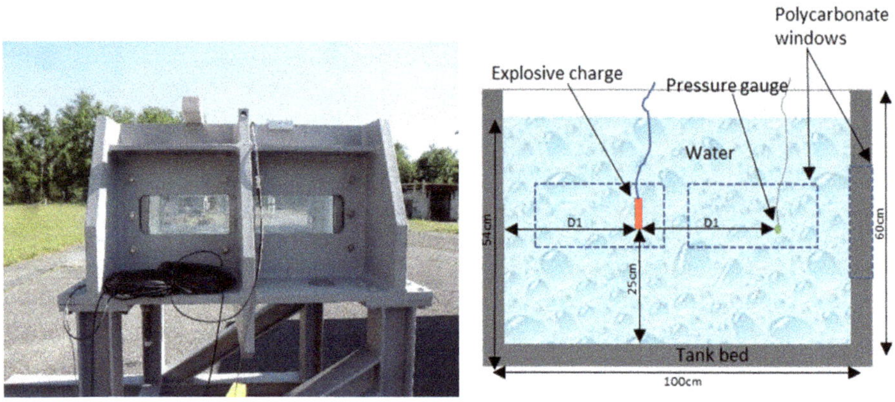

Figure 6: 310.5 litre steel blasting tank.

3.2 Experimental conditions

During the experimental period, a series of 20 underwater explosions using four condensed explosive charges of TNT-equivalent weights ranging from 0.598 g to 10 g (illustrated in Fig. 7) were detonated.

Some experiments were reproduced several times to assess repeatability. The largest charge consists of a plastic explosive bullet (SEMTEX made of RDX and PETN, seen in Fig. 7(d)) stacked together to a bridgewire detonator (namely, a RP501 detonator from Teledyne Defense containing 227 mg of HMX and 136 mg of PETN, see Fig. 7(a)). All the explosive (SEMTEX, RDX, HMX and PETN) weights have been converted into TNT-equivalent weights by applying a TNT-equivalent ratio comprised between 1.45 and 1.72. These ratios, calculated using detonation velocities (according to Cooper methodology [6]), are consistent with the fact that a TNT-equivalent ratio for underwater conditions is higher than for air conditions, as suggested in Cole [1]. However, due to uncertainties on the TNT equivalency for the reduced explosive charges, a ± 10% uncertainty factor is to be considered on the weight of the TNT-equivalent charges.

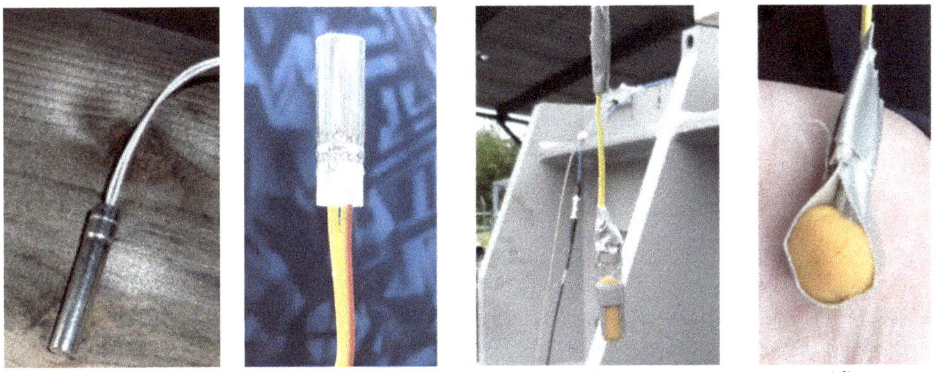

(a) (b) (c) (d)

Figure 7: Explosive charges. (a) RP83; (b) RP501; (c) RP501 and Semtex (5 g eq. TNT); and (d) RP501 and Semtex (10 g eq. TNT).

The blast wave generated by the detonation of the explosive charge produces a pressure wave measured with piezoelectric sensors in the steel tank. Piezoelectric sensors are PCB M 102A, 102B104 and Neptune Sonar T11 sensors (respectively connected to PCB 482C and 5018 charge amplifiers). Data are recorded and filtered (5th order Bessel filter, 500 kHz) by a 2 MHz Transcom Recorder.

Detonation tests characteristics, including the charge weight, the range distance, the charge depth and the water depth, are presented in Table 1.

Table 1: Characteristics of the explosive tests in steel tank.

Explosive charge	Weight W in TNT eq.[a] (g)	Range D_1 (m)	Test iterations	Water depth (m)	Charge depth H (m)
RP501	0.598	0.35	5	0.54	0.29
RP83	1.89	0.35	3		
RP501 (inversed)	0.598	0.35	1		
RP501	0.598	0.35	5		
RP83	1.89	0.40	1		
RP501	0.598	0.45	2		
RP83	1.89	0.45	1		
RP501 + Semtex	5	0.45	1		
RP501 + Semtex	10	0.60	1		

4 EXPERIMENTAL RESULTS

4.1 Free field results

Fig. 8 presents a comparison of maximal overpressure results from underwater explosions obtained with both sensor technologies (Neptune Sonar, denoted as NS, in Fig. 8(a), and PCB in Fig. 8(b)) with overpressures predicted by Cole (thick line) as a function of the charge weight W (kg) and as a function of the scaled range $D_1/W^{1/3}$ (m/kg$^{1/3}$) where D_1 is the distance from the explosion. Symbols represent experimental data, each point being the mean value

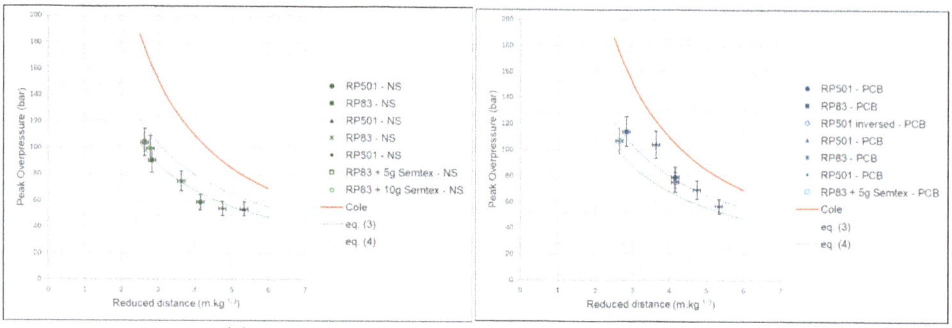

(a) (b)

Figure 8: Variation of the peak overpressure recorded by the various pressure transducers as a function of the scaled range for all explosive charges, detonated in the steel tank.

of a series of tests in the case they have been reproduced. Vertical error bars are based on uncertainties on the maximal peak pressure of ± 10% due to the sensor sensitivity and to the digitizer resolution, and on the standard deviation of the repeated series, which is defined as:

$$\sigma = \sqrt{E[(X - E(X))^2]}, \quad (1)$$

where X represents a series of experimental values.

Horizontal uncertainties on the scaled range are due to the misestimation of ± 10% of the TNT equivalency of the charge as mentioned earlier and to the charge positioning error of ± 0.01 m.

From Fig. 8, it can be observed that the peak pressure decreases with the scaled distance. Moreover, whatever the type of pressure gauge, measurements do not fit the curve for the peak pressure P_{max} (in bar) given by the empirical law (eqn (2)) from Cole reported in the literature (e.g., [1] and [2]) and that applies for shallow charges in deep water:

$$P_{max} = 524\,(D_1/W^{1/3})^{-1.13}, \quad (2)$$

where D_1 is the distance (in m) from the explosion, and W is the charge weight (in kg TNT-equivalent). Instead, for instance, for the PCB gauges, measurements rather follow the curve resulting from the power law fit:

$$P_{max} = 278\,(D_1/W^{1/3})^{-0.9}. \quad (3)$$

For the NS gauges, measurements are well described by the curve resulting from the following law:

$$P_{max} = 233.1\,(D_1/W^{1/3})^{-0.89}. \quad (4)$$

Eqns (3) and (4) significantly differ from eqn (2). In contrast to the empirical law that is defined for an ‘open’ (deep water) environment, eqns (3) and (4) implicitly account for the environment characteristics (essentially, the volume, the small scale and the impedance contrast between water and the tankbed and/or water–air interface). It could be interesting in the future to see whether the location of the detonation (namely, on the tankbed or close to the water surface) does or not affect the peak pressure.

4.2 Bubble dynamics results

Fig. 9 shows different images of the burnt gas bubble for a RP501 placed at 0.35 m from a PCB sensor in free field, in correlation with recorded pressure–time evolution We observe that, after the detonation, the bubble forms itself and expands (Figs 9(a) and 9(b)) until reaching its maximum radius at 14 ms (Fig. 9(c)); it consequently contracts downwards (Figs 9(d) and 9(e)) before collapsing (Fig. 9(f)).

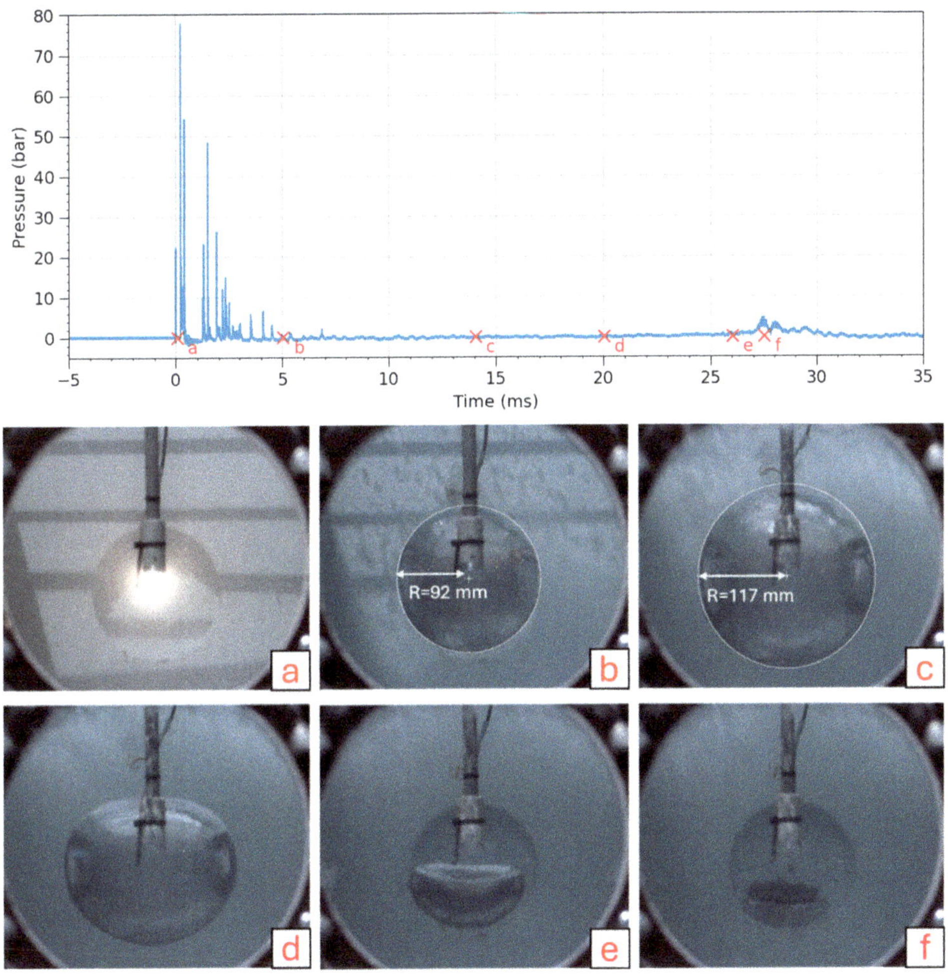

Figure 9: Burnt gas bubble characteristic times and bubble radius *R* evolution (in mm).

At 27.5 ms (Fig. 9(f)), the sensor records an overpressure of approximately 4.5 bar, whereas no significant pressure peak is detected at Figs 9(c), 9(d) and 9(e). This suggests that the overpressure measured at instant f corresponds to the shock wave generated by the bubble collapse.

The empirical law (eqn (5)) from Cole estimates the first pseudo-period T_p (in ms) (as a function of W, the explosive weight in kg, and H, the explosion depth in m) of the bubble (first reaching a minimum size of the bubble) to be 25.5 ms, which aligns quite well with our recordings of the pressure–time signal and the high-speed videos reported in Fig. 9.

$$T_p = 2.11\ (W^{1/3}/(H+10)^{5/6}). \tag{5}$$

To measure the evolution of the bubble radius over time, the images are post-processed to enhance bubble edges using the MorphoLibJ library [12], followed by a circular Hough Transform analysis in Fiji [13]. This approach provides useful data to describe the bubble during most of its expansion phase, except at the very beginning due to processing issues (as the bubble is quite dark, and when cavitation clouds the window). However, it does not convey the potential asymmetry of the bubble and becomes totally irrelevant during the contraction phase, as the bubble shape transitions from circular to a more jellyfish-like form. To assess the distortion of the bubble, some additional manual measurements of horizontal and vertical radius of the bubble, respectively R_x and R_y in Fig.10, were done. It showcases that even during the expansion phase the bubble tends to grow longer than larger which could be due to interactions with tank walls. The maximum radius $R_{\max}$ for this test is estimated to be 117 mm (seen in Fig. 9) which fits well Cole empirical law (136 mm from (eqn (6)) within a 14% difference and considering an approximate uncertainty of 20%.

$$R_{\max} = 3.5\ (W/H+10)^{1/3}. \tag{6}$$

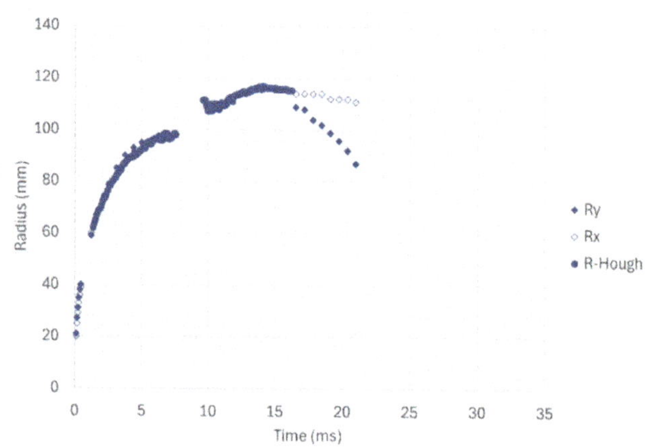

Figure 10: Burnt gas bubble radius over time.

5 CONCLUSION

The study conducted by DEND highlights the importance of understanding the effects of underwater explosions on critical infrastructure. Through reduced-scale experiments, the research focuses on capturing and analysing shockwave dynamics generated by small-scale underwater detonations and their interaction with obstacles. The results reveal valuable insights into the propagation of underwater shockwaves and the behaviour of the burnt gas volume as it collapses on itself. The data collected from pressure measurements and high-speed imaging has provided a clearer understanding of shockwave characteristics, such as peak pressure and wave propagation (Fig. 10), as well as the impact of detonation distance and charge weight. The findings underscore the significance of environmental factors, such as the volume of the testing area and impedance contrast, which influence the shockwave's behaviour compared to open-water conditions. Moreover, evolution of detonation gas bubble has been observed in detail [11], offering insights into the dynamic response of submerged structures. These experiments form the foundation for developing predictive models and enhancing the design of infrastructure to withstand underwater blast impacts. The next phase of the project will involve testing additional diagnostic methods and comparing experimental results with numerical simulations to further refine these models. This work will contribute to improving safety measures for nuclear infrastructure (such as spent fuel pools notably) exposed to underwater explosive threats.

REFERENCES

[1] Cole, R.H., *Underwater Explosions*, 1st ed., Princeton University Press, 1948.

[2] Swisdak, M., Explosion effects and properties. Part II. Explosion effects in water. Report No. NSWC/WOL/TR-76-116, Defense Technical Information Center, 1978. https://doi.org/10.21236/ada056694.

[3] Kinney, G.F. & Graham, K.J., *Explosives Shocks in Air*, 2nd ed., Springer-Verlag: Berlin, 1985.

[4] Legifrance, Arrêté du 20 avril 2007 fixant les règles relatives à l'évaluation des risques et à la prévention des accidents dans les établissements pyrotechniques. Article 11, 2007. https://www.legifrance.gouv.fr/loda/id/LEGISCTA000006111540.

[5] Costanzo, F.A., Underwater explosion phenomena and shock physics. *Structural Dynamics*, **28**(3), pp. 917–938, 2011.

[6] Cooper, P.W., *Explosives Engineering*, Wiley-VCH: New York, 1996.

[7] Ben-Dor, G., *Shock Wave Reflection Phenomena*, 2nd ed., Springer, 1991.

[8] Gavart, R., Mesure des combinaisons, focalisation et diffraction d'ondes de choc autour de structures type. PhD thesis, University of Orléans, France, 2023.

[9] Kaca, J., An interferometric investigation of the diffraction of a planar shock wave over a semicircular cylinder. Technical Note No. 269, Institute for Aerospace Studies, University of Toronto, 1988.

[10] Zółtak, J. & Drikakis, D., Hybrid upwind methods for the simulation of unsteady shock-wave diffraction over a cylinder. *Computer Methods in Applied Mechanics and Engineering*, **162**(1–4), pp. 165–185, 1998.

[11] Trélat, S. & Sturtzer, M.-O., Blast effects of an underwater explosion: a small-scale approach using high-speed imaging. Presented at *EUROPYRO International Conference*, Saint-Malo, France, 2023.

[12] Legland, D., Arganda-Carreras, I. & Andrey, P., MorphoLibJ: Integrated library and plugins for mathematical morphology with ImageJ. *Bioinformatics*, **32**(22), pp. 3532–3534, 2016. https://doi.org/10.1093/bioinformatics/btw413.

[13] Schindelin, J. et al., Fiji: An open-source platform for biological-image analysis. *Nat. Methods*, **9**, pp. 676–682, 2012. https://doi.org/10.1038/nmeth.2019.

URBEX: A VALIDATED, FAST-RUNNING CODE TO COMPUTE BLAST EFFECTS IN URBAN CONFIGURATIONS

MAROUA HEDHLI[1], EMMANUEL LAPÉBIE[1], ERNHOLLA MARINASY[1] & ISABELLE SOCHET[2]
[1]Apex Solutions, France
[2]PRISME Laboratory, France

ABSTRACT

This paper presents the URBEX fast-running code, one of the outputs of the research project URB(EX)[3], co-funded by the French National Research Agency (grant #ANR-21-CE39-0016). The URBEX code is able to compute all urban effects following an explosion in urban configurations (multiple diffractions, regular and Mach reflections, channelling and urban canopy bypassing) in less than 2 minutes on a standard laptop, without the need of specific modelling knowledge so it could be used by first responders in emergency situations. Urban configurations are retrieved from widely available geographic data (OpenStreetMap, the IGN Institute in France or other sources). The URBEX code relies on a breakthrough approach, far above the international state-of-the-art, and its underlying mathematical method has been patented. This method will be detailed and discussed in the paper, together with its current limitations and the planned future improvements. A number of comparisons with high-quality, small-scale experiments done in the PRISME laboratory will also be presented, as well as with results from numerical codes and forensic analysis of past bombings. Finally, current actions aimed to extend the URBEX method will be outlined, in particular the development of fast-running codes for explosions in complex buildings and in tunnel networks (MALBEC project, cofounded by the French Innovation Defense Lab).
Keywords: fast-running code, explosion, blast effects, urban configurations.

1 INTRODUCTION

Explosion risks, whether accidental or intentional, pose a major threat to people and property in various environments, including industrial, public, and military areas. Past disasters like those in Seveso (Italy) 1976, Toulouse (France) 2001, and Beirut (Lebanon) 2020 highlight the severe consequences. Understanding explosion dynamics is essential for developing a rigorous fast-running model to simulate explosions, particularly in urban and confined spaces, where shock wave reflections and diffractions add complexity to the analysis.

Currently, only costly three-dimensional computational fluid dynamics (CFD) methods can accurately predict blast loads. Tools like BlastFoam [1] and Viper::Blast [2] have shown potential in modelling complex blast scenarios. Unfortunately, these techniques are not suitable for certain applications where a short execution time is required. Based on the authors' knowledge, limited efforts have been made to develop fast-running models. One promising approach involves the use of neural networks [3], [4]. This method requires a significant amount of reliable data to train the network, and its ability to generalize to other configurations has yet to be fully established. Another, more relevant method is the use of composite approaches. In fact, tools like 3D Blast from ARA (Applied Research Associates) [5], provide simplified solutions for blast load calculations on isolated buildings. Fi-Blast [6], [7] and VAPO [8] offer more comprehensive approaches, although no real configuration data has been published. In France, the FLASH code [9]–[12], developed at CEA/DAM, is the most advanced, using a shortest path method, diffraction model, and discretized volume approach for channelling. The URB(EX)[3] research project, including a detailed review of

doi:10.2495/SUSI250051

existing approaches, URBEX code requirements, and comparison of results, was presented in Chereau et al. [13].

This study aims to develop an innovative, fast-running, meshless model for evaluating blast consequences in urban environments. The proposed modelling approach utilizes advanced computational methods to simulate blast effects efficiently. The performance of the new model is compared with experimental data, numerical simulations, and real-world event data to assess its accuracy and reliability.

2 PRINCIPLE OF ANALYTICAL CODE

2.1 A new approach to presenting the problem

The URBEX model is based on a spatial vision of the propagation of the air shock wave (Fig. 1(a)), rather than temporal evolution as in numerical approaches (Fig. 1(b)). It describes the spatial domains of a succession of unit waves generated by various physical phenomena (diffractions, regular and Mach reflections, channelling and urban canopy bypassing), all resulting from the initial wave emitted at the explosion point. The characteristics of each unit wave in its spatial domain are known analytically based on a functional approach, no meshing being required. The model developed is very innovative and has never been described in the literature (the patent [14] has already been filed).

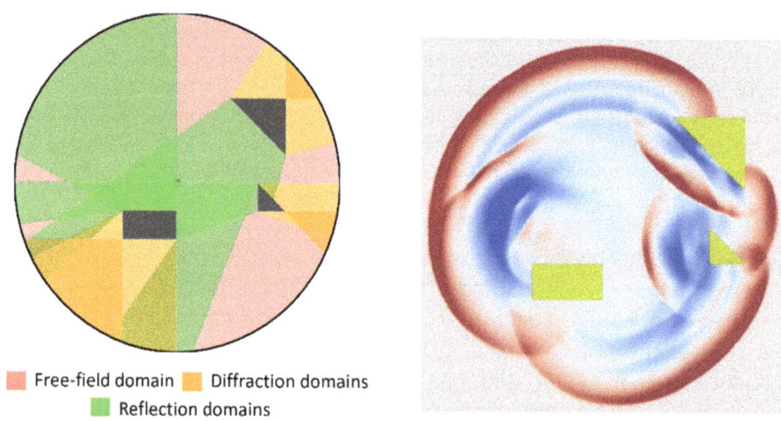

(a) (b)

Figure 1: Comparison of spatial vs temporal evolution of a wave. (a) Spatial domains of the URBEX Model, Level 1; and (b) Temporal evolution of a wave Viper::Blast.

The model describes a tree of unitary waves, where level 0 corresponds to the initial 'free-field' wave, and level $i + 1$ includes the unitary waves resulting from the reflections and diffractions of the waves from generation i. Each wave of level i can be treated as a new initial wave, which in turn generates diffracted and reflected waves of level $i + 1$, and so on. This recursive approach is depicted in Fig. 2. Each unitary wave of level i is described by:

- The reference to its parent wave of level $i - 1$.
- An influence zone (visibility polygon, apex, angular range, supporting segment).
- An analytical function that provides the various parameters of the Friedlander function.

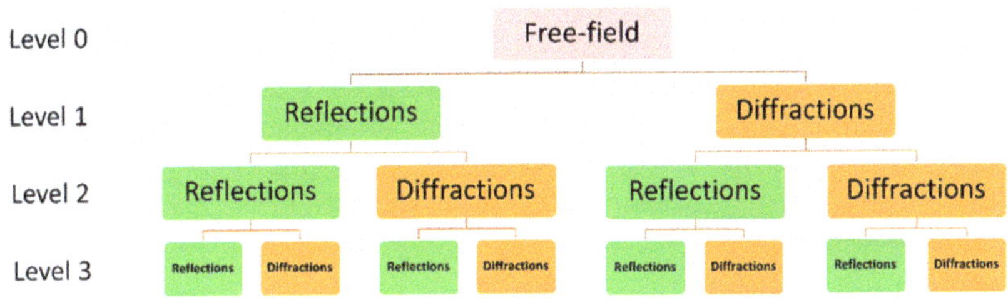

Figure 2: Recursive tree structure of unitary waves.

For each point of interest, all relevant zones it belongs to are identified. Then, the contribution of each wave is calculated by tracing back through the recursion tree to the direct wave.

Using a simple example with a charge and three obstacles, Fig. 1 shows the visibility polygon of the free-field wave in pink. This is level 0 in the tree structure. From this parent wave, we have six diffractions, represented by six diffraction visibility polygons in orange, and three reflections, represented by three reflection visibility polygons. These diffractions and reflections represent level 1 in the tree structure.

2.2 Unit waves superposition

It is assumed that at each point in the domain, the received pressure signal is the sum of unit signals. Fig. 3 demonstrates a simple example of unit wave superposition. The free-field wave (blue path) of level 0 produces a reflection (orange path) and a diffraction (green path) wave of level 1 at the point of interest. The final signal at this point consists of three peaks: the first corresponds to the free-field wave, the second to the reflected wave, and the third to the diffracted wave. The strength of our new approach lies in calculating N waves instead of just calculating the wave that passes through the shortest path [9]–[12].

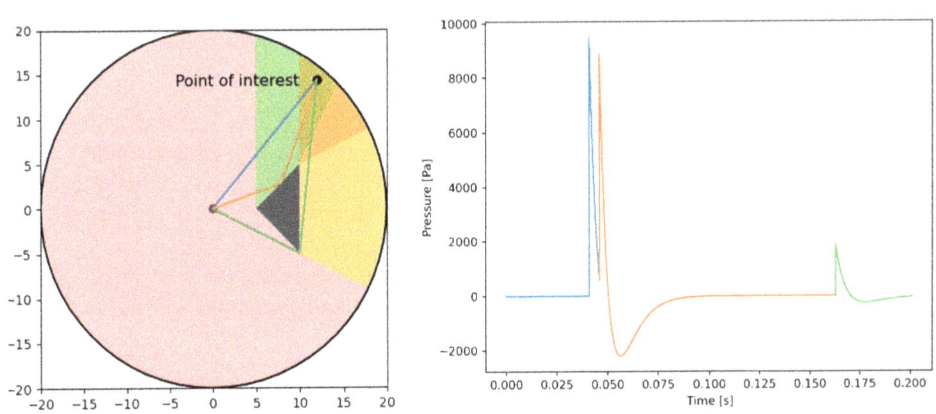

(a) (b)

Figure 3: Principle of unit waves superposition. (a) Simple case, different paths to reach the point of interest. Level 1; and (b) Sum of unit signals calculated for each path.

2.3 URBEX model: Free-field processing

In this study, new fits are developed as several limitations in the previous approaches are identified [15]. The old fits were often redundant, sometimes inconsistent, and lacked physical coherence. Furthermore, they required numerous coefficients, which led to frequent errors, and some were non-invertible, making the calculation of the reduced distance more complex. It may be recalled that the reduced distance is defined as follows:

$$Z = \frac{R}{\sqrt[3]{m}}, \tag{1}$$

where m is the explosive hemispherical mass and R is the true distance in metres.

To solve these issues, a new approach was adopted. Redundancies were eliminated and a lower bound for reduced distance was set. Additionally, reversible fits were chosen, allowing for a more accurate and consistent calculation of the reduced distance.

2.3.1 Lower bound for reduced distance

Z can be related to the non-dimensional radius ratio $\frac{R}{R_0}$, R_0 being the spherical charge radius, as follows:

$$Z = \frac{R}{\sqrt[3]{m}} = \frac{R}{\sqrt[3]{\frac{4}{3}\pi R_0{}^3.\rho_0}} = \frac{1}{\sqrt[3]{\frac{4}{3}\pi.\rho_0}}\frac{R}{R_0}, \tag{2}$$

where ρ_0 is the high explosive density.

It should be noted that the fireball radius R_{fireball} is 25 to 35 times the charge radius R_0 for most solid high-explosives. A commonly agreed value for TNT is $30.R_0$, which corresponds to $Z \sim 1.6$. Inside the fireball, phenomena strongly differ from free-field shock propagation, and we consider that it is useless to try modelling waveform evolution below this threshold.

Moreover, $Z = 3$ (for TNT) is considered to be the upper threshold for which blast consequences are extremely severe, both for humans and structures. It is thus not important to be precise below $Z = 3$ since everything will be destroyed/everybody will be killed.

2.3.2 URBEX free-field fits for TNT

We aim to obtain the evolution of the different parameters of the Friedlander as a function of the reduced distance for each unit wave. Indeed, regardless of the phenomenon being studied (diffraction or reflection), it is reduced to a free-field equivalence. We will then recalculate the reduced distance differently using the appropriate function and then use our free-field fits to find the different parameters.

Instead of providing a model for the maximum overpressure ΔP_+, the equivalent option of modelling the shock Mach number M (wave speed divided by sound speed) was chosen instead. Both variables are related through:

$$\frac{\Delta P_+}{P_0} = \frac{2\gamma}{\gamma+1}\,(M^2 - 1), \tag{3}$$

with γ the polytropic coefficient (ratio of specific heats) and P_0 the atmospheric pressure.

The mathematical expression proposed for reduced Mach variation is:

$$M^2(Z) = \left(a_0 + a_1 \frac{Z}{Z_0} + a_2 \left(\frac{Z}{Z_0}\right)^2\right)^{-\frac{1}{n}} + 1, \tag{4}$$

with $Z_0 = 1 \frac{\text{m}}{\sqrt[3]{\text{kg}}}$ to have non-dimensional coefficients.

Since overpressure and Mach number are related, knowing the evolution of Mach with Z, the expression for the reduced overpressure can then be derived; this is given by the following formula:

$$\Delta P_+(Z) = P_0 \times \frac{2\gamma}{\gamma+1} \times (M^2(Z) - 1). \tag{5}$$

The expression of the reduced time of arrival is as follows:

$$\overline{t_a}\,(Z) = \overline{t_{a,\text{ref}}}\,(Z_{\text{ref}}) + \int_{Z_{\text{ref}}}^{Z} \frac{dZ}{a_0 . M(Z)}, \tag{6}$$

in which M is the shock Mach number, a_0 is the speed of sound and $\overline{t_{a,ref}}$ is the time of arrival at the reference reduced distance Z_{ref}.

A fit for the reduced positive phase duration $\overline{\Delta t_+}$ is also needed; the following was chosen:

$$\overline{\Delta t_+} = b_0 + b_k \left(\frac{Z - Z_{\text{min}}}{Z_0}\right)^k. \tag{7}$$

The impulse I_+ can be computed from overpressure ΔP_+, positive phase duration Δt_+ and exponential factor α. The corresponding equation is:

$$I_+ = \frac{\Delta P_+ \Delta t_+}{\alpha^2}(e^{-\alpha} + \alpha - 1). \tag{8}$$

2.4 Functional approach

The core reasoning of the URBEX model is based on determining the visibility polygons for each unit wave and on the fact that within each visibility polygon, pressure and impulse can be calculated analytically, with no meshing required. Each unit wave is associated with a function that computes, across the entire spatial domain, the different Friedlander parameters as a function of the reduced distance.

2.4.1 Regular reflections

The initial wave (level 0) propagates directly to every point in the study area that is visible from the initial explosion point. The corresponding signal is described at any point in this zone by the free-field functions presented previously. Each reflective surface generates a reflective unit wave. This unit wave seems to be emitted from a virtual source, see Fig. 4, symmetrical to the initial point in relation to the segment. It propagates within an angular sector defined by the segment. The corresponding signal, at any point in this area, is described by the free field functions $\Delta P_+(Z)$ derived from the virtual source.

As shown in Fig. 4(b), at the point of interest, the signal exhibits two main peaks: one peak corresponds to the wave arriving directly, and the other to the wave that will reflect and come from a virtual source. The second peak arrives slightly later, as it travels a longer path.

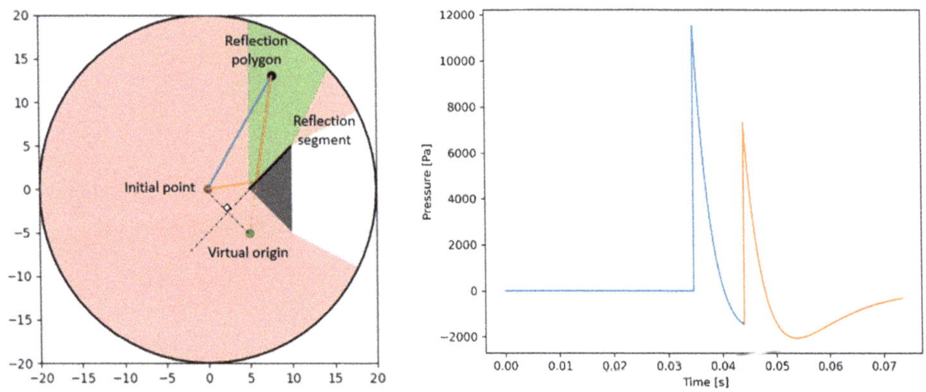

(a) (b)

Figure 4: Principle of the mirror-image approach. (a) Path of the direct wave (in blue) and path of the reflected wave (in orange); and (b) Signal with two main peaks.

The mirror-image approach for handling regular reflections is strictly valid for sound waves. In the case of intense shock waves, it is important to quantify the error introduced by the sonic approximation, especially with regard to the reflection coefficient at the reflective surface, which is currently set to 2.0. A correction factor must be applied to the overpressure in the region affected by the regular reflected wave (excluding the Mach domain). In this paper, we only present the simple version of URBEX, which does not include the new model of regular reflection.

2.4.2 Diffractions approach

As shown in Fig. 5, each corner of a building meeting certain criteria generates a diffracted wave. This wave propagates from the diffraction apex *A*, in an angular sector (diffraction polygon) limited by a building surface (reference segment) and a segment of the free-field zone which we call line of sight. The corresponding signal is described at every point of this diffraction polygon by the Bazhenova et al. [16] functions at the apex and by the free-field decay when moving away from the apex.

The approaches of Whitham [17], [18] (theoretical relationship between the section of a shock tube and the Mach number) and Bazhenova et al. [16] (empirical, based on the

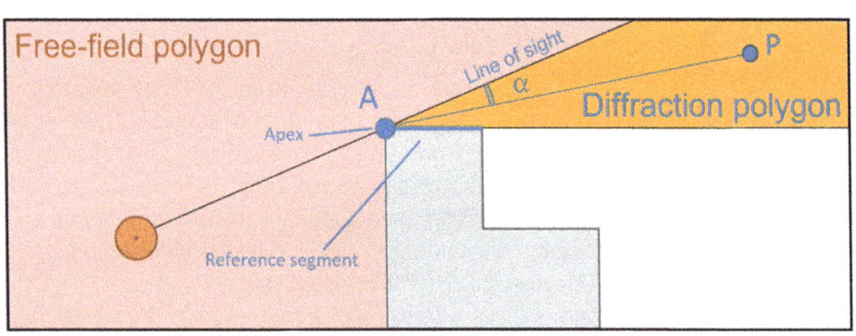

Figure 5: Simple diffraction model.

experimental results of skews [19], [20]) relate the Mach number after diffraction to the Mach number before diffraction and the diffraction angle. This approach is similar to the concept of a 'singular pressure drop' in incompressible fluid mechanics. As shown in Fig. 6, the diffraction point (apex) causes a sudden drop in pressure (or Mach number), except in the region near the apex, where the behaviour becomes more complex. This phenomenon can also be modelled by an instantaneous increase in the reduced distance Z, referencing the free-field curves.

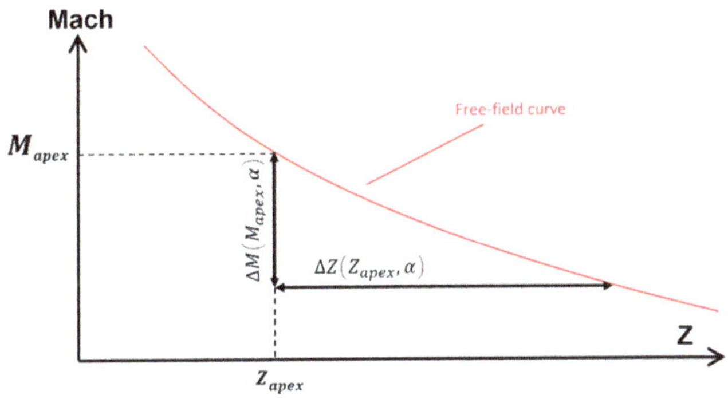

Figure 6: Diffraction apex influence on the evolution of the Mach number, with reference to the free-field curve.

The Bazhenova model consists in determining the reduced distance Z of any point P within the diffraction polygon, at a distance AP from the apex, and observed at an angle α relative to the line of sight, by summing:

- The reduced distance at the apex Z_{apex}.
- A sudden increase $\Delta Z(Z_{\text{apex}}, \alpha)$ resulting from the decrease $\Delta M(M_{\text{apex}}, \alpha)$ in the Mach number between the incident Mach number $M_{\text{apex},0}$ and the Mach number calculated for the α angle, $M_{\text{apex},\alpha}$.
- The reduced distance corresponding to the account of the distance AP, ΔZ_{AP}.

The evolution of the Mach number at the apex according to Bazhenova is given by the following formula:

$$M_{\text{apex},\alpha} = 1 + (M_{\text{apex},0} - 1)\left(1 - \sin\left(\frac{\alpha}{2}\right)\right), \tag{9}$$

from which we can simply derive:

$$\Delta M(M_{\text{apex}}, \alpha) = -(M_{\text{apex},0} - 1)\sin\left(\frac{\alpha}{2}\right). \tag{10}$$

For a point P in the diffraction polygon at a distance ΔZ_{AP} from the apex, the reduced distance is expressed as:

$$Z = Z_{\text{apex}} + \Delta Z(Z_{\text{apex}}, \alpha) + \Delta Z_{AP}. \tag{11}$$

The diffraction approach employed in the basic URBEX model is relatively conservative, as it results in an overestimation of the Mach number after diffraction due to the expansion zone defined by the angle beta. Specifically, the Bazhenova model tends to overestimate the overpressure when the point of interest lies within the expansion zone, assuming that the overpressure remains equivalent to that in the free-field, since the point falls within the visibility polygon of level 0. This paper presents the basic version of URBEX, excluding the newly developed diffraction model that incorporates the angle beta and the initiation of diffraction within the expansion zone.

2.5 Other urban effects

2.5.1 Mach reflection

When a wave of level $i-1$ generates, from the same reflecting segment, both a regular wave and a Mach wave, a Boolean operation on the representative polygons of these waves can be used to avoid double-counting the reflected wave. The positioning of the contact point will determine the function to prioritize in order to calculate the different parameters of the unit wave. Fig. 7 shows the overlap of regular reflection areas with Mach reflection areas. This paper presents only the basic version of URBEX, which excludes Mach reflection.

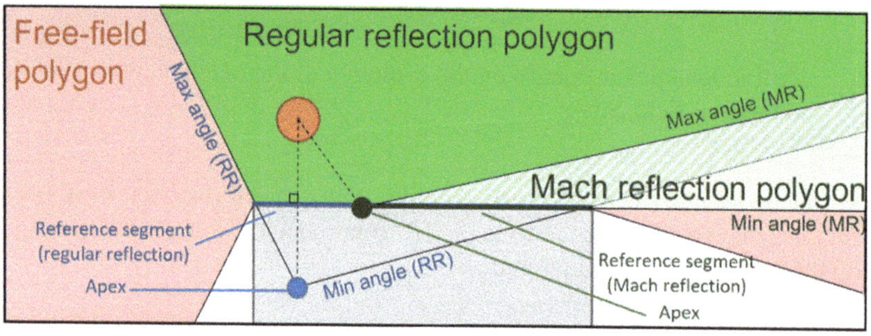

Figure 7: Overlap of regular reflection areas and Mach reflection areas.

2.5.2 Channelling

Urban canyon effects lead to a gradual transition from a 3D free-field wave to a 2D wave. The wave initially propagates spherically down the street, with reflections off the walls that result in Mach reflections. At a certain distance from the source, the propagation evolves into a cylindrical shape, concentrating the energy along the street and increasing the wave's intensity. This results in a reduced pressure drop compared to free-field conditions, potentially leading to larger danger zones.

2.5.3 Urban canopy bypassing

What is currently being calculated is the behaviour on the ground, but another possibility is that the wave could bypass the building from above. This wave will be highly diffracted, so the expected overpressures are not very high. This allows for overpressures to be mapped in the case of a building with an inner courtyard. The wave inside arrives with low intensity and is significantly delayed compared to the other waves. This paper discusses only the basic version of URBEX, omitting the urban canopy wave.

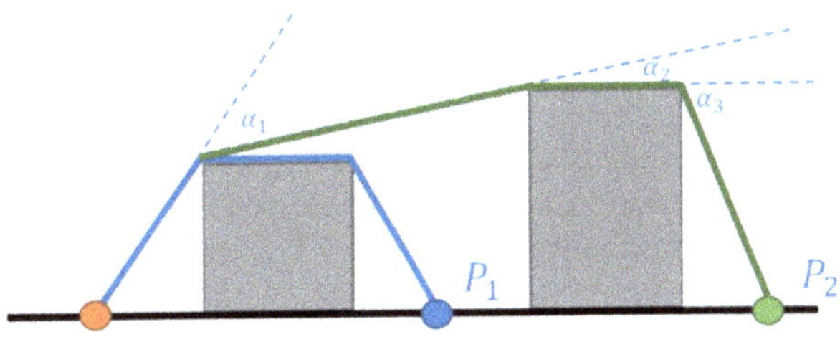

Figure 8: Example of an 'urban canopy' wave.

3 URBEX APPLICATIONS AND RESULTS

The URBEX model can calculate a complex wave at every point in the area of interest, taking into account all urban effects: regular and Mach reflections, diffractions, shock wave channelling, and bypassing by the urban canopy. In this section, we will present the results obtained with the basic version of the model, which does not include the improved diffraction model, the new regular reflection model, Mach reflections, or the urban canopy wave.

3.1 Comparison for urban configurations

A first comparison is conducted with the DEMOCRITE platform (image courtesy of IT Link), which uses the CEA/DAM FLASH code [9]–[12]. The scenario involves the explosion of a 'small vehicle' (227 kg TNT, following FEMA standards [21]) in a large square in the city of Cahors (France). Fig. 9 compares the FLASH results with the numerical simulation results from the Viper::Blast software and an URBEX simulation at precision level 1. The node spacing (FLASH) is 2 m, while the pixel size (URBEX) is 300^2. The output displays danger zones based on overpressure thresholds in accordance with French regulations, where Zone 1 (red) is 0.430 mbar, Zone 2 (orange) is 0.2 mbar, Zone 3 (yellow) is 0.140 mbar, Zone 4 (green) is 0.050 mbar, and Zone 5 (blue) is 0.02 mbar. The coloured circles represent the corresponding free-field danger zones.

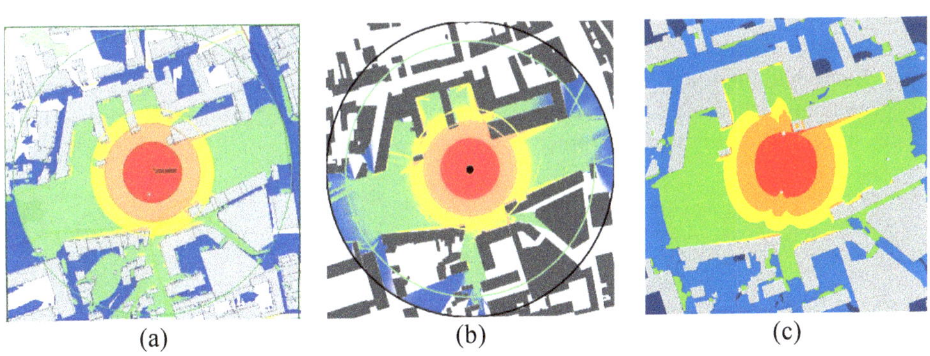

Figure 9: Danger zones calculated using (a) FLASH (courtesy of IT Link), runtime = 41 s; (b) URBEX: precision level 1, runtime =18 s; and (c) Viper::blast, runtime = 34 min.

At precision level 1, URBEX outperforms FLASH by delivering nearly identical results with almost half the execution time. It also shows performance comparable to the

Viper::Blast software, which requires 34 minutes for execution, although URBEX achieves similar results in a significantly shorter time. The free-field danger zones are comparable across all three methods.

A second comparison is made with Viper::Blast and the experimental results of INSA-CVL using a charge of propane/oxygen mixture at small scale. The event involves an explosion equivalent to a hemispherical charge of 1,777 kg TNT occurring in the Butte-aux-Cailles district of Paris (France). As illustrated in Fig. 10, the results obtained with URBEX at level 1 are similar to those found with the Viper::Blast simulation software as well as the experimental results, which proves that our model is reliable and accurate in simulating an explosion in an urban environment.

(a) (b) (c)

Figure 10: Danger zones calculated using (a) Viper::Blast; (b) experimental results of INSA-CVL; and (c) URBEX with precision level 1.

3.2 Application to the Oslo bombing

Christensen and Hjort [22] provide a forensic analysis of the Oslo bombing occurred on 22 July 2011, including field surveys to determine the percentage of broken windows on various façades near the explosion. Fig. 11 compares this analysis with URBEX results. It uses the Green Book consequence model for glass breakage [23], applied to the signal's maximum peaks. Some buildings were manually added to the OSM data (with a few still missing), as they were demolished after the event. Glass breakage is strongly influenced by diffraction effects, making it highly non-isotropic. The red zone generated by URBEX, indicating over

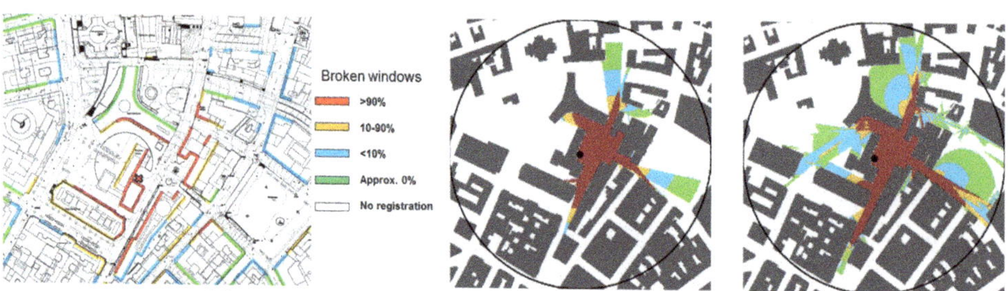

(a) (b) (c)

Figure 11: Percentage of broken windows from the Oslo bombing. (a) Ground truth [20]; URBEX results: (b) Precision level 1; and (c) Precision level 2.

90% of broken windows, closely matches the field data. Other areas provide more conservative estimates due to the simple Bazhenova diffraction model.

4 CONCLUSION AND FUTURE WORK

In this paper, we presented the basic model of URBEX and demonstrated its effectiveness through various results. These findings show that URBEX delivers reliable and accurate simulations, particularly in urban environments. The model proves to be efficient, with significant potential for future developments and applications. It is readily usable by first-responders in crisis situations (potential VB-IED for instance). Furthermore, URBEX's efficiency, especially at higher precision levels, highlights its value for applications requiring precise danger zone analysis. The URBEX code is set to define a new benchmark for fast and accurate computations of explosion consequences in urban environments. A separate paper will be devoted to presenting the new regular reflection model, the Mach reflection model, a more rigorous diffraction model, and urban canopy bypassing.

Extensions of this work are currently underway. The CONVEX project, co-funded by the MAIF foundation, is investigating the explosion risks associated with the thermal runaway of electric vehicle batteries in confined spaces such as underground car parks, tunnels, and ferries. The MALBEC project, co-funded by the French Defense Innovation Agency, is addressing the challenges of explosive breaching by Special Forces. It is developing a fast-running tool to assist in preparing such operations, which requires a 3D extension of the URBEX approach and modelling of quasi-static pressure generation and propagation.

ACKNOWLEDGEMENTS

The authors would like to thank their partners, INSA-CVL and Blast Solutions & Associates, for their support and collaboration. They also wish to express their gratitude to the French National Research Agency for selecting and co-funding the URB(EX)^3 project. Furthermore, the authors appreciate the valuable discussions with the project's advisory board members: SAFE cluster, IRCGN, LCPP, CETID, PIAM, 17th RGP, IPE, IRSN, INERIS, EURENCO, BS&A, SPIEZ laboratory (CH), SSSL (UK), and other organisations that supported the project, including DGA, CEA/DAM, RisCrises, and CERL (CA).

REFERENCES

[1] Synthetik Applied Technologies, BlastFoam: An OpenFOAM solver for compressible multi-fluid flow with application to high-explosive detonation, 2019.

[2] Stirling, C.G., Viper::Blast user manual. Stirling Simulation Services Limited: Glasgow, 2020.

[3] Dennis, A.A., Stirling, C. & Rigby, S.E., Towards the development of machine learning tools for blast load prediction. *Proceedings of the 6th International Conference on Protective Structures (ICPS6)*, International Association of Protective Structures, 2023.

[4] Dennis, A.A & Rigby, S.E., The direction-encoded neural network: A machine learning approach to rapidly predict blast loading in obstructed environments. *International Journal of Protective Structures*, 20414196231177364, 2023.

[5] 3D Blast, https://www.ara.com/3d-blast/. Accessed on: 24 Feb. 2025.

[6] Fi-Blast, Fast analysis of blast and fragment loaded structures. https://numerics.de/products/fi-blast/. Accessed on: 24 Feb. 2025.

[7] Tank, A., Greulich, B. & Rottenkolber, E., Rapid calculation of urban blast propagation using FI-BLAST. *Proceeding of 14th International Symposium on Interaction of the Effects of Munitions with Structures (ISIEMS14)*, Seattle, WA, USA, 2011.
[8] Nichols, J.F. & Doyle, G., Current engineering models and capabilities in the vulnerability assessment and protection option (VAPO) software. *Structures Congress*, 2014.
[9] Lapébie, E., Soulié, R. & Youinou, L., FLASH: Fast Lethality Assessment for Structures and Humans. *Proceeding of the 24th International Symposium on Military Aspects of Blast and Shock (MABS24)*, Halifax, Canada, 2016.
[10] Lapébie, E. et al., Fast-running models for blast and QSP in complex geometries. *Proceedings of the 17th International Symposium on Interaction of the Effects of Munitions with Structures (ISIEMS17)*, Bad-Neuenahr, Germany, 2018.
[11] Lapébie, E. et al., Fast-running models for blast and QSP in complex geometries. *Improved Explosives and Munitions Risk Management (IEMRM)*, MSIAC workshop, Granada, Spain, 2018.
[12] Osmont, A. & Lapébie, E., Models for rapid calculation of breath and quasi-static pressure in complex environments. *Proceeding in French Congress of Mechanics*, Brest, France, 26–30 Aug. 2019.
[13] Chereau, G., Lapébie, E. & Sochet, I., URBEX: A fast yet accurate model to compute blast consequences in urban geometries. *Proceeding of the 26th International Symposium on Military Aspects of Blast and Shock (MABS26)*, Wollongong, Australia, 2023.
[14] Lapébie, E. (inventor), *Procédé de gestion d'une explosion dans un environnement réel*. France Patent FR2400950. To be published.
[15] Swisdak, J.M., Simplified Kingery airblast calculations. *Proceedings of the 26th DoD Explosives Safety Seminar*, Aug. 1994.
[16] Bazhenova, V., Gvozdeva, L.G. & Nettleton, M.A., Unsteady interactions of shock waves. *Aerospace Sciences*, vol. 21, Pergamond Press, pp. 249–331, 1984.
[17] Whitham, G.B., A new approach to problems of shock dynamics: Part I Two dimensional problems. *Journal of Fluid Mechanics*, **2**(2), pp. 145–171, 1957.
[18] Whitham, G.B., *Linear and Nonlinear Waves*. John Wiley & Sons, 1974.
[19] Skews, B.W., The perturbed region behind a diffracting shock wave. *J. Fluid Mech.*, **29**, pp. 705–719.
[20] Skews, B.W., The shape of the diffracting shock wave. *J. Fluid Mech.*, **29**, pp. 297–304.
[21] Joint Counterterrorism Assessment Team (JCAT), Bomb threat standoff distances. https://www.dni.gov/nctc/jcat/references.html. Accessed on: 5 Mar. 2025.
[22] Christensen, S.O. & Hjort, O.J.S., Back calculation of the Oslo bombing on July 22nd, using windows breakage. *Proceeding of the 22nd International Symposium on Military Aspects of Blast and Shock (MABS22)*, Bourges, France, 2012.
[23] TNO, *Green Book – CPR 16E – Methods for the Determination of Possible Damage*, 1992.

DEFORMATION OF BURIED PIPELINES BY SUBSOIL DETONATION

ALEXANDER J. M. SCHMETS[1], JASPER T. BLOM[1,2] & LAMBERTUS KOENE[3]
[1]Faculty of Military Sciences, Netherlands Defence Academy, The Netherlands
[2]Engineers Regiment, Prinses Margrietkazerne, The Netherlands
[3]Knowledge Centre Weapon Systems and Munitions, Material and IT Command, The Netherlands

ABSTRACT

Transport pipelines are critical for the distribution of energy commodities, refined products and essential materials like water and hydrogen. While often buried underground for protection from environmental factors, they remain vulnerable to distant threats such as intentional or accidental explosions. This study investigates the mechanical response of steel pipelines to seismic excitation from nearby buried explosive charges. An experimental approach was employed to measure the stresses and deformation of a buried, straight steel pipeline segment using high-frequency strain gauges attached to its surface before burial in damp, sandy soil. The gauges were strategically placed to capture various deformation modes, such as ovalisation and axial strain. Seismic loading was applied using buried plastic explosive charges weighing up to 16 kg TNT-equivalent, detonated at distances up to 8 m from the pipeline centre. Seismic ground shock characteristics were evaluated using triaxial MEMS accelerometers. The results establish a relationship between the explosion intensity, measured ground shock characteristics (e.g., peak particle velocity) and pipeline deformation. For the experimental scenarios, pipeline deformation remained within the elastic limit, with maximum strains measured in the order of $100 - 200 \times 10^{-6}$. The ground shock was found to be significantly damped between the detonation point and the pipeline. Thus, when no other loads are present, transport pipelines buried in dry, sandy soil are predicted to deform elastically for subsoil detonations up to 100 kg TNT-equivalent at a 14.5-m distance.

Keywords: subsoil detonation, gas pipelines, UXOs, soft soil, safety distances, steel deformation.

1 INTRODUCTION

The Netherlands was heavily bombed during World War II (see Fig. 1). Additionally, the country served as a flight route for Allied bombers heading to Germany. On their return flights, bombers often dropped unexploded bombs in so-called safe areas to save fuel. Many of these bombs never detonated (named UXOs (UneXploded Ordnance), also known as duds) and remain in the Dutch soil. These duds are frequently discovered, particularly during construction works, by specialized detection companies. Once these unexploded remnants of war have been localised, the Dutch Army Explosive Ordnance Disposal Regiment (EOD) is called in to neutralize the threat in the safest possible way.

Although the safest possible way of neutralising the UXO is aimed for, for adequate safety measures to be taken during the dismounting intervention, one must consider the possibility that the UXO unfortunately will detonate. Therefore, evacuations of people and feedstock may be necessary, and possible damage to neighbouring structures must be considered.

The damage due to undesired detonation of such UXO may be caused by fragments, blast and seismic effects of the detonation, the so-called ground shock. The first two damage causes are in The Netherlands' EOD practice mitigated by covering the UXO by a temporary large heap of sand, which will on the other hand increase the coupling of the UXO to the soil, hence increasing the ground shock effect. For the ground shock effects safety distance tables are used, which consider the size of the charge, the local soil characteristics and the ground water level, amongst others. However, for these safety distances of underground detonations

WIT Transactions on The Built Environment, Vol 215, © 2025 WIT Press
www.witpress.com, ISSN 1743-3509 (on-line)
doi:10.2495/SUSI250061

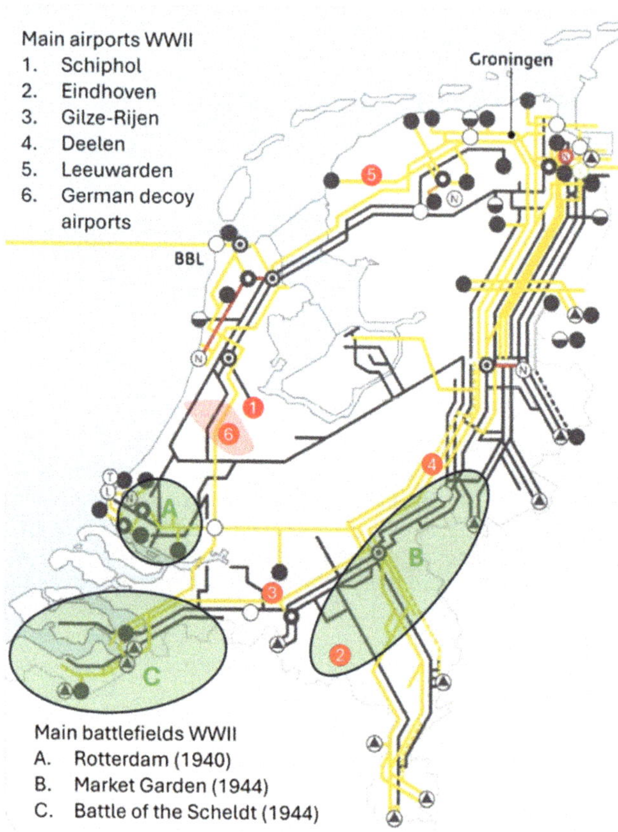

Figure 1: The high-pressure gas network in The Netherlands. Also shown are the main airports and battlefields in World War II. *(Source: Gasunie Transport Services and Dutch Army Explosive Ordnance Disposal Regiment.)*

different criteria exist, depending on whether it is the owner of some sort of (critical) infrastructure or the Dutch EOD. Discrepancy in these safety distances arises partly due to a lack of empirical knowledge about the impact of ground shock on pipelines, particularly in Dutch soil. The Netherlands has highly variable soil conditions, ranging from dry sands to saturated clays, which further complicates the assessment. Research is needed to better understand these effects and therefore the development of experimentally validated prediction methods for determining safety distances. Such knowledge could also benefit the design of underground infrastructure near industrial facilities where explosion risks exist, such as fuel or hydrogen storage terminals. Relatively few similar studies in this field exist, for example [1]–[3].

For instance, pipeline owners have an interest in a validated method, as their current approach – dating back to the 1980s – is necessarily conservative. This may lead to unnecessary protective measures, such as depressurizing pipelines, installing sheet piles, or digging trenches, all of which are costly. Additionally, shutting down a pipeline temporarily can have significant societal impacts. A scientifically grounded safety distance would eliminate unnecessary precautions while ensuring public safety.

To acquire empirical knowledge, ten detonation experiments have been performed, see Table 1. In all these experiments 10-inch straight steel tubes without internal pressure – representing pipelines – were used, in one type of soil. The ground shock characteristics are measured using triaxial MEMS sensors and the deformation of the pipeline was measured using strain gauges. For all experiments, a common plastic military explosive was used. The explosive material was used without any casing. The latter could decrease the effect of the shock wave upon the pipeline.

Table 1: Overview of the performed experiments. For each experiment, the mass of the plastic explosive, its distance from the steel pipeline and its location of the detonation is given.

Detonation	**Mass explosive (kg)**	**Distance (m)**	**Location of detonation**
1	2.5	5	Perpendicular; middle pipe
2	4	8	Perpendicular; middle pipe
3	2.5	4	Extension of pipeline
4	4	4	Extension of pipeline
5	3.5	5	Perpendicular; middle pipe
6	3.5	5	Perpendicular; end pipe
7	5	5	Perpendicular; middle pipe
8	5	5	Perpendicular; middle pipe
9	10	5	Perpendicular; end pipe
10	10	5	Perpendicular; middle pipe

The structure of this paper is as follows. First, some theoretical background is given. Next, the experimental set-up and the materials used are described. Then, results of the experiments, i.e., the mechanical response of the pipeline, is discussed. This is followed by conclusions and recommendations for future research in this topic.

2 THEORETICAL BACKGROUND

2.1 Deformation in longitudinal direction

The characteristics of the ground shock wave originating from the explosive charge will be treated as the loading of the pipeline segment. Details of this wave, i.e., whether it has a surface wave character etc. [4], are not considered here. Only the measured strength of the wave, as measured close to the pipe segment, is considered as the load. Compression and shear wave components of the ground shock introduce strain in both the longitudinal (L, axial) and circumferential (C) directions of the buried transport pipeline. Shear waves, and possibly compression waves, depending on the angle of impact on the pipeline, result in axial strain in the pipeline. This leads to the development of tensile and compressive stresses within the pipeline. The axial strain can be calculated using:

$$\varepsilon_{aL} = \frac{u}{c_p} \cos\theta. \tag{1}$$

In this equation u is the peak particle velocity, c_p the propagation velocity of the compression wave, and θ the angle between de location on the pipeline and the detonation [5].

Additionally, shear waves, and possibly compression waves, will cause bending strain due to difference in arrival of the shear and compression waves over the length of the pipeline. Assuming the bending of the pipe follows the deformation of the surrounding soil, the strain caused by bending can be calculated as follows:

$$\varepsilon_{bL} = \frac{u(2\pi f)d}{2c_s^2}\cos\theta. \tag{2}$$

In this equation d is the outer diameter of the pipeline, f is the dominant frequency of the shear wave, and c_s the propagation velocity of the shear wave [5], [6].

2.2 Deformation in circumferential direction: Ovalisation

The circumferential strain occurs because of loading by the compression wave component of the ground shock. The maximum circumferential strain can be estimated using the maximum strain caused by the compression wave, plus a small additional strain due to bending caused by the ovalisation – i.e., the deformation that leads to change of shape of the pipe segment's cross section – of the pipeline. The circular strain can be obtained from:

$$\varepsilon_h = \frac{u}{c_p}\left(1 + \frac{3t}{d}\right)\sin\theta, \tag{3}$$

in which ε_h is the circular strain and t is the thickness of the pipe [5].

2.3 Damping characteristics of the ground shock

The Bornitz equation [7] describes the two most important factors of the attenuation of intensity of the ground shock with distance: material damping and geometric damping. Here we formulate the Bornitz equation for a chosen strength parameter, the peak particle velocity (PPV) of the ground shock, at a distance r from the detonation location

$$v(r) = v(r_0)\left(\frac{r_0}{r}\right)^n e^{-\alpha(r-r_0)}. \tag{4}$$

In this equation $v(r)$ and $v(r_0)$ are the peak particle velocities at distance r and a reference distance r_0 from the source. α is the decay or attenuation coefficient. This coefficient is dependent on material characteristics of the soil. In addition, it is dependent on the dominant frequency f of the vibration:

$$\alpha = \frac{2\pi D f}{c}. \tag{5}$$

In this equation c is the propagation velocity of the waves, and D is the so-called damping ratio. Finally, the parameter n determines the geometric damping (e.g., $n = \frac{1}{2}$ for surface waves). Thus, by measuring the PPV at a reference distance r_0, the expected PPV can be calculated for any distance r, provided that α and n are known.

Often seismic loading is given in terms of the local acceleration as determined by vibration sensors. However, in Dutch norms [8] iso-damage distances are related to peak particle velocities, which explains the choice for this metric for the seismic loading.

3 EXPERIMENTAL

To investigate the response from a buried pipeline to transient loading from an underground detonation, ten detonation experiments have been performed, see Table 1. In all these experiments a straight pipeline segment was used. Fig. 2 provides details of the experimental set-up that has been used.

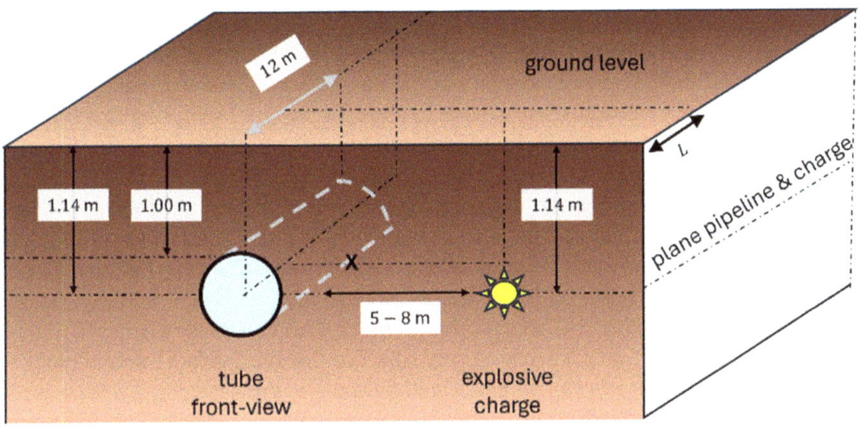

Figure 2: Overview of the experimental set-up: pipeline and explosive charge buried in damp sandy soil. The X shows the location of the MEMS-seismic sensor whose data are shown in Fig. 5(a) (the sensors are buried at 1.14 m below ground level).

3.1 Buried steel pipeline

The buried, steel (L245MB) pipeline segment has a total length of 12 m and a circular cross section with an outer diameter of 273.0 mm, and a nominal wall thickness of 6.3 mm. After the strain gauges were placed, the two parts were welded together. The specified minimum yield strength of the pipeline steel is 245 MPa and its tensile strength is 466 MPa, as obtained from tests on a 20 mm wide specimen [9]. For transport and placement two lifting eyes were welded to the pipe. While hanging in free air and excitation of the pipe with a steel hammer, the only observable eigenfrequency of the pipeline segment was found about 10 Hz.

3.2 Seismic loading

For the experiments a plastic explosive consisting of 83% penthrite was used with a mass density of 1,770 kg/m^3 and a detonation velocity of 8,350 m/s. The specified explosion temperature is 4,230°C, and its mass specific energy 5,800 kJ/kg. The relative effectiveness factor of this form of penthrite is close to 1.25 [10]. Usually, one performs this type of experiments at constant depth of burst, i.e., at constant $d/\sqrt[3]{w}$, with d the depth relative to ground level, and w the TNT-equivalent mass of the explosive charge [9]. However, here it's chosen to have the pipeline and explosive charge buried at the same depth. After burying the charge, the hole is filled with sand, and finally all is covered with a heap of sand to achieve optimal coupling between the charge and the soil. It also should be noted, especially when detonations take place at the same location, that experiments will locally compact the soil, which leads to different soil properties – hence, shock propagation properties – in subsequent tests.

3.3 Ground shock seismic properties and pipeline response

Triaxial MEMS accelerometers or vibration sensors (Convergence Instruments) with a dynamic range of $8g$ are used to measure the PPV of the ground shock that originates from the detonation. The sensors record the particle velocity of the ground wave along three orthogonal axes. The convention was adopted that the sensor's x-axis was oriented from the point of detonation perpendicular to the pipeline, its y-axis parallel to the longitudinal direction of the pipeline, and the z-axis is perpendicular to the ground level. The sensors are buried in the ground such that their xy-plane coincides with the plane defined by pipeline and explosive charge. The vibration sensors were operated at a signal sampling rate of 1000 Hz, and were set to measure in the velocity mode, providing the particle velocities for all three directions. From this, the PPV is readily determined.

Six vibration sensors are used in this experiment. Three of these sensors were placed in front of the pipeline, positioned close to it. One sensor was placed at the rear of the pipeline, also positioned close to it. Finally, two additional sensors were placed behind the detonation site. For the experiments reported here, the sensor indicated with X – between detonation location and pipeline (see Fig. 2) – was used. Moreover, the sensors are also used to measure the ground shock propagation properties of the soil. For this the sensors internal clocks need to get mutually synchronised prior to experimentation. For the sandy soil as present at the experiment location ('t Harde, the Netherlands), a seismic velocity of 426 m/s and wavelength of about 85 m (frequency of 5 Hz) was found.

Precision strain gauges (Micro Measurements, USA) are used to measure the strain in both the longitudinal (L) and circumferential (C) directions of the pipeline. They are carefully attached to the pipeline, so that any deformation of the pipeline is transferred to the strain gauges as well. The strain gauges are connected to a power source, and when the strain gauges deform, their resistance changes accordingly. Finally, the local strain as function of time is obtained by measuring the changes in electrical resistance at 1 kHz. The measured strain can be converted into mechanical stress, allowing for the determination of the stress occurring in the pipeline. The strain gauges used in this experiment are so-called T-rosettes. They consist of two strain gauges positioned at a 90° angle to each other.

Fig. 3 shows the strain gauges used in this experiment and the way they are mounted to the pipeline. In Fig. 4, the locations of the eight strain gauges attached to the pipeline segment are indicated.

4 MECHANICAL LOADING AND RESPONSE

Now the most important experimental results are given. Here we focus on extreme values found for the measured strains, as they are important to obtain objective estimates for safety distances. Fig. 5 shows the mechanical loading and strain response at various locations, for scenario 10 as presented in Table 1. Table 2 presents the experimentally obtained loading results from the vibration sensors, i.e., the peak particle velocities resulting from the ground shock, measured in the x-, y-, and z-directions as described in Section 3. In Fig. 5(a) it is shown how the PPV is obtained, and in Table 2 the PPVs are given for the seven accelerometers installed. Sensor 1 has the highest value of PPV. This is the sensor represented by an X in Fig. 2, and which PPV is being used in the remainder of this paper. The wobbly pattern observed in Fig. 5(b) is observed in all T-rosettes and happens to have a frequency of about 50 Hz. It is considered as a measurement artifact, that will not be further elaborated here.

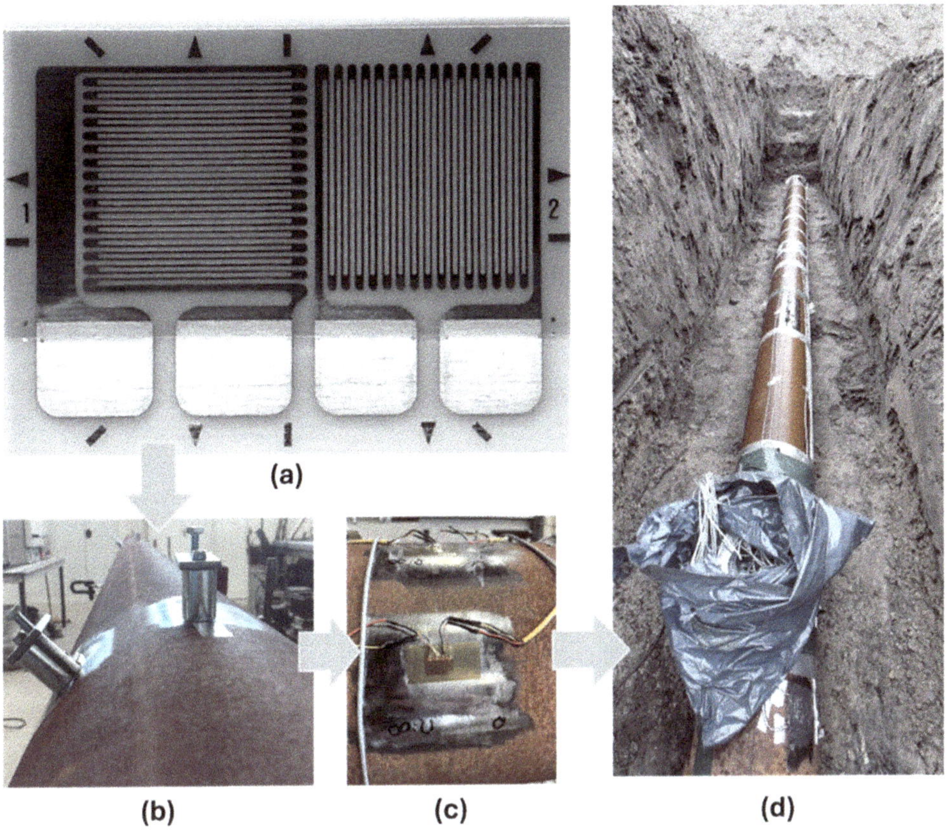

Figure 3: (a) General purpose strain gauge (Tee Rosette 250UT with resistance $(350 \pm 0.4\%)\Omega$; (b) Mounted to a polished area of the pipeline, coated and press-dried; (c) Wired pipe-gauges system; and (d) Finally at the test site (with protective cover).

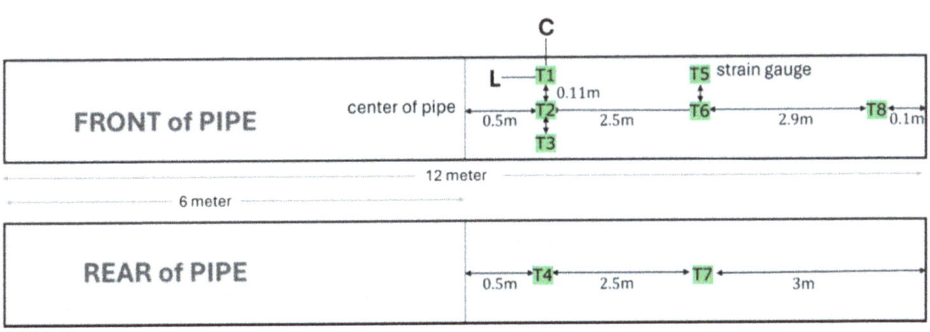

Figure 4: Location of strain gauges (T1–T9) at the front (directed to incoming ground shock) and rear of the pipeline. The bold L (length of the pipe direction) and C (circumferential direction) identify the strain measuring directions of the gauges. Front of pipe means the face of the pipe directed towards the detonation location.

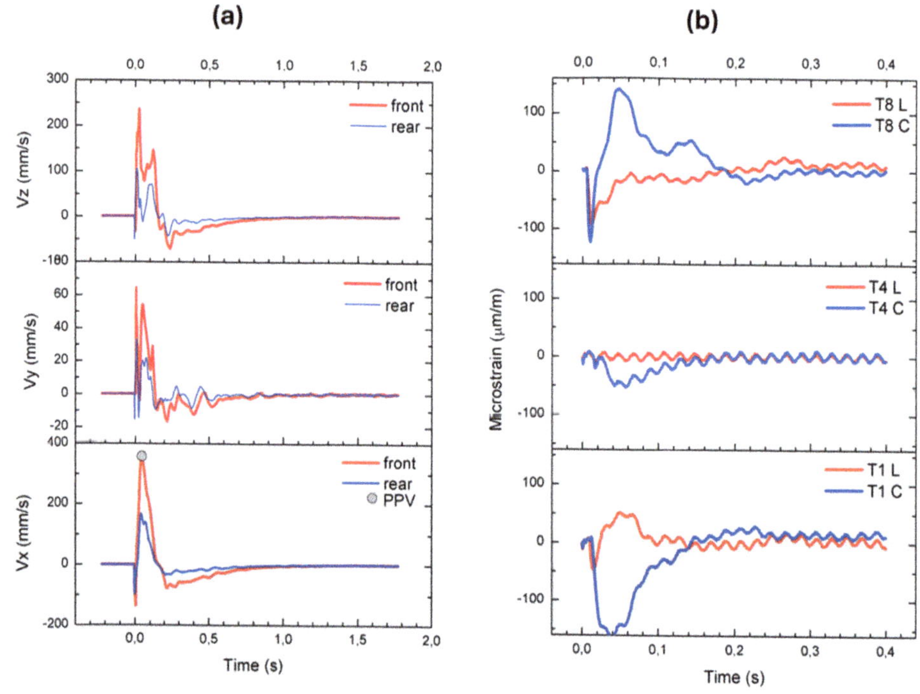

Figure 5: Loading and strain response of a 10 kg charge. (a) Measured seismic particle velocities just in front and behind (rear) the pipeline; and (b) Axial (L) and radial (O) deformation (strain) as measured at various locations (T1, T4, T8).

Table 2: Peak particle velocity of the vibration sensors for Detonation 10. For each sensor the distance to the detonation is given.

Sensor	Peak particle velocity *x*-direction (mm/s)	Peak particle velocity *y*-direction (mm/s)	Peak particle velocity *z*-direction (mm/s)	Distance to detonation (m)
1	378.4	62.55	251.9	5.0
2	228.8	147.9	118.2	5.6
3	48.17	79.51	102.4	7.4
4	174.7	33.75	109.7	5.5
6	29,44	15.78	73.76	8.0
7	34.18	15.23	36.04	17.25

In Figure 6, the observed strains and the corresponding calculated stresses, from Hooke's law and Equations (1) to (3), are presented. During the experiment, strain measurements were taken at select number of points on the pipeline, as shown in Figure 4. Therefore, the strain in the remaining sections is an estimation. The strain is given in micro strain, i.e. 10^{-6}, and the stress is measured in MPa. Strain is treated as a positive value, as the focus is primarily on the magnitude rather than the direction in which it occurs. However, for stress, also the negative strain values were considered, and the stress values shown in the figure represent the absolute values.

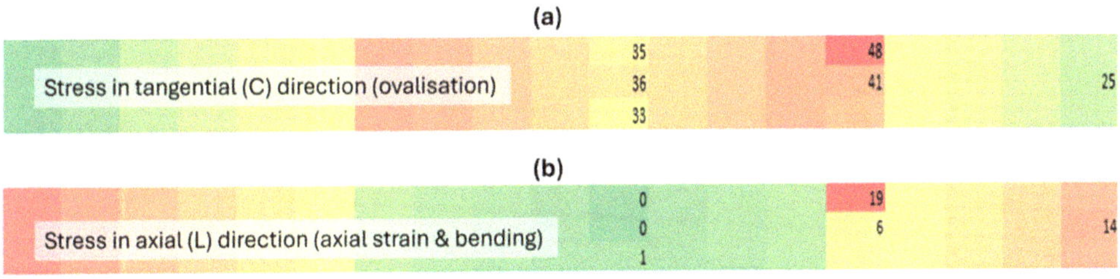

Figure 6: Schematic view of the stresses in the pipeline for Detonation 10. The values that have been measured are shown. For other locations the values are estimated.

Table 3: Peak particle velocity and maximum strain for Detonation 10. Strain in longitudinal direction is indicated with an L and strain in the circumferential direction is indicated with a C.

Sensor & strain gauge	Peak particle velocity x-direction (mm/s)	Peak particle velocity y-direction (mm/s)	Peak particle velocity z-direction (mm/s)	Maximum strain L (µ)	Maximum strain C (µ)	Distance to detonation (m)
1 & T2	378.4	62.55	251.9	53	−174	5.0
2 & T6	228.8	147.9	118.2	87	−205	5.6
3 & T8	48.17	79.51	102.4	−102	139	7.4
4 & T4	174.7	33.75	109.7	−17	−60	5.5

It is interesting to compare the peak particle velocity of the seismic waves with the deformation in the pipeline, as this provides a better understanding of the response of the tube (see Table 3).

For Detonation 10 (Table 1), the largest deformation on average occurs in the circumferential direction at strain gauge T5. However, this does not automatically mean that the highest stress also occurs at the location of strain gauge T5. The largest deformations and, consequently, stresses may also have occurred in sections of the pipeline where no strain gauges were placed.

For strain gauges T5 and T7, it is possible to plot a linear relationship between the peak particle velocity at vibration sensor 1 (located at X in Fig. 2) and the maximum occurring strain in both the longitudinal and circumferential directions. Results are shown in Fig. 7. From these graphs, it can be observed that for strain gauge T5, the longitudinal strain increases in the positive strain direction as a function of peak particle velocity, while the circumferential strain increases in the negative strain direction. For strain gauge T7, both the circumferential and longitudinal strains increase in the negative strain direction as a function of peak particle velocity.

So, two possible stress states of the pipeline are observed: one where the strain in the circumferential and longitudinal directions essentially compensate each other, shown in the lower graph in Fig. 7, and one where the strain in the longitudinal and circumferential directions reinforce each other, shown in the upper graph in Fig. 7.

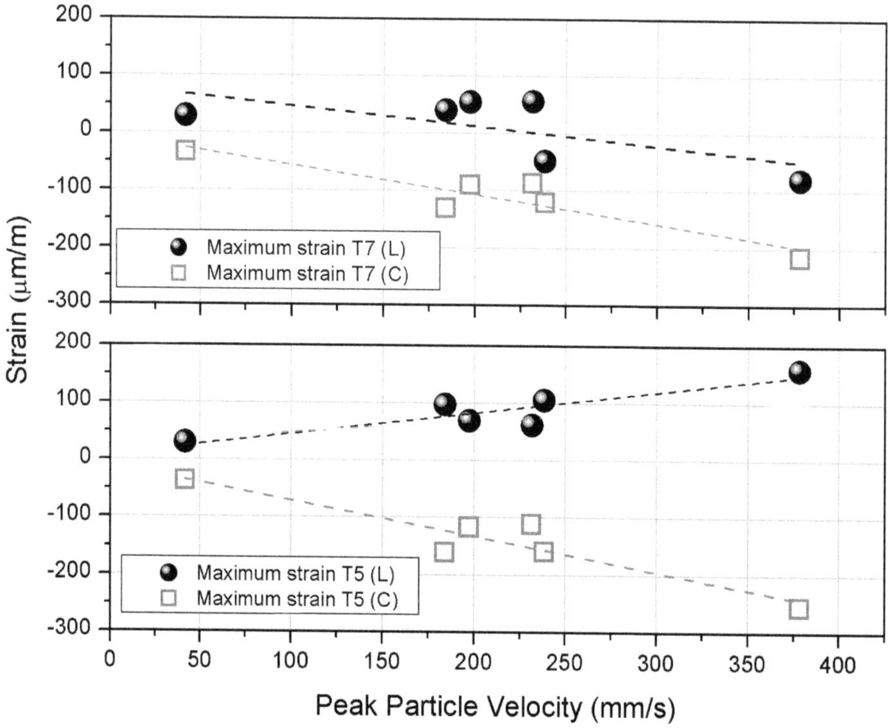

Figure 7: Maximum strain in longitudinal (L) and circumferential/tangential (C) direction as a function of the PPV measured by vibration sensor 1.

5 CONCLUSIONS

This research investigates the relationship between ground waves caused by underground detonations and the resulting stresses in buried transport pipelines. Based on the literature reviewed and experimental results, the study confirms that such a relationship exists. Two types of ground waves are generated by underground detonations: compression waves and shear waves. Among these, compression waves cause the most significant deformations and, consequently, the highest stresses in pipelines. Shear waves can amplify this effect. The most critical scenario occurs when a detonation charge is positioned perpendicular to the pipeline, and the explosive charge is highest

The study identifies three main deformation modes leading to stress in the pipeline: bending, axial tension and compression, and ovalisation. There are also two different stress scenarios. The highest stress occurs when both circumferential and longitudinal strains are either positive or negative. When one strain is positive and the other negative, the resulting stress is lower.

For the specific pipeline studied, a correlation was found between peak particle velocity of the ground waves and the induced stress. The highest stress was observed at strain gauge T7, where both circumferential and longitudinal strains were negative (see Fig. 7). At a peak particle velocity of 378 mm/s, a maximum stress of 55 MPA was recorded, which remains well below the pipe's yield strength. To exceed the yield limit of 245 MPa, a detonation would need to generate a peak particle velocity of 1,877 mm/s, equivalent to a 100 kg TNT explosion at 14.5 m from the pipeline, assuming similar soil conditions.

In conclusion, while a relationship between detonation-induced ground shock waves and pipeline stress is evident, it remains difficult to quantify precisely due to the problem's complexity. This research treated it as a static issue, although it is inherently dynamic. Since strain varies along the pipeline and strain waves may propagate along the pipeline segment. Moreover, deformations can occur in multiple ways, which makes a direct correlation to be challenging. However, the study confirms that higher peak particle velocities lead to greater deformations and higher stress levels, However, when a maximum allowed stress level is defined, e.g., as a percentage of the yield stress, a good estimate of safety distances can in principle be derived from the experiments presented. However, it should be noted that these experimental results provide a lower limit to the deformation of the tube. The experiments were conducted in dry sand (high seismic damping), while the tube had a high thickness-to-diameter ratio. Smaller ratios, occurring for large-diameter transport lines, will probably lead to larger deformations.

Therefore, additional experiments are planned in the future. Moreover, finite element simulations may further clarify the relation between seismic waves and the stresses and deformations they cause in buried pipelines.

ACKNOWLEDGEMENTS

We would like to thank our colleagues at the Knowledge Centre Weapon Systems & Munitions (KCW&M) of the Material and IT Command (COMMIT) in 't Harde for performing the experiments. Moreover, the authors thank Gasunie, the Faculty of Military Sciences (FMS) of the Netherlands Defence Academy (NLDA), Material and IT Command (COMMIT) for funding and facilitating this research. Last but not least, the authors would like to thank Mr Paul Wesselius (Gasunie) and Mr Dennis Krabbenborg (FMS/NLDA).

REFERENCES

[1] Nourzadeh, D., Bargi, K. & Takada, S., Evaluation of gas distribution pipeline network subjected to seismic wave propagation in greater than Tehran area, Iran. *10th International Conference on Urban Earthquake Engineering*, p. 6, 2013.

[2] Esparza, E.D., *Pipeline Response to Blasting in Rock*, PRCI (Catalogue No. L51661, Contract PR-15-712), 1991.

[3] Olarewaju, A.J., Kameswara, R.N.S.V. & Mannan, M.A., Response of underground pipes to blast loads. *Earthquake-Resistant Structures: Design, Assessment and Rehabilitation,* ed. A. Moustafa, InTechOpen, 2012.

[4] Bulson, P.S., *Explosive Loading of Engineering Structures*, CRC Press: London, 1997.

[5] Nyman, D.J., Howding, C.H. & Oriard, L.L., Evaluation of close-in blasting effects on welded steel pipelines. *7th International Pipeline Conference*, Calgary, Alberta, Canada, pp. 123–132, 2008.

[6] Carlos Santamarina, J., Klein, K.A. & Fam, M.A., *Soils and Waves: Particulate Materials Behavior, Characterization and Process Monitoring*, Wiley & Sons: Chichester, 2001.

[7] Bornitz, G., *Über die Ausbreitung der von Großkolbenmaschinen erzeugten Bodenschwingungen in die Tiefe* (in German), Springer: Berlin, 1931.

[8] SBRCURnet, *SBR Trillingsrichtlijn A: Schade aan bouwwerken: 2017* (in Dutch), Delft, 2017.

[9] Salzgitter Mannesmann Line Pipe GmbH, Inspection certificate 3.2, No. 45161227, Steel line pipe, HFW-HFI-longitudinally welded, 2025.

[10] Akhavan, J., *The Chemistry of Explosives*, 3rd ed., Royal Society of Chemistry: Cambridge, 2011.

NUMERICAL ASSESSMENT OF THE EFFECTIVENESS OF TRADITIONAL RESILIENT MOUNTING SYSTEMS IN MITIGATING VIBRATIONS OF NAVAL EQUIPMENT SUBJECTED TO UNDERWATER EXPLOSIONS

JACOPO BARDIANI, CLAUDIO SBARUFATTI & ANDREA MANES
Dipartimento di Meccanica, Politecnico di Milano, Milan, Italy

ABSTRACT

Underwater explosions (UNDEX) generate shock waves and bubble oscillations that propagate through marine structures, inducing vibrations that may compromise onboard equipment performance and reliability. The dynamic response of marine structures to these events presents significant challenges to operational safety, particularly in non-contact scenarios. While traditional resilient mounting systems are commonly employed to mitigate these effects, their efficiency is often limited by design constraints such as stiffness, target frequency range and equipment mass. This study investigates the effectiveness of resilient mounting systems in reducing structure-borne vibrations in a patrol vessel subjected to UNDEX events. A coupled acoustic–structural numerical approach, implemented in ABAQUS CAE, is used to predict the transient response and frequency behaviour of a foundation-resilient-mass subsystem supporting onboard equipment. The system is located on the ship's double bottom at mid-hull length, with an explosive charge positioned directly beneath the structure to simulate a realistic non-contact UNDEX scenario. Numerical simulations evaluate the ability of resilient mountings to attenuate high-frequency shock-induced vibrations and lower-frequency oscillations caused by bubble dynamics. Results highlight performance limitations and provide insights into their effectiveness in maintaining operational reliability under extreme loading conditions. This study advances vibration control strategies in marine structures, offering a reference for optimizing anti-shock and anti-vibration solutions in naval engineering applications.

Keywords: underwater explosion, acoustic–structural simulation, resilient, vibration control, equipment safeguard.

1 INTRODUCTION

In naval shipbuilding and offshore engineering, comprehending the dynamic response of submerged and floating structures under pressure loads is crucial for ensuring both safety and structural integrity. Among the various types of pressure loads encountered, underwater explosions (UNDEX) are particularly significant due to their substantial effects on the structural resilience and operational functionality of marine systems.

UNDEX events are typically categorized as either contact or non-contact, depending on the proximity of the detonation to the structure [1]–[3]. Contact explosions occur in direct or near-direct contact with the hull, often resulting in severe structural damage such as ruptures. Conversely, non-contact explosions take place at a certain distance from the structure but still generate intense shock waves and oscillating gas bubbles, leading to both elastic and plastic deformations. Although these events are less likely to cause immediate hull breaches, they present considerable risks to onboard equipment and overall operational efficiency.

In non-contact UNDEX scenarios, two primary loading mechanisms dominate: high-frequency shock waves and low-frequency gas bubble pulsations [2], [4]. Shock waves induce significant structural vibrations, exciting the ship's hull across a broad spectrum of frequencies. These vibrations can propagate throughout the vessel, potentially affecting equipment and components far from the initial point of impact [5]. Meanwhile, gas bubble

www.witpress.com, ISSN 1743-3509 (on-line)
doi:10.2495/SUSI250071

oscillations, despite their lower frequency, can generate large-scale global motions such as whipping, further intensifying the risks to onboard systems [6].

To mitigate the adverse effects of these vibrations on onboard equipment, resilient mounting systems are commonly employed. These systems, designed to absorb and dissipate vibrational energy, can be categorized as either elastic or elastic–plastic, depending on their ability to recover post-shock. However, designing effective resilient mountings is a complex task influenced by multiple factors, including the mass of the equipment, system stiffness, target frequency range, and mounting location [7].

This study investigates the effectiveness of resilient mounting systems in mitigating vibrations resulting from non-contact UNDEX events. A patrol vessel is selected as the case study, and a high-fidelity numerical model is developed using a coupled acoustic–structural approach within ABAQUS CAE. This modelling technique effectively captures the dynamic behaviour of the ship's structure and onboard equipment, allowing for a comparative assessment of different resilient mounting configurations in vibration reduction and equipment protection.

The paper is structured as follows: Section 2 introduces the case study vessel and outlines the numerical modelling approach. Section 3 discusses the traditional resilient mounting system as a solution for vibration control, detailing its design, implementation, and expected performance. Section 4 presents the comparative analysis results, and Section 5 concludes with key findings and recommendations for future research.

2 NUMERICAL MODELLING SET-UP

This section outlines the computational framework employed to investigate the structural response of a patrol vessel subjected to far-field underwater explosions (UNDEX). A representative geometry of the Swedish patrol vessel KBV 202 (Fig. 1(a) and (b)), belonging to the KBV 201 class, is utilized to validate the proposed methodology. The vessel has a displacement of 476 tons, a length of 52 m, a beam of 8.6 m and a designed draught of approximately 3.2 m. Its structural configuration follows a conventional mixed design, incorporating both longitudinal and transverse reinforcements. A view of the structure is depicted in Fig. 1(c).

The numerical model is developed to accurately capture the vessel's behaviour under extreme loading while maintaining computational efficiency. Implemented within ABAQUS CAE, it employs a coupled acoustic–structural approach to account for fluid–structure interactions. Due to the lack of detailed information regarding internal arrangements and secondary stiffeners, a simplified structural representation is adopted. This model derived using Smith's method and the MAESTRO software for ship hull strength evaluation [8], assumes a uniform plating thickness of 0.04 m.

The structural domain is discretized using S4 shell elements with an average mesh size of 0.1 m. This resolution is determined based on established guidelines for acoustic analyses [9], ensuring that the modal behaviour of the vessel is accurately represented while optimizing computational resources. The maximum frequency of interest is set at 100 Hz.

The surrounding fluid domain is meshed with 1,374,293 AC3D8R acoustic brick elements. The mesh size is chosen following a criterion that guarantees at least 12 internodal divisions per shortest acoustic wavelength. Considering a maximum frequency of 300 Hz, the upper limit for element size is set at 0.96 m. Non-reflecting boundary conditions are applied at the outer domain edges, while a free-surface boundary condition is enforced at the top to accurately capture wave propagation. The domain extends to a radius of 15 m, meeting the necessary conditions for effective wave absorption. Furthermore, cavitation effects –

(a)

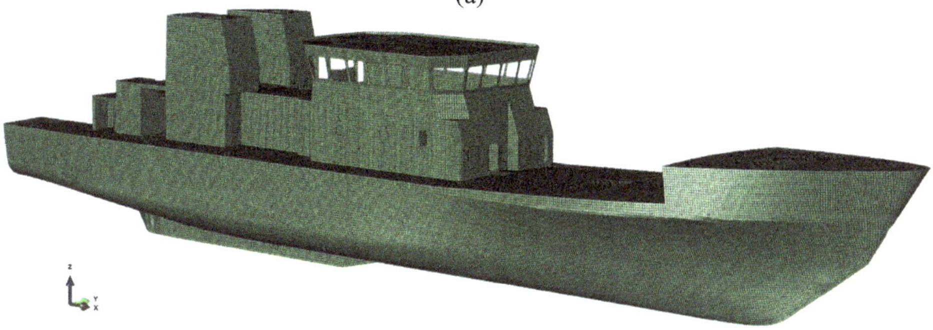

(b)

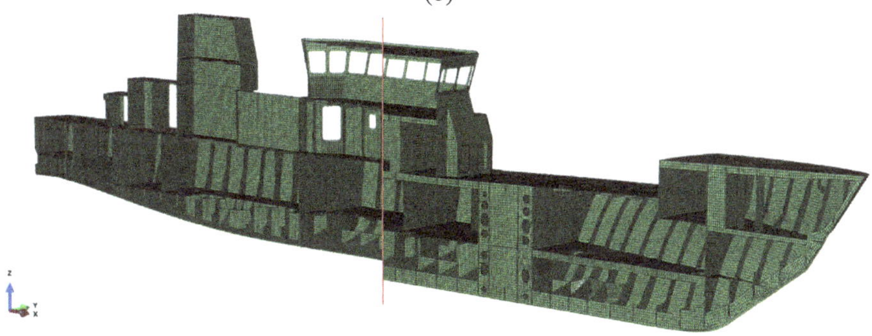

(c)

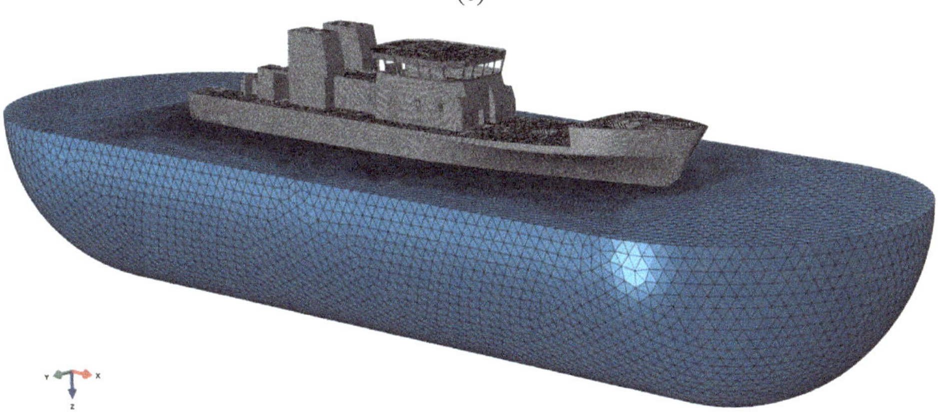

(d)

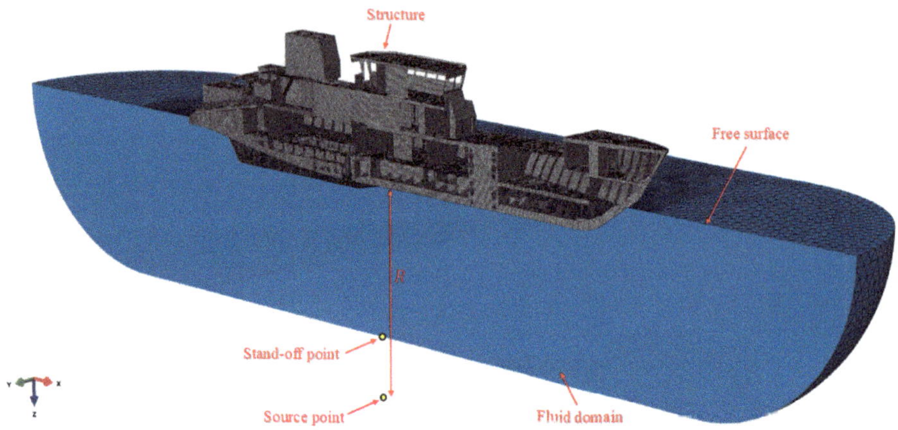

(e)

Figure 1: (a) CAD geometry of the vessel created in Rhinoceros suite; (b) Numerical model of the ship structure; (c) Internal structural arrangement; (d) Global FEM model of the ship portion surrounded by the external fluid; and (e) Vertical section of the model presented in (d).

including both local and bulk cavitation – are incorporated using the total wave formulation in the transient analysis. The coupled acoustic–structural numerical model is illustrated in Fig. 1(d) and (e).

The vessel's primary structural material is naval mild steel, characterized by a Young's modulus of 210 GPa, a Poisson's ratio of 0.3, and a density of 7850 kg/m^3. Plastic behaviour is modelled using the Johnson-Cook material law, with parameters sourced from Lin et al. [10]. However, damage modelling is excluded as deformations remain within the elastic regime. The water properties are defined with a bulk modulus of 2.03 GPa, a density of 1024 kg/m^3, and a sound velocity of 1500 m/s. Fluid–structure interaction is enforced through surface-based tie constraints along the wetted interface.

To simulate a torpedo detonation, a 516 kg TNT charge is placed at a depth of 31.6 m and a stand-off distance of 28.4 m. The total simulation duration is set to 1 s to capture gas bubble pulsations, which are modelled using the Geers and Hunter approach in ABAQUS. The secondary shockwave arrival time is estimated at 0.771 s, following Cole's empirical formulae [1].

3 TRADITIONAL RESILIENT MOUNTING SYSTEM AS SOLUTION TO VIBRATION CONTROL

To assess the vibrational effects on onboard equipment, a representative subsystem is included in the model. A schematic of this configuration is presented in Fig. 2. The subsystem comprises a foundation (explicitly modelled), resilient elements (spring-type connectors), and a cylindrical mass representing a generic equipment to be protected. The resilient supports, selected from the Vulkan MG series [11] (Fig. 3), feature a stiffness of 1.5×10^5 N/m. The cylindrical mass is assumed to be made of steel (density 7850 kg/m^3) and weighs 270 kg, with height constraints dictated by spatial limitations within the compartment.

This subsystem is positioned within a compartment at the extreme aft section of the vessel, near the primary shock arrival location, as shown in Fig. 4. The foundation is meshed with shell elements and affixed to the deck using tie constraints, while the spring connectors are kinematically coupled to the mass. Since ABAQUS Explicit does not allow spring elements to function in a single direction, additional kinematic constraints are introduced to restrict in-plane motion.

The selection of this type of resilient mount from the catalogue (Fig. 3) was based on the specific application requirements and the nominal load of the equipment, ensuring an appropriate balance between vibration isolation efficiency and structural compatibility.

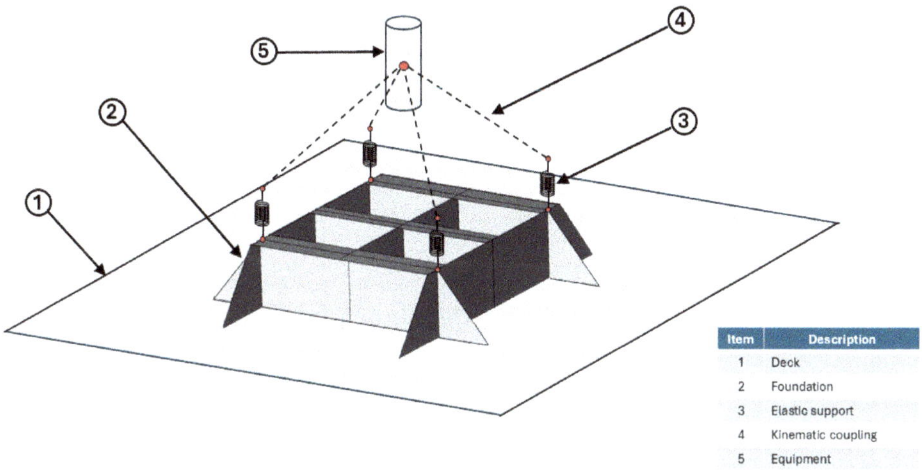

Figure 2: Schematization of the subsystem composed of the foundation, resilient elements, and equipment.

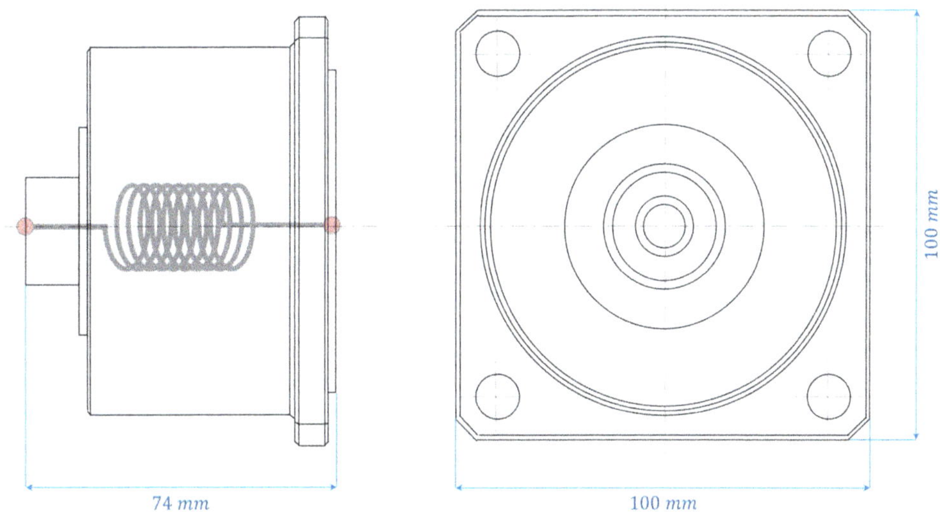

Figure 3: The actual geometry of the elastic supports modelled as spring elements and inspired by the Vulkan Group's MG Series [11].

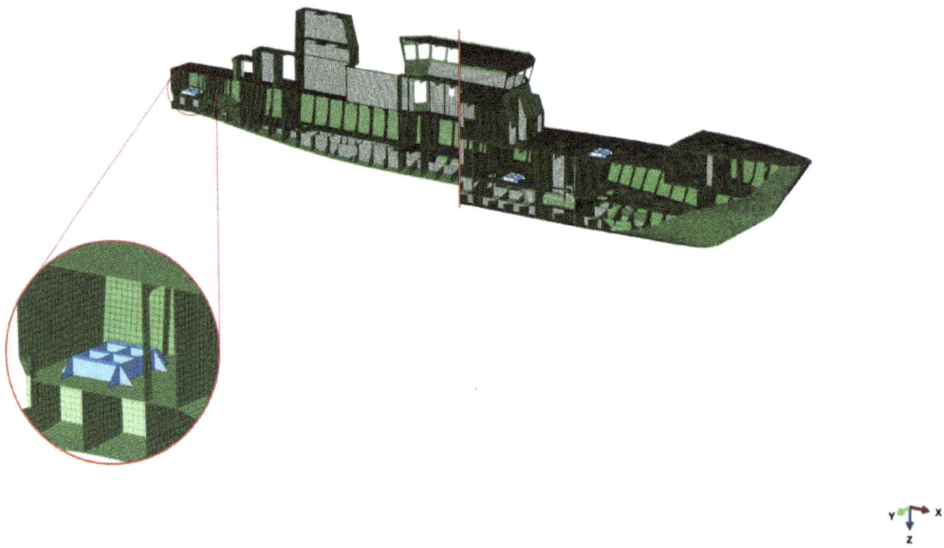

Figure 4: Installation of the foundation-resilient-equipment subsystem inside the patrol vessel.

4 RESULTS AND DISCUSSION

This section provides a description of the response of the foundation-resilient-equipment subsystem subjected to the UNDEX event outlined in the previous sections. As an example, Fig. 5 documents the dynamic acoustic pressure (POR) in a longitudinal section of the entire numerical model, at the time instant where the primary shock wave reaches the structure. The recorded POR values are consistent with those provided by analytical formulations available in the literature, demonstrating the numerical model's reliability and accuracy in predictions.

Fig. 6 presents the simulation results for the foundation-resilient-mass subsystem over a 1-second interval. The plot compares the vertical acceleration recorded on the mass (black) with that measured at the central point of the foundation's upper surface (red). The system's dynamic response to the underwater explosion follows three distinct phases: an initial sharp peak upon the arrival of the primary shock, a subsequent phase of oscillations, and a final phase characterized by the first gas bubble pulsation.

The foundation's vertical acceleration initially peaks at -1000 m/s^2 (upward), while the mass responds with a much lower peak of ± 50 m/s^2. In the next phase, the foundation's oscillations gradually diminish, with the mass showing a smoother response. Around 0.8 s, distinct peaks in the mass's acceleration correspond to the first gas bubble pulsation, while the foundation continues exhibiting sharper oscillations.

The results of Fig. 6 confirm that the resilient mounting system – featuring the Vulkan material – effectively reduces the transmission of high-frequency vibrations from the foundation to the mounted equipment. The comparison between the two acceleration curves clearly demonstrates the system's ability to decouple the mass from the more severe and higher-frequency dynamic content of the foundation, thus protecting the mounted equipment. Nonetheless, the response to low-frequency excitations, such as those associated with bubble pulsation, remains partially transmitted due to the limited damping effectiveness in that frequency range.

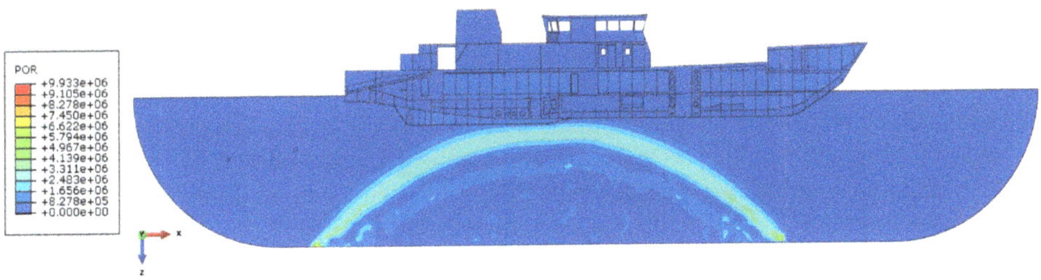

(a)

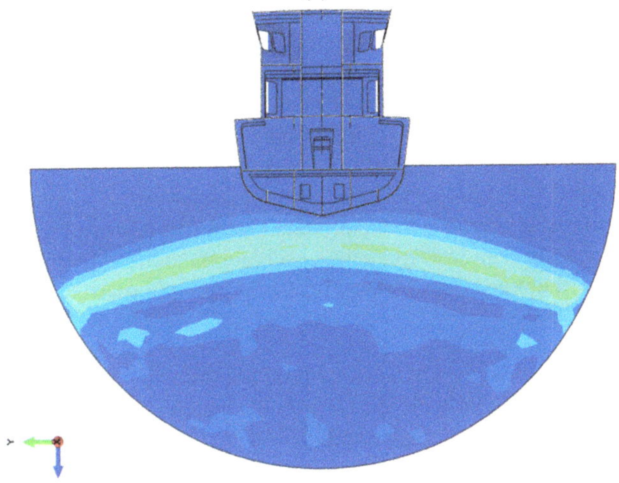

(b)

Figure 5: Total dynamic acoustic pressure (POR) field (in Pa) at the time instant where the shock hits the structure: (a) $t = 0.007$ s (longitudinal section); and (b) $t = 0.007$ s (transversal section).

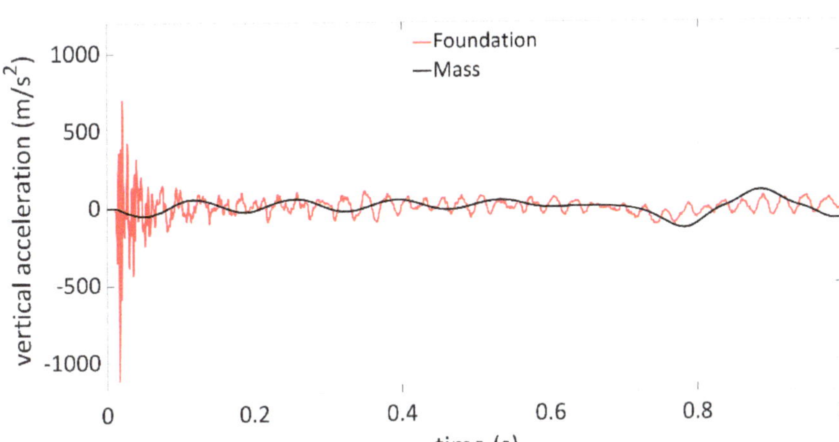

Figure 6: Comparison between the vertical acceleration versus time plots of the masses under study and its respective foundations accelerations.

5 CONCLUSIONS

The dynamic response of marine structures to underwater explosions (UNDEX) poses a significant challenge to operational safety and structural integrity, especially in non-contact scenarios. Shock waves and bubble oscillations from UNDEX events generate vibrations that propagate through the hull, adversely affecting onboard equipment and its performance. While traditional resilient mountings are commonly used to mitigate these effects, their efficiency is often constrained by design limitations, such as stiffness, target frequency range, and equipment mass.

This study investigates the effectiveness of traditional resilient mountings in mitigating vibrations induced by non-contact UNDEX events and explores the potential of advanced metamaterials as a future enhancement. A patrol vessel served as the reference military ship model for this study. A foundation-resilient-mass subsystem, designed to support onboard equipment, was positioned inside a compartment located at the aft section of the hull. Its dynamic response was assessed during a non-contact UNDEX event, with the explosive charge placed directly beneath the hull.

The results showed that traditional resilient mountings provided a degree of vibration attenuation, particularly in response to the primary shock wave. However, their effectiveness was limited when dealing with low-frequency oscillations induced by the gas bubble. These findings emphasize the need for further research into alternative vibration mitigation strategies [12]–[15].

ACKNOWLEDGEMENTS

This work is part of the dTHOR project, which has received funding from the European Defence Fund (EDF) under Grant agreement 101103257 – dTHOR – EDF-2021-NAVAL-R-2.

REFERENCES

[1] Cole, R.H. & Weller, R., Underwater explosions. *Phys Today*, **1**, p. 35, 1948. https://doi.org/10.1063/1.3066176.

[2] Aman, Z., Weixing, Z., Shiping, W. & Linhan, F., Dynamic response of the non-contact underwater explosions on naval equipment. *Marine Structures*, **24**, pp. 396–411, 2011. https://doi.org/10.1016/j.marstruc.2011.05.005.

[3] Bardiani, J., Sbarufatti, C. & Manes, A., Transfer learning with deep neural network toward the prediction of the mass of the charge in underwater explosion events. *Journal of Marine Science and Engineering*, **13**(2), p. 190, 2025. https://doi.org/10.3390/jmse13020190.

[4] Tran, P., Wu, C., Saleh, M., Bortolan Neto, L., Nguyen-Xuan, H. & Ferreira, A.J.M., Composite structures subjected to underwater explosive loadings: A comprehensive review. *Compos. Struct.*, **263**, 2021. https://doi.org/10.1016/j.compstruct.2021.113684.

[5] Ming, F.R., Zhang, A.M., Xue, Y.Z. & Wang, S.P., Damage characteristics of ship structures subjected to shockwaves of underwater contact explosions. *Ocean Engineering*, **117**, pp. 359–382, 2016. https://doi.org/10.1016/j.oceaneng.2016.03.040.

[6] Didoszak, J.M. & Kwon, Y.W., Failure assessment of shipboard equipment subjected to underwater explosions. *Multiscale and Multidisciplinary Modeling, Experiments and Design*, **8**, 2025. https://doi.org/10.1007/s41939-024-00637-7.

[7] Du, J.-Y., Du, Z.-P., Li, Y. & Shao, K., The progress of naval ship explosion protection technology. *Binggong Xuebao/Acta Armamentarii*, **36**, pp. 391–400, 2015.

[8] Smith, C.S., Influence of local compressive failure on ultimate longitudinal strength of a ship's, hull. *Proceedings of the International Symposium on Practical Design in Shipbuilding (PRADS)*, Tokyo, 17–21 Oct., pp. 73–79, 1977.

[9] Kyaw Oo, D., D'Amore, G., Morgut, M., Biot, M., Mauro, F. & Kašpar, J., Integration and optimization of the after-treatments systems to reduce the acoustic footprint of the ships. *Applied Acoustics*, **213**, 2023. https://doi.org/10.1016/j.apacoust.2023.109625.

[10] Lin, L., Zhi, X.-D., Fan, F., Meng, S.-J. & Su, J.-J., Determination of parameters of Johnson-Cook models of Q235B steel. *Zhendong Yu Chongji/Journal of Vibration and Shock*, **33**, pp. 153–158+172, 2014. https://doi.org/10.13465/j.cnki.jvs.2014.09.028.

[11] Vulkan Group, MG series resilient mounts catalogue. https://www.vulkan.com/. Accessed on: 28 Jan. 2025.

[12] Ji, J.C., Luo, Q. & Ye, K., Vibration control based metamaterials and origami structures: A state-of-the-art review. *Mech. Syst. Signal Process*, **161**, 2021. https://doi.org/10.1016/j.ymssp.2021.107945.

[13] Pelat, A., Gautier, F., Conlon, S.C. & Semperlotti, F., The acoustic black hole: A review of theory and applications. *J. Sound Vib.*, **476**, 2020. https://doi.org/10.1016/j.jsv.2020.115316.

[14] Krylov, V., Acoustic black holes: Recent developments in the theory and applications. *IEEE Transactions on Ultrasonics, Ferroelectrics, and Frequency Control*, **61**, pp. 1296–1306, 2014. https://doi.org/10.1109/TUFFC.2014.3036.

[15] D'Amore, G., Rognoni, G., Biot, M. & Mauro, F., Acoustic black holes: The new frontier for soundproofing on board ships. *Advances in the Analysis and Design of Marine Structures*, eds J.W. Ringsherg & C. Guedes Soares, CRC Press: London, pp. 179–184, 2023. https://doi.org/10.1201/9781003399759-20.

THE INCEPTION OF A RAPID PREDICTOR FOR BLAST WAVE SHIELDING: THINKING RADI(C)ALLY

JAY KARLSEN[1], ADAM A. DENNIS[1] & SAMUEL E. RIGBY[1,2]
[1]School of Mechanical, Aerospace & Civil Engineering, University of Sheffield, Sheffield, UK
[2]Arup Resilience, Security & Risk, Manchester, UK

ABSTRACT

A rapid and accurate predictor of blast effects in a geometrically complex urban environment has considerable practical value. Therefore, it is imperative to understand the behaviours that arise from blast–obstacle interaction; in particular the 'primary' behaviours, like shielding and channelling, which govern the magnitude and distribution of blast loading globally. However, these are entangled with highly-localised, second-order behaviours, like clearing and incidence effects, whose sensitivity and non-linearity have impeded the development of a generalised, first-order predictor for the primary behaviours. Consequently, this work begins to decouple them to permit a study of shielding in isolation. It is demonstrated that many of these secondary effects stem from the mismatch in geometry between the curved blast wave and the flat faces of the rectilinear obstacles commonly studied. Therefore, by curving the obstacle to match the wavefront, a planar interaction is compelled at all stand-off distances, and these secondary behaviours are eliminated. With the effects of shielding isolated, a parametric study of 273 curved wall geometries is undertaken in *Viper::Blast*, varying the footprint dimensions of an infinitely tall, rigid structure, and its stand-off from a hemispherical surface charge. Full-field peak specific impulse is measured at ground level and the data is used to demonstrate a fundamental consistency in the distribution of the loading across the full extent of the shielded region. A conservative, reduced-order representation for the distribution of peak specific impulse within the shielded region is developed and it is shown to be generalisable for annular sector obstacles of any stand-off distance, size or scale. Future work shall leverage this in the development of a fast-running engineering model for the prediction of the blast loading in the wake of an obstacle.

Keywords: blast, shielding, blast–obstacle interaction, parametric, numerical modelling, polar.

1 BACKGROUND

Blast wave shielding is the behaviour that describes the mitigation of blast loading in the wake of an obstacle [1], [2]. Shielding can thus be leveraged by blast protection engineers [2] and, more generally, it (along with channelling) fundamentally dictates blast loading parameters in an urban environment [1]. Understanding, quantifying, and predicting the mechanisms involved in shielding is therefore of immense practical significance, and as a consequence, blast wave shielding has been extensively studied. There are many independent numerical and experimental investigations that parametrically explore the relationship between an obstacle's dimensions and peak pressure and impulse reduction, often relative to the unobstructed case (e.g., [3]–[9]). Obstacles are shown to have considerable mitigative effects, reducing peak pressure by up to 80% [4], [6], peak specific impulse by up 60% [4], and sustaining the loading reduction to significant distances behind the obstacle [2].

From these investigations, a diverse array of predictive tools have been produced for the estimation of the reduced loading caused by shielding. There are largely two loading cases for which there are predictors: the reflected loading on a structure protected by a leading obstacle (e.g., [3]–[6]), and the full-field incident effects in the wake of a lone obstacle (e.g., [7]–[10]). The latter is of particular interest herein, as estimating full-field conditions rapidly and accurately is a prerequisite to the post-blast inverse characterisation of explosive events,

WIT Transactions on The Built Environment, Vol 215, © 2025 WIT Press
www.witpress.com, ISSN 1743-3509 (on-line)
doi:10.2495/SUSI250081

which has implications for the coordination of emergency life-saving response efforts [11]. It is observed that, in all cases, these incident predictors are developed from datasets measured at an arbitrary lateral extent, e.g., measurements constrained to ground level loading, or elevations below the obstacle's height (or some multiple of it). However, the lateral extent of an obstacle's influence grows considerably with the distance behind it [12]. Consequently, none of these predictive methodologies can estimate the full extent of an obstacle's shielding, nor the magnitude and distribution of the loading between this boundary and the spatial limits of their measurements.

Therefore, this work seeks to expand on prior parametric studies to characterise the influence of an obstacle's dimensions and stand-off distance on its mitigation of blast loading. In particular, the full lateral extent of the shielded zone shall be identified, with the view to support its prediction in forthcoming work. Notably, the rectilinear blast wall, on which prior studies have focused, is posited to be somewhat limiting in the insight it offers into shielding as a fundamental blast–obstacle interaction behaviour. Consequently, an alternative geometry is adopted – that of an annular sector – to eliminate the interference of several second-order behaviours that otherwise superpose with shielding, distorting the measurable effects by an unquantifiable amount. It is shown that the shielded region, once its full extent is isolated, exhibits an extremely consistent pattern of behaviour, irrespective of the distance behind the obstacle at which the effects are measured, and irrespective of the dimensions, scale and stand-off distance of the obstacle. Finally, a physically rationalised, reduced-order model is developed for this consistent behavioural regime.

2 DESIGNING THE OBSTACLE

An investigation of fundamental blast–obstacle interaction behaviour reasonably necessitates the study of lone obstacles of relatively simple geometry. Published literature seemingly conforms with this as a rectilinear, planar blast wall (often of infinite length) is commonly employed in the study of shielding effects [3]–[9]. The simplicity of the rectilinear case may have been assured during the nuclear-scale investigations of the mid-twentieth century [2], when 'an atomic bomb [was] the most likely form of attack against surface structures' [13, p. 239]. Considering explosive yields on the order of kilotonnes and megatonnes [14], the planar shock assumption is approximately valid [15] and thus impingement on a rectilinear obstacle would be uniform and perfectly orthogonal. However, modern threats have net explosive quantities between three and six orders of magnitude lower [16], meaning the assumption of a planar shock ceases to be appropriate and a blast wave's curvature must be considered [2].

Fig. 1(a)–(d) illustratively contrasts the interaction of a planar and a curved blast wave with a rectilinear obstacle. In both cases, the mechanisms of reflection, clearing and shielding are present. These shall be termed 'primary' behaviours, as they are guaranteed to manifest from the interaction of a blast wave with any finitely sized obstacle. In the case of the curved wave's impingement, an array of additional 'secondary' behaviours arise from the geometric mismatch between the spherically expanding shock and the flat wall, e.g., variability in incident loading parameters across the obstacle's face due to differing stand-off distances point-to-point [17]; oblique reflection with variation in incident angle [10]; the potential for Mach reflection [14], [18]; local shielding in the shadow of the front corners [12], [13], [19]. These secondary behaviours produce unquantifiable interference with the flow-field, superposing with the shielding (and other primary behaviours) in a way that cannot be decoupled.

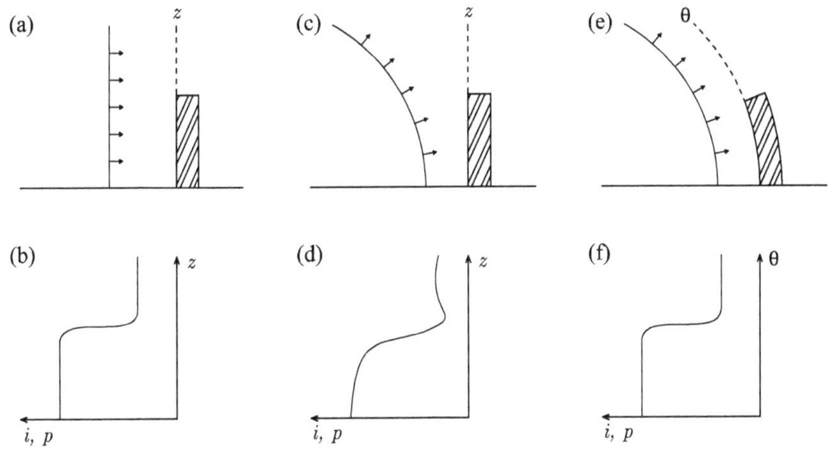

Figure 1: An illustrative contrast of planar and oblique blast–obstacle interaction. (a), (c) and (e) Radial cross-section through a blast wave's propagation; (b), (d) and (f) Representative peak specific impulse (i) and pressure (p) distribution along and beyond the reflecting plane for the scenarios depicted in (a), (c) and (e), respectively. Note the disrupting effects of the superposing secondary behaviours in (d) compared to planar interaction of (b) and (f).

Ultimately, the data-driven analysis of such a scenario is inappropriate when trying to understand any singular interaction behaviour. As such, despite the apparent simplicity of the rectilinear obstacle, an alternative geometric configuration is required to compel the manifestation of only the primary behaviours and thereby isolate 'pure' shielding.

A planar shock provides the necessary behavioural simplifications to enable a direct and isolated study of shielding, however its conventional representation is impractical to assume. Fig. 1(e) and (f) offers an alternative to this: rather than 'flattening' the shock to match the obstacle, the obstacle is instead curved to match the shock. In so doing, the analytical benefits of the planar shock assumption are obtained whilst guaranteeing physical validity for obstacles of any stand-off distance, size or scale.

The interacting geometry studied herein shall be annular sectors that match exactly the plan curvature of a free-field blast wave's spherical expansion. Using these, a data-driven, numerical investigation is undertaken to identify the extent of a single obstacle's shielding in the absence of localised, sensitive and interfering secondary behaviours.

3 MODELLING AND POST-PROCESSING

3.1 Parametric geometry

The magnitude and distribution of peak specific impulse in the wake of an obstacle is to be parametrically investigated with respect to the geometry of an annular sector. The variables considered are illustrated in Fig. 2 and defined below:

- Stand-off distance to rear face, R (m).
- Projected angle, θ (°), as a measure of lateral extent.
- Obstacle depth, d (m), measured radially.

The charge is a hemispherical ground burst (0.5 kg), and the longitudinal dimension of the obstacle is infinite – as is typical of prior investigations. Note, these two specifications permit the subsequent findings to be applicable to either a blast wall of infinite length or a building of infinite height.

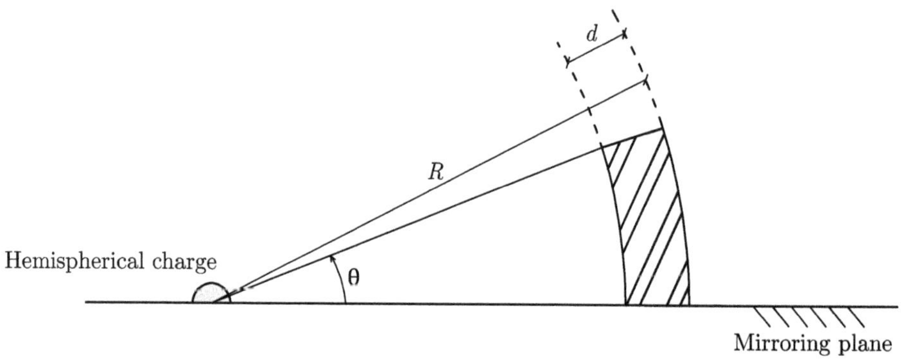

Figure 2: Plan elevation of the annular sector's geometry, with the parametric variables indicated.

The values sampled for each variable are as follows: R (m) = {4, 5, … 9, 10}; θ (°) = {10, 15, … 35, 40}; d (m) = {0.04, 0.2, 0.5, 1, 2, 3}.

To eliminate the potential for near-field blast–obstacle interaction and the associated complexities [20], variable combinations where the obstacle's front face stand-off distance is less than 3 m/kg$^{1/3}$ were neglected from the study, i.e., $R - d \geq 3$ m/kg$^{1/3}$. This reduces the number of unique geometry combinations available to study from 294 to 273.

3.2 Simulation set-up

Viper::Blast is a commercially available computational fluid dynamics software that '[predicts] blast loads on structures of non-trivial geometry' [21, p. 6]. Its predictive accuracy has been experimentally validated for blasts in both unobstructed and obstructed environments [21], [22], including cases of shielding [23]. Consequently, *Viper::Blast* is appropriate for use in the subsequent numerical investigation of shielding.

Beyond a scaled distance of 2 m/kg$^{1/3}$, the load predicted by both the isothermal burst and multi-material representations of a charge in *Viper::Blast* are shown to have converged [21]. Therefore, when modelling far-field interactions, the more computationally efficient isothermal burst method is commonly adopted [21]–[23]; this will be repeated herein.

Each geometric configuration is therefore simulated independently within *Viper::Blast*, importing the annular sector obstacle geometries as a triangulated surface mesh from a stereolithography file. The simulations leverage 1D to 3D remapping, and quarter symmetry about the ground plane and obstacle centreline to improve computational efficiency; eighth symmetry was considered, but interfering reflections from the fictitious second obstacle occurred with small values of R. All boundaries are therefore transmissive, besides the symmetry planes which are rigid reflecting. To produce the necessary 0.5 kg hemispherical ground burst, a unit mass sphere of TNT is initialised.

A polar arrangement of probes is used to record the spatial and temporal variation in pressure at ground level, at stand-off distances between 1 m and 15 m. The probes are radially separated by two cell lengths and circumferentially separated by 1°. From these, peak specific impulse at each measurement point is approximated via numerical integration of the pressure–time history. It was confirmed for all obstacle geometries that a 70° arc of probes is sufficient to capture the full lateral extent of the shielded region.

Following a systematic domain study, as outlined by Chester et al. [23], it was confirmed that the dimensions of the domain are such that rarefaction waves from the ambient transmit boundaries arrive too late to interfere with the peak specific impulse, though a 2 m 'buffer' is required [22] (reducing the extent of the effective data collection zone to 13 m). A subsequent mesh refinement study identified sufficient convergence in peak specific impulse for unit cells of length 0.04 m.

3.3 Post-processing

Throughout, peak specific impulse is reported relative to the unobstructed case; $i_{p,r}$ is the measured specific impulse as a percentage proportion of the free-field value.

Fig. 3(a) depicts a spatial map of the post-processed relative peak specific impulse for a representative wall geometry. The colour map highlights three distinct and intuitive behavioural regions: relative enhancement, due to the front face reflection; relative reduction where the obstacle's redistribution of energy means that the loading is locally lessened; and a region of no significant change, where (beyond some distance) the obstacle causes the peak impulse to be no more or less severe than the unobstructed case. For clarity, Fig. 3(b) extracts the circumferential distribution of $i_{p,r}$ at the common radius highlighted in Fig. 3(a).

Herein, a point is considered 'shielded' if its relative peak specific impulse is 97.5% or less, i.e., the obstacle must reduce peak impulse by at least 2.5% from the free-field event. All data points that do not meet this criterion, e.g., the solid black lines in both Fig. 3(a) and (b), are then eliminated from consideration, as illustrated by the shaded region in Fig. 3(b).

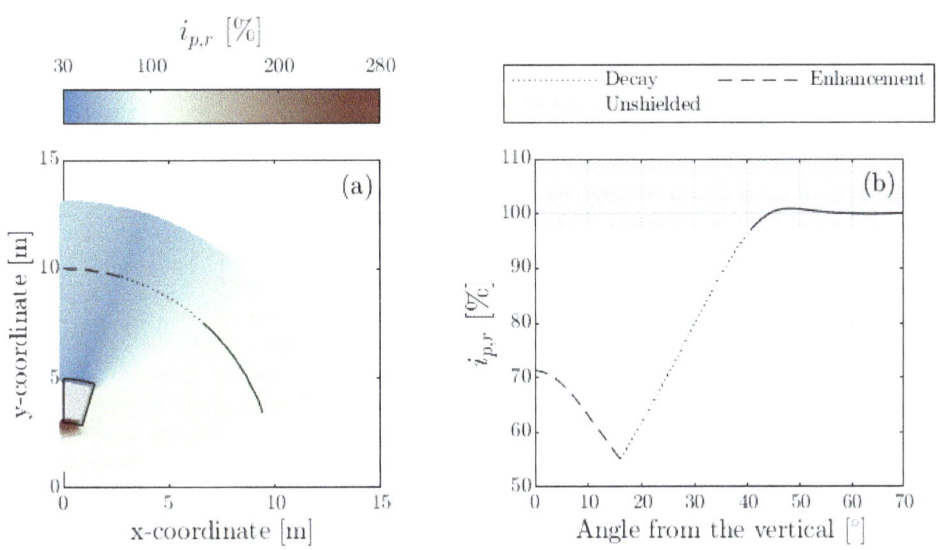

Figure 3: Spatial distribution of relative peak specific impulse for a representative obstacle (R = 5 m, θ = 35°, d = 2 m). (a) Plan elevation, with the charge positioned at the origin, and the obstacle is indicated by the grey block; and (b) Circumferential distribution at a constant radial distance.

4 DATA INTERROGATION AND INSIGHTS

4.1 Behaviours in the shielded region

Using Fig. 3(b), it is apparent that it is dichotomised between two regimes, termed 'decay' and 'enhancement'. These are physically rationalised as follows.

Consider the blast wave as it propagates perfectly incidentally and one-dimensionally (besides some amount of interference from the reflecting face) alongside the obstacle. At the rear corner, the rigid obstacle is replaced by atmospheric air, resulting in a pressure differential that drives the otherwise radially propagating blast wave to expand circumferentially into the obstacle's shadow. This loads the region above atmospheric conditions, but relatively less than had the obstacle not been present. As the shock travels deeper into the shadow, the loading lessens with increasing distance [17]. It is this process that rationalises the initial decay of the shielded region (depicted in Fig. 3(b), as the angle from the vertical reduces from 40° to 30°). Note, this expansion induces a rarefaction that propagates outwards from the obstacle into the otherwise free-field loading conditions, extending the load reduction effects beyond the obstacle's immediate shadow.

As the shock continues to diffract circumferentially, it reaches a distinct minimum value in measured impulse (17° from the vertical in Fig. 3(b)). Thereafter, the loading increases due to superposition with the wave that diffracted around the opposite side of the obstacle. The effect of this interference is demonstrably maximal along the wall's centreline, where the travelled distance for both waves is least, and their loading period is aligned. The subsequent wave, which propagates in the opposite direction to the first, excites points that have already been loaded, enhancing the measured peak impulse. As the interfering wave travels out from the centreline, not only does it continue to expand, but it arrives later and later through the first wave's loading period. Ultimately its impulse enhancement lessens until it ceases to exceed the peak of the primary wave's positive phase, hence the peak impulse in the decay region is unaffected by the superposition.

4.2 Behavioural consistency

The decay–enhancement dichotomy previously described arises exclusively and without exception for all 273 configurations simulated. Fig. 4 demonstrates this consistency in the behaviour of the shielded region. Depicted is the $i_{p,r}$ distribution as before (now with unshielded data eliminated), but for obstacles whose geometry is at the approximate upper and lower quartiles of the range of variable values considered in this investigation.

Whilst there are differences in the magnitudes of the relative peak specific impulse values across this range of problems, the pattern formed by the distribution of the loading is identical: a decay from the unshielded boundary to some minimum, followed by a rise to some maximum at the obstacle's centreline. This behavioural pattern is also true at all radial distances behind an obstacle, as demonstrated in Fig. 5. At small distances behind the wall, the decay is exaggerated further by vortex shedding [13].

Ultimately, this consistency is a product of the generalised planar shock assumption being compelled in all instances. No other behaviours can physically arise because secondary behaviours are eliminated, and the self-similarity of the blast–obstacle interaction ensures that the primary behaviours always arise in the same way.

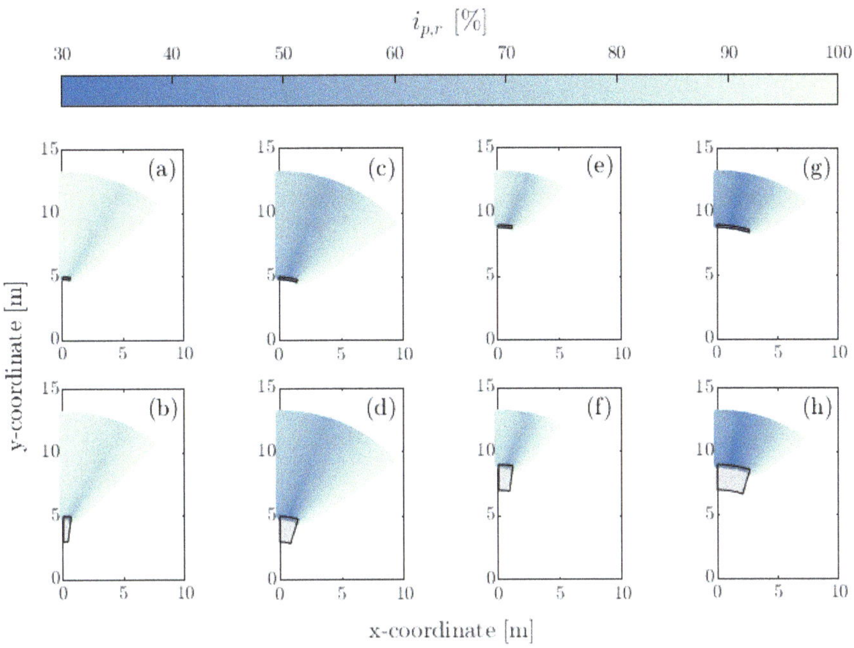

Figure 4: Spatial distribution of relative peak specific impulse in the shielded region of several obstacles. The obstacles are represented by the grey blocks, and the charge is positioned at the origin in all cases. (a) R = 5 m, θ = 15°, d = 0.2 m; (b) R = 5 m, θ = 15°, d = 2 m; (c) R = 5 m, θ = 35°, d = 0.2 m; (d) R = 5 m, θ = 35°, d = 2 m; (e) R = 9 m, θ = 15°, d = 0.2 m; (f) R = 9 m, θ = 15°, d = 2 m; (g) R = 9 m, θ = 35°, d = 0.2 m; (h) R = 9 m, θ = 35°, d =2 m.

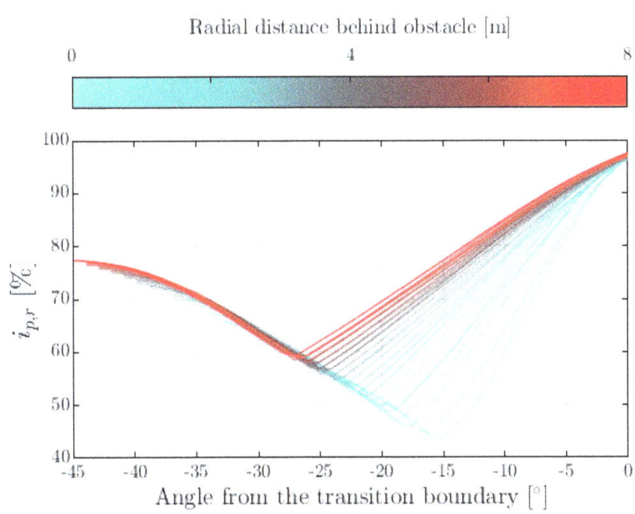

Figure 5: Circumferential distribution of relative peak specific impulse, plotted at each radial distance behind a representative obstacle (R = 5 m, θ = 35°, d = 0.2 m).

5 SELF-SIMILARITY AND PREDICTABILITY

5.1 An exploitable pattern

For several obstacles of differing geometry, Fig. 6 isolates and plots the spatial position of the transition boundary that delineates the shielded and unshielded regions. In all instances arc angle is variable whilst the obstacles' stand-off distance is constant (6 m). In Fig. 6(a), the obstacle depth is also constant (0.2 m), and it is qualitatively apparent that despite a variable arc angle, the transition boundary for all seven obstacles is approximately coincident.

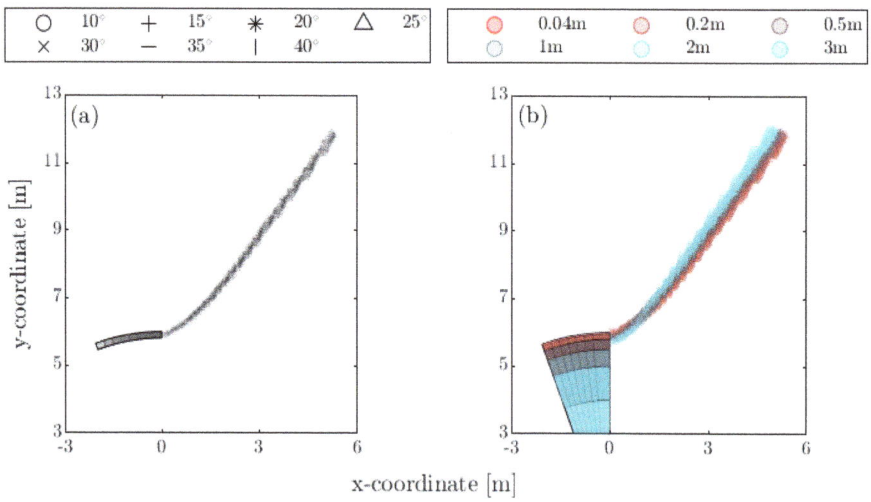

Figure 6: The spatial position of the shielded–unshielded transition boundaries of several obstacles. $R = 6$ m in all cases. (a) Obstacle θ is varied and d constant; and (b) Both θ and d are varied.

This is a consequence of two phenomena previously discussed. Firstly, the transition boundary is in the decay region beyond the minimum stream of measured peak impulse. The peak value of specific impulse at points in this region is exclusively the product of the primary wave; superposition with the mirrored diffracting shock is irrelevant here. Secondly, the initial conditions of the primary shock at the instant it begins to diffract around the rear corner of the obstacle are a function only of the stand-off distance (due to approximately one-dimensional incident propagation alongside the obstacle) and the obstacle's depth (which affects interference from the front face reflection). As such, the pressure differentials and momentum vectors that drive flow in the shielded region are identical for all these obstacles, given their common R and d. The impulse–time histories at the transition boundary are therefore identical in all these cases, diverging only upon the commencement of superposition (which occurs physically later or earlier depending on arc angle), by which time the peak value has already been reached.

Fig. 6(b) permits both arc angle and obstacle depth to vary (though only a single marker style is needed because the transition boundary has proven nominally independent of θ). It is shown that varying depth alters the spatial position of the transition boundary somewhat. However, considering that the depth has been varied by two orders of magnitude, the change in the transition boundary's position is extremely small relatively, especially given that minor variability is to be expected having used a Cartesian mesh for this radial analysis [12]. This indicates another nominal independency (a conclusion supported by previous work [8]).

The transition boundary is simply the limiting case of the decay region. Consequently, the loading at all locations between the points of minimum impulse and the unshielded transition boundary must, by definition, be nominally independent of θ and d. Fig. 7(a) confirms this to be true, plotting the circumferential distribution of $i_{p,r}$ at a constant radial distance for the same array of obstacles as in Fig. 6(b).

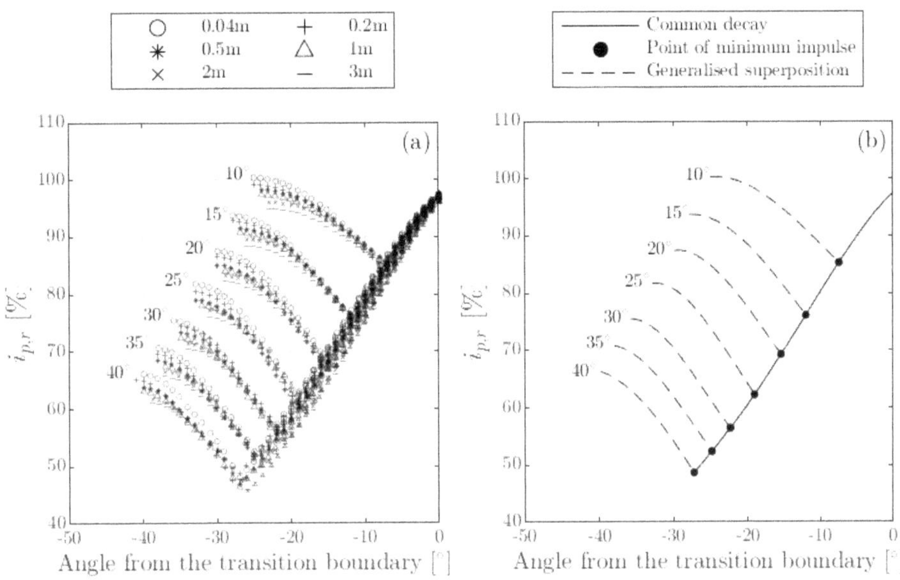

Figure 7: Circumferential distribution of relative peak specific impulse at a constant radial distance for obstacles of a constant R (6 m) and varied θ and d. (a) Raw data; and (b) Reduced-order representation.

5.2 A framework for prediction

Fig. 7(a) demonstrates that obstacles of differing arc angle and depth share a very similar decay from their transition boundary to their respective minimum, which has been physically rationalised. Branching from this common decay regime is the enhancement of impulse due to superposition. For obstacles of differing lateral extent, θ, the superposition occurs physically later or earlier, though this point of inflection occurs at approximately the same location for obstacles with the same lateral extent. Finally, though the superposition demonstrates some sensitivity to d, it is conservative in all instances to assume the obstacle to be infinitely thin.

From these observations, a reduced-order model of the impulse distribution within the shielded region can be constructed. Fig. 7(b) depicts this.

The behavioural consistency that has been demonstrated throughout this work means that the general form of this simplified model (though not its explicit magnitudes) describes the circumferential distribution of peak specific impulse at any stand-off distance in the shielded region of an annular sector obstacle of any geometry. Furthermore, because all secondary interferences were eliminated, this pattern of behaviour represents the blast–obstacle interaction behaviour that is shielding, fundamentally and in isolation. Understanding and predicting the features highlighted in Fig. 7(b) would therefore be foundational to a fast-running engineering model (FREM) for the prediction of blast wave shielding that is generalisable and extensible to any scenario.

Direct polynomial assembly (e.g., [8], [9]) seems appropriate for said predictor, given the apparent simplicity of the decay profile, though the considerable data acquired in this study would also support the application of machine learning (ML). ML has successfully been leveraged in the prediction of urban blast loading, e.g., [22], [24], [25], though, once trained, such a tool is often described as a 'black box'. Not only does this prevent the user from learning about any novel physical relationships the model may have identified, but also it means that the model may not be conforming to the physical laws that we know to be true of the real-life problem. Past work has shown that ML failed to identify physics as well understood and simple as impulse's monotonic decay with increasing distance (in free-field conditions) [26]. The ML predictor attained a 'significant performance premium' [26, p. 148] once the researcher directly implemented this physics, demonstrating that an engineer's understanding of the physics governing a problem continues to add considerable value to such analyses [27].

6 SUMMARY AND OUTLOOK

Shielding is a blast–obstacle interaction behaviour of significant practical interest. Despite extensive prior study, however, the full lateral extent of the mitigation has yet to be mapped and reliably predicted. This work therefore extends prior parametric studies, exploring the relationship between an obstacle's dimensions and the shielding in its wake. 273 unique obstacle configurations were trialled in *Viper::Blast*, capturing the full lateral extent of their mitigation, as well as the magnitude and distribution of peak specific impulse within it. Critically, rather than a more typical rectilinear form, the obstacles employed in this study were annular sectors; by matching the curvature of the spherically expanding shock, all secondary interferences that may superpose with the shielding were eliminated, permitting its directed and isolated study.

The regime of behaviours within the shielded region was demonstrated to be consistent for obstacles of any size, scale, and stand-off distance. Several physically rationalised consequences of this self-similarity were identified, particularly that the decay in impulse during the diffraction into an obstacle's shadow was nominally independent of its lateral extent and its depth. From these principles, a simplified, physics-informed representation of the impulse distribution in an obstacle's shielded region was derived. Given the elimination of all secondary behaviours, this model is representative of shielding in isolation and therefore as a fundamental blast–obstacle interaction behaviour. Notably, this reduced-order model could form the basis for a framework for predicting the loading distribution within the wake of any obstacle, extensible beyond annular sectors.

A machine learning predictor for shielding could begin to be trained from the data acquired in this work. Furthermore, the direct interrogation and manipulation of the data identified several simple physical relationships that could be used to support its training. Yet, upon reflection, it is striking that had machine learning been applied without this critical appraisal of the data, several unique and elegant manifestations of the physics may never have been noticed, appreciated, and understood by the authors. They would instead have been lost within its black box, if it ever identified them at all.

ACKNOWLEDGEMENTS

J.K. would like to thank Baldan Devecioğlu for her assistance with the drawing package used to create some of the figures in this paper, and Dr. Dain G. Farrimond and Ross Waddoups for their formative insights. This work was performed as part of an iCASE PhD studentship funded by EPSRC and Synthetik Applied Technologies.

REFERENCES

[1] Smith, P.D., Rose, T.A. & Ng, S.H., The influence of areal density on the shielding and channelling of blast by buildings. *Proceedings of the Eighteenth International Symposium on Military Aspects of Blast and Shock*, 2004.

[2] Isaac, O.S., Alshammari, O.G., Pickering, E.G., Clarke, S.D. & Rigby, S.E., Blast wave interaction with structures: An overview. *International Journal of Protective Structures*, **14**(4), pp. 584–630, 2023. https://doi.org/10.1177/20414196221118595.

[3] Beyer, M.E., Blast loads behind vertical walls. *Proceedings of the 22nd Department of Defense Explosives Safety Seminar*, pp. 767–797, 1986.

[4] Rose, T.A., Smith, P.D. & Mays, G.C., Design charts relating to protection of structures against airblast from high explosives. *Proceedings of the Institution of Civil Engineers – Structures and Buildings*, **122**(2), pp. 186–192, 1997. https://doi.org/10.1680/istbu.1997.29307.

[5] Geng, J. & Thomas, J.K., Diffracted blast loads behind structures. *Proceedings of the ASME 2020 Pressure Vessels and Piping Conference*, pp. V004T04A008-1–V004T04A008-7, 2020.

[6] Gautier, A., Sochet, I. & Lapebie, E., Analysis of 3D interaction of a blast wave with a finite wall. *Shock Waves*, **32**(3), pp. 273–282, 2022. https://doi.org/10.1007/s00193-022-01081-7.

[7] Chapman, T.C., Rose, T.A. & Smith, P.D., Reflected blast wave resultants behind cantilever walls: A new prediction technique. *International Journal of Impact Engineering*, **16**(3), pp. 397–403, 1995. https://doi.org/10.1016/0734-743X(94)00051-W.

[8] Zhou, X.Q. & Hao, H., Prediction of airblast loads on structures behind a protective barrier. *International Journal of Impact Engineering*, **35**(5), pp. 363–375, 2008. https://doi.org/10.1016/j.ijimpeng.2007.03.003.

[9] Sung, S.-H. & Chong, J.-W., A fast-running method for blast load prediction shielding by a protective barrier. *Defence Technology*, **16**(2), pp. 308–315, 2020. https://doi.org/10.1016/j.dt.2019.07.011.

[10] United States Department of Defense, Structures to resist the effects of accidental explosions, UFC-3-340-02, United Facilities Criteria: Washington D.C., USA, 2008.

[11] Karlsen, J., Farrimond, D.G., Lodge, T.J., Rigby, S.E., Tyas, A., Clarke, S.D. & Brewer, T.R., Trilateration of blast wave arrival time: An inverse method for determining explosive yield and position. *Philosophical Transactions of the Royal Society A: Mathematical, Physical and Engineering Sciences*, in press.

[12] Payne, T., Williams, A., Worfolk, T. & Rigby, S., Numerical investigation into the influence of cubicle positioning in large-scale explosive arena trials. *International Journal of Protective Structures*, **7**(4), pp. 547–560, 2016. https://doi.org/10.1177/2041419616676438.

[13] Norris, C.H., Hansen, R.J., Myle, J.H. Jr., Biggs, J.M., Namyet, S. & Minami, J.K., *Structural Design for Dynamic Loads*, 1st ed., McGraw-Hill: New York, USA, 1959.

[14] Beshara, F.B.A., Modelling of blast loading on aboveground structures: I. General phenomenology and external blast. *Computers and Structures*, **51**(5), pp. 585–596, 1994. https://doi.org/10.1016/0045-7949(94)90066-3.
[15] Glasstone, S. & Dolan, P.J., *The Effects of Nuclear Weapons*, 3rd ed., United States Department of Defense; Energy Research and Development Administration: Washington D.C., USA, 1977.
[16] United States Department of Homeland Security. IED attack fact sheet: Improvised explosive devices. https://www.dhs.gov/publication/ied-attack-fact-sheet. Accessed on: 17 Apr. 2023.
[17] Taylor, G.I., The formation of a blast wave by a very intense explosion I. Theoretical discussion. *Proceedings of the Royal Society of London. Series A. Mathematical and Physical Sciences*, **201**(1065), pp. 159–174, 1950. https://doi.org/10.1098/rspa.1950.0049.
[18] Ben-Dor, G., *Shock Wave Reflection Phenomena*, 2nd ed., Springer: Berlin, Germany, 2007.
[19] Needham, C.E., *Blast Waves*, 1st ed., Springer: Berlin, Germany, 2010.
[20] Farrimond, D.G., Rigby, S.E., Clarke, S.D. & Tyas, A., Time of arrival as a diagnostic for far-field high explosive blast waves. *International Journal of Protective Structures*, **13**(2), pp. 379–402, 2022. https://doi.org/10.1177/20414196211062923.
[21] Stirling, G.C., *Viper::Blast User Manual and Examples for Version 1.20.0*, Stirling Simulation Services Limited: Glasgow, UK, 2021.
[22] Dennis, A.A. & Rigby, S.E., The direction-encoded neural network: A machine learning approach to rapidly predict blast loading in obstructed environments. *International Journal of Protective Structures*, **15**(3), pp. 455–483, 2024. https://doi.org/10.1177/20414196231177364.
[23] Chester, A., Hazael, R. & Critchley, R., Comparative analysis of blast prediction software for far-field shock wave effects behind a blast wall. *International Journal of Protective Structures*, **0**(0), pp. 1–28, 2024. https://doi.org/10.1177/20414196241284299.
[24] Dennis, A. & Rigby, S., Prediction of blast loads using machine learning approaches. *Proceedings of the 2023 Society for Earthquake and Civil Engineering Dynamics Conference*, 2023.
[25] Dennis, A.A., A direction-encoded machine learning approach for peak overpressure prediction in urban environments. *Proceedings of the Nineteenth International Symposium on Interaction of the Effects of Munitions with Structures*, 2024.
[26] Pannell, J.J., Surrogate modelling strategies for the prediction of near-field blast impulse. PhD thesis, University of Sheffield, 2022.
[27] Karlsen, J. & Rigby, S.E., The role of AI in engineering: Towards rapid inverse blast analysis. *Proceedings of the Fourth International Conference on Structural Safety Under Fire & Blast Loading*, 2024.

DAMAGE ASSESSMENT OF H-SECTION STEEL COLUMNS UNDER VEHICLE EXPLOSIONS

SIQI CHEN & WEIFANG XIAO
College of Civil Engineering, Tongji University, China

ABSTRACT

To evaluate the damage levels of H-section steel columns in steel frame structures under vehicle explosions, this study adopts a blast scenario similar to the Oklahoma City bombing (passenger/cargo van with a TNT equivalent mass of 1,814 kg). Five standoff distances (R = 6, 7, 8, 9 and 10 m) and three angles of incidence (α = 0°, 45° and 90°) are considered to analyse the damage modes and residual load-bearing capacity of a full-scale steel column. Numerical simulations are conducted to capture the dynamic response and failure mechanisms of the column under blast loading. The results reveal that the columns may exhibit various failure modes, including tearing at flange–web connections, shear fracture at the base, global bending and torsion, and local buckling, depending on the standoff distance and angle of incidence. The study also quantifies the residual load-bearing capacity corresponding to each damage mode, providing a basis for evaluating the post-blast performance of steel columns. These findings offer valuable insights for the design and retrofitting of steel frame structures to enhance their resilience against vehicle explosions.

Keywords: steel column, damage mode, damage assessment, vehicle explosion.

1 INTRODUCTION

In steel frame structures, columns are one of the primary load-bearing components. The loss of load-bearing capacity and stiffness softening of steel columns under blast loading can lead to the collapse of the entire structure. For instance, the 1995 Oklahoma City bombing of the Murrah Federal Building, which resulted in 168 fatalities, demonstrated the catastrophic consequences of vehicle explosions. This incident has spurred increasing research interest in the blast resistance of structures. Inspired by this event, Magallanes et al. [1] replaced the concrete columns of the Murrah Building with steel columns and conducted full-scale blast tests to study the response of steel columns under similar explosion conditions. Their results showed no significant local fractures, highlighting the excellent ductility of structural steel. Similarly, Morrill et al. [2] conducted full-scale blast tests on steel columns subjected to vehicle-borne and satchel charge explosions, observing localized fractures in the columns. Due to the high cost and logistical challenges of full-scale tests, many researchers have turned to scaled experiments to study the blast performance of steel columns. For example, Grisaro et al. [3]–[5] performed close-in blast tests and numerical simulation of steel wide-sections to examine their dynamic response under shock waves along the strong and weak axes. They analysed deformation modes, including global bending and local deformation under certain conditions. Remennikov and Uy [6] investigated the behaviour of square hollow steel tubular columns under contact and near-field blast scenarios, comparing the failure mechanisms of hollow and concrete-filled tubes. Additionally, Gangolu and Grisaro [7] studied the influence of axial load ratio (ALR) on the residual load-bearing capacity of steel columns, focusing on the P-δ effect.

In this study, a numerical model is developed using LS-DYNA to investigate the effects of standoff distance and angle of incidence on the damage modes and residual load-bearing capacity of a full-scale steel column under an explosion with the TNT equivalent mass of 1,814 kg, representing a passenger/cargo van explosion.

WIT Transactions on The Built Environment, Vol 215, © 2025 WIT Press
www.witpress.com, ISSN 1743-3509 (on-line)
doi:10.2495/SUSI250091

2 NUMERICAL MODELS AND VALIDATION

A three-dimensional finite element model, incorporating high explosive, air, and steel column, was developed using LS-DYNA. To validate the model and methodology, a series of experiments conducted by Nassr et al. [8], [9] and Zhao et al. [10] were simulated.

2.1 Establishment of numerical model

The S-ALE (Structured Arbitrary Lagrangian–Eulerian) method is employed in this study, where the air domain is not explicitly modelled but defined using *ALE_STRUCTURED_MESH to specify the air mesh and corresponding coordinates. The detonation location and geometry of the explosive are defined using *INITIAL_DETONATION and *ALE_STRUCTURED_MESH_VOLUME_FILLING. The fluid–structure interaction (FSI) process is established through the keyword *ALE_STRUCTURED_FSI. The air boundaries are set using *BOUNDARY_SALE_MESH_FACE. In this study, the steel column is positioned vertically on the ground, and ground effects are considered in the analysis.

TNT is selected as the high explosive, with its material properties and equation of state defined using *MAT_HIGH_EXPLOSIVE_BURN and *EOS_JWL. Air is modelled as an ideal gas using *MAT_NULL and *EOS_LINEAR_POLYNOMIAL. The steel column incorporates the Cowper–Symonds material model to account for strain rate effects, with its material properties defined by *MAT_PLASTIC_KINEMATIC, a yield strength of 459 MPa, and parameters $C = 6{,}844\ s^{-1}$ and $P = 3.91$. Given the wide range of standoff distances studied (6–10 m), a fixed air mesh size is unsuitable. Therefore, the mesh convergence is determined by comparing the reflected impulse for air mesh sizes of $R/50$, $R/100$, $R/150$, and $R/200$ at $R = 7.0$ m and $\alpha = 0°$, as shown in Fig. 1. Since the results for $R/150$ and $R/200$ differ by only 3.9%, $R/150$ is selected as the air mesh size. For the steel column mesh, a size of 2 cm is chosen based on comparisons with experimental responses.

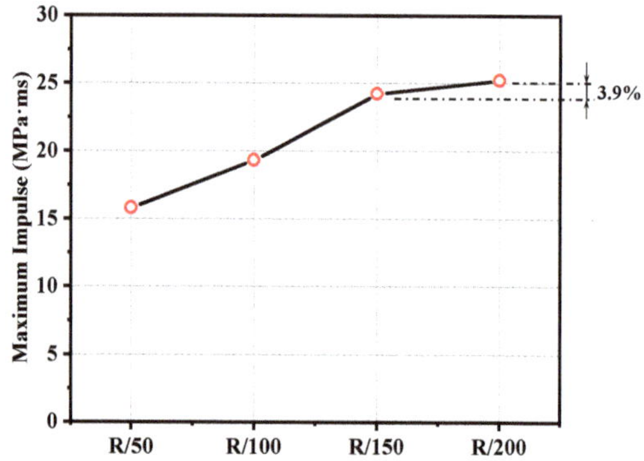

Figure 1: Mesh convergence for maximum impulses versus mesh sizes under scenario $R = 7.0$ m, $\alpha = 0°$.

This study employs a four-stage simulation method, as illustrated in Fig. 2. The four stages include:

- Step I: preloading of the axial force on the steel columns;
- Step II: shock wave propagation and its interaction with the steel columns;

- Step III: free oscillation of the steel columns;
- Step IV: axial loading phase for the residual capacity of steel columns after the blast.

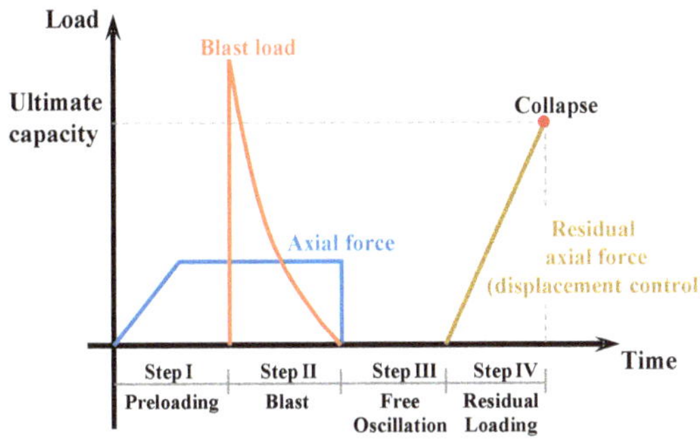

Figure 2: Numerical algorithm.

2.2 Validation of numerical model

To the best of the authors' knowledge, no tests have been conducted on steel columns that comprise all four stages mentioned above. Therefore, two separate experiments were used to simulate these stages individually.

The first experiment involves the blast tests on H-section steel columns conducted by Nassr et al. [8], [9]. The experimental setup and the quarter-symmetric finite element model are shown in Fig. 3. In the validated scenario, 100 kg of ANFO explosive was placed 10.3 m from the structure. Since the scaled distance exceeds 1.0 m/kg$^{1/3}$, the shape effect of the explosive can be neglected, and it is modelled as a sphere. The geometric dimensions of the steel column are $b \times h \times h_w \times h_f = 102 \times 160 \times 7 \times 10$ mm, where h and b are the sectional height and width, and h_f and h_w are the thicknesses of the flange and web. The column is supported by a hinge at the top and a roller support at the bottom. Plywood panels in between two neighboured column prevents shockwave diffraction and is modelled as a flexible body with a failure strain of 0.25 to control material erosion. The displacement–time history curve at the midpoint of column 2C1 and the simulated residual deformation of the three columns after the blast are shown in Fig. 4 and Fig. 5, exhibiting a good agreement with the experimental data.

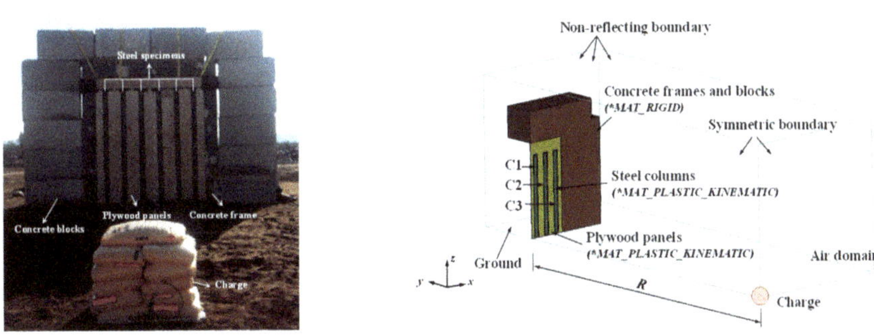

(a) (b)

Figure 3: Nassr's test: (a) Blast test setup [8], [9]; and (b) corresponding FE model.

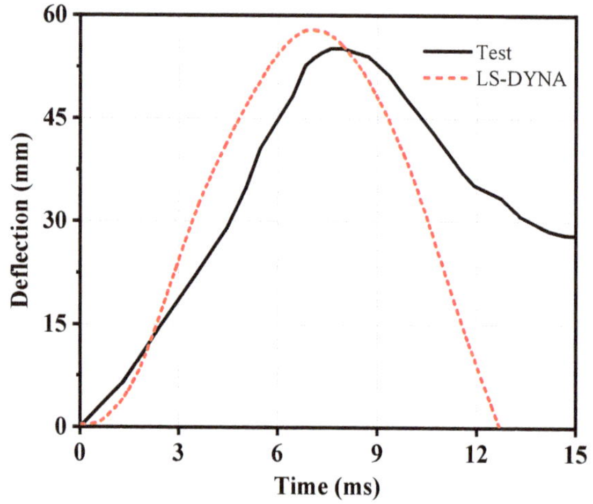

Figure 4: Deflection–time history of column 2C1.

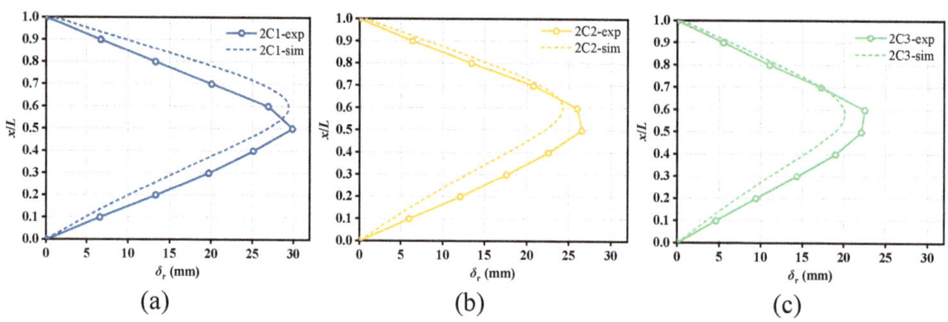

Figure 5: Residual deflection profile δ_r of 2C1, 2C2 and 2C3. (Note: x is the elevation of the gauge porin from the ground; L is the height of the columns.)

The second experiment involves the load-bearing capacity tests on steel columns conducted by Zhao et al. [10]. The experimental setup is shown in Fig. 6. Displacement-controlled loading was applied at rates ranging from 0.1 to 10 cm/s, and the load-lateral displacement curves and load-bearing capacities at different loading rates are presented in Fig. 7. At a loading rate of 1 cm/s, the simulated load-bearing capacity is 605.2 kN, which shows a relative error of 4.83% compared to the experimentally measured capacity of 577.3 kN. Therefore, a loading rate of 1 cm/s is selected for evaluating the post-blast residual load-bearing capacity.

3 PARAMETRIC STUDY

3.1 Geometry of the designed steel columns and scenarios

This study focuses on H-section steel columns with dimensions HW 350 × 350 ($b \times h \times h_w \times h_f$ = 350 × 350 × 12 × 19 mm), which were used as ground-floor columns in an actual steel

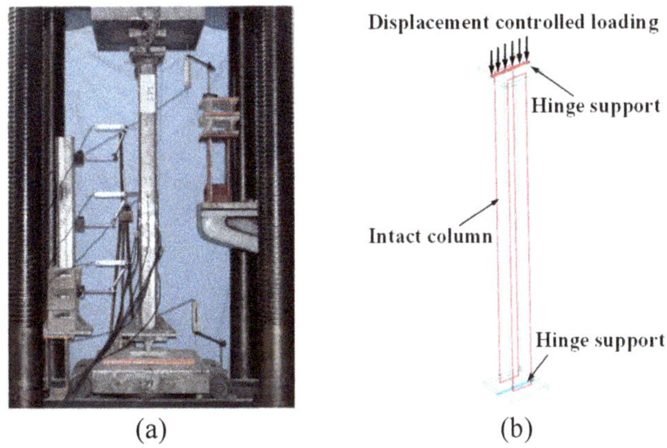

Figure 6: Test of residual bearing capacity of the steel column. (a) Test setup; and (b) Numerical model.

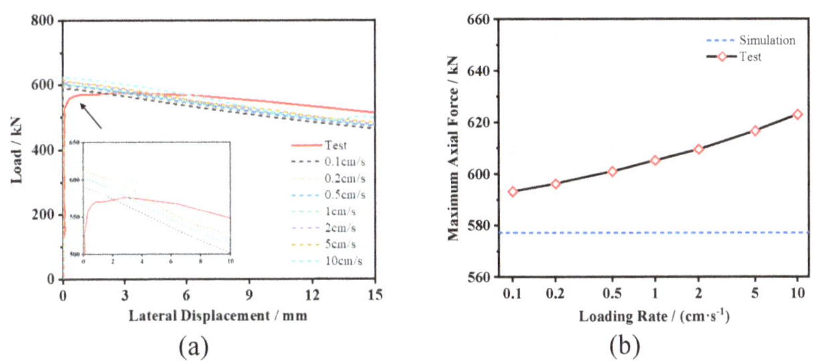

Figure 7: Numerical results. (a) Axial load-lateral displacement curves; and (b) Maximum axial forces versus the loading rate.

frame structure in China. The column height is 3.72 m, corresponding to the net height of the ground floor. The column is fixed at the base, and a 50 mm end plate is added at the top to ensure uniform load distribution during axial force application. The initial axial force is designed to be 500 kN. To represent explosions caused by a passenger/cargo van, as outlined in FEMA 428 [11], a TNT equivalent mass of 1,814 kg is used as the explosive charge. Fig. 8 illustrates the explosion scenarios, where the damage levels of the steel columns are evaluated for various combinations of standoff distance (R = 6–10 m) and angle of incidence (α = 0°, 45° and 90°) under the passenger/cargo van explosion. Here, H = 1.37 m represents the height of burst. The angle of incidence α is defined as the angle between the line connecting the charge centre and the mid-span centre of the column and the strong axis of the column (Fig. 9).

Fig. 10 shows the numerical model of the steel column under scenario with R = 7 m and α = 45°. The FE model, including S-ALE, Lagrangian mesh sizes, material models, and numerical methods all employ the validated methods from Section 2 and will not be repeated

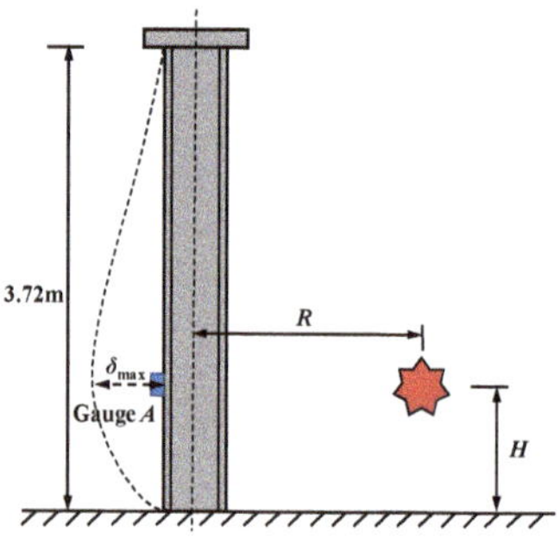

Figure 8: Diagram of explosion scenarios.

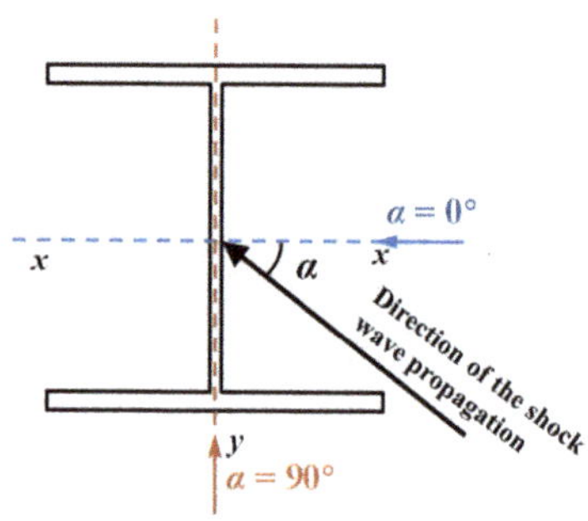

Figure 9: Definition of angle of incidence.

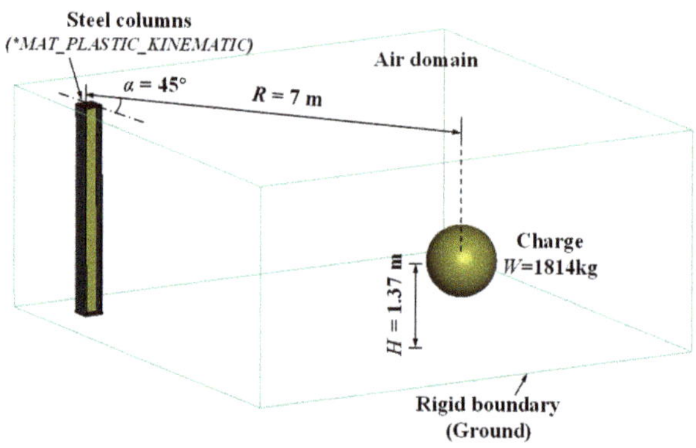

Figure 10: Numerical model of the scenario with $R = 7$ m, $\alpha = 45°$.

here. It should be noted that this study ignores the influence of charge shape, i.e. all explosives are modelled as spherical charges. Regarding the boundary conditions, except for the rigid boundary representing the ground as shown in Fig. 10, the other five faces of the air domain are all set as non-reflecting boundaries. All translations and rotations are constrained for the nodes on the top face of the endplate and the bottom face of the column to realize a total fixed support ($\delta_x = \delta_y = \delta_z = 0$, $\theta_x = \theta_y = \theta_z = 0$).

3.2 Results and discussion

Using the numerical model mentioned above, the damage modes of the steel columns under different explosive scenarios will be discussed first in this section. Then the damage evaluation will be conducted based on residual capacity.

3.2.1 Damage modes

When the standoff distance is relatively small, i.e. $R = 6$ m, at an angle of incidence $\alpha = 0°$, the shock wave primarily acts on the web, causing the column to bend about its weak axis. Additionally, multiple reflections of the shock wave within the flange–web–flange region result in the outward opening of the two half-flanges and backward deformation of the web along the shockwave propagation direction, accompanied by tearing at the flange–web connections (Fig. 11(a)). At $\alpha = 45°$, when the shock wave reaches the steel column, it splits into two parts at the edge of the blast-facing flange: one part propagates along the flange, while the other enters the flange–web–flange region, leading to multiple reflections (Fig. 11(b)). Under such shock wave impacts, the steel column experiences tearing at the flange–web connections and shear failure at the base of the web. At $\alpha = 90°$, the shock wave impinges the flange facing the charge, resulting in shear fractures in the web at two locations: near the base and slightly below the burst height. Additionally, significant backward bending of the two half-flanges and separation at the flange–web connection are observed.

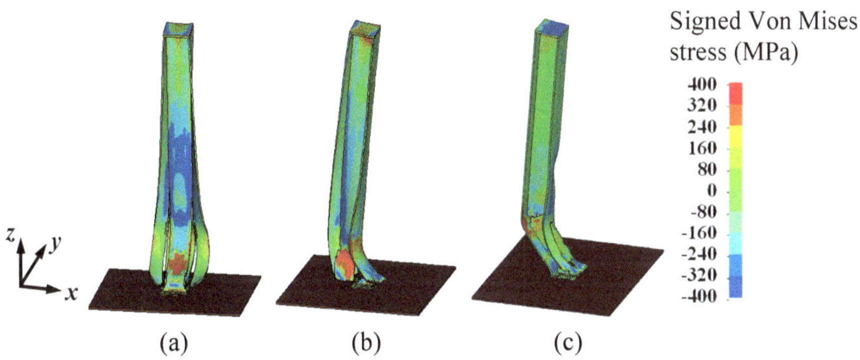

Figure 11: Damage modes of steel columns after the blast under $R = 6$ m. (a) $\alpha = 0°$; (b) $\alpha = 45°$; and (c) $\alpha = 90°$.

At relatively large standoff distances, i.e. $R = 7$–10 m, the steel columns generally do not fracture after the blast. Instead, damage primarily manifests as a combination of overall bending, torsion, and local buckling. Fig. 12 illustrates the damage modes of the steel column at $R = 8$ m and $\alpha = 0°$, 45° and 90°. At the scenario with $\alpha = 0°$, the column primarily bends about its weak axis, with a significant hogging moment at the base leading to high stress concentrations. Additionally, substantial shear forces induced by the shockwave result in large stresses at the flange–web connections (Fig. 12(a)). At $\alpha = 45°$, the column experiences biaxial bending, torsion, and local buckling. The web exhibits diagonal shear deformation along the dashed line at the column base indicated in Fig. 12(b). At $\alpha = 90°$, the column primarily bends about its strong axis, with significant shear forces at the base primarily borne by the web (Fig. 12(c)).

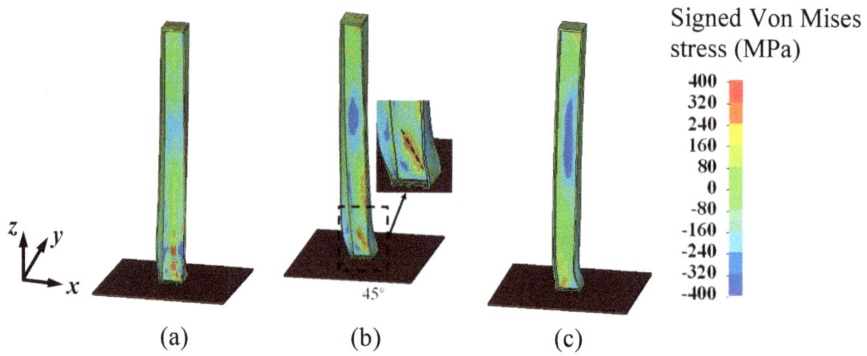

Figure 12: Damage modes of steel columns after the blast under R = 8 m: (a) α = 0°; (b) α = 45°; and (c) α = 90°

3.2.2 Damage assessment

After determining the damage modes of the steel columns, an axial load is applied to the top endplate using displacement control until the column fails completely. The resulting load-displacement curves for steel columns under various standoff distances and angles of incidence are shown in Fig. 13. The corresponding damage index (DI) for each scenario is presented in Fig. 14.

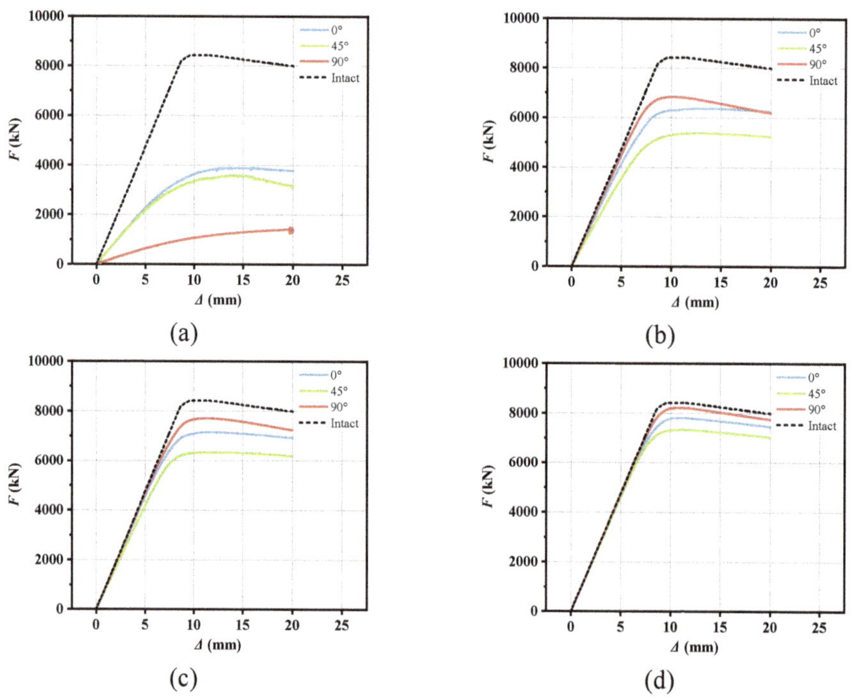

Figure 13: Force-displacement curves of the columns at: (a) R = 7 m; (b) R = 8 m; (c) R = 9 m; and (d) R = 10 m.

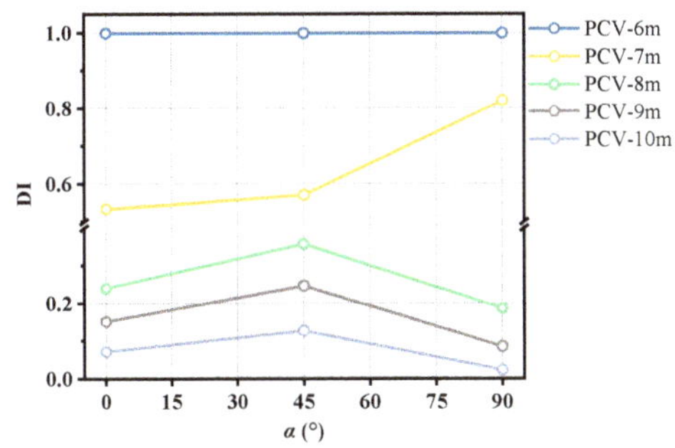

Figure 14: Damage index of the columns versus angles of incidence at diverse standoff distances.

At a standoff distance of $R = 6$ m for all angles of incidence, the steel columns lose their axial load-bearing capacity due to tearing and shear failure. For $R = 7$ m at $\alpha = 0°$ and 45°, the columns primarily experience bending or a combination of bending and torsion, which do not compromise the cross-sectional area or overall integrity, corresponding to DI values of 0.54 and 0.57. At $\alpha = 90°$, the load-displacement curves show a significant reduction in capacity and stiffness, attributed to pronounced shear failure at the base, resulting in a weakened section composed only of the two flanges, corresponding to DI = 0.82. For $R = 8$–10 m, the load-displacement curves for all angles of incidence resemble those of undamaged columns, with no significant reduction in post-blast stiffness, corresponding to DI values of 0.02–0.25, indicating low levels of damage.

4 CONCLUSION

In this study, the damage modes and residual load-bearing capacity of a full-scale H-section steel column subjected to a vehicle explosion of passenger/ cargo van are investigated. The numerical models for the blast phase and residual capacity phase are validated using blast tests by Nassr et al. [8], [9] and axial compression tests by Zhao et al. [10], respectively. The standoff distance and angle of incidence are varied to observe the column's response and residual capacity under blast loading.

The results show that at smaller standoff distances, the shock wave primarily impacts the flange–web–flange region at $\alpha = 0°$, causing tearing failure at the flange–web connections. At $\alpha = 90°$, the shock wave mainly impacts the blast-facing flange, leading to shear fracture at the column base. At $\alpha = 45°$, both tearing and shear fractures occur simultaneously.

At larger standoff distances, the column exhibits no significant fractures but experiences overall bending or torsion with local buckling. The post-blast damage modes can be used to evaluate damage levels. When significant tearing or shear fracture occurs, the damage index (DI) exceeds 0.8, indicating potential collapse and substantial stiffness loss. When only bending or torsion occurs without significant fractures, the stiffness remains relatively unchanged, with a slight reduction in load-bearing capacity, and the DI is less than 0.6.

ACKNOWLEDGEMENTS

The authors would like to thank the National Natural Science Foundation of China (grant number: 52278521) and the Fundamental Research Funds for the Central Universities for financial support to carry out this research.

REFERENCES

[1] Magallanes, J.M., Martinez, R. & Koenig J.W., *Experimental Results of the AISC Full-Scale Column Blast*, The American Institute of Steel Construction: USA, 2006.
[2] Morrill, K.B., Crawford, J.E., Magallanes, J.M. & Choi, H.J., Development of simplified tools to predict blast response of steel beam-column connections. *Structural Engineering Research Frontiers*, 2007.
[3] Grisaro, H.Y., Packer, J.A. & Seica, M.V., Experimental and numerical investigation of bare and strengthened wide-flange sections subject to near-field blast loading. *Journal of Structural Engineering*, **147**(9), 04021139, 2021.
[4] Grisaro, H.Y., Seica, M.V., Packer, J.A. & Li, W., Strong-axis response of steel I-sections subjected to close-in detonations. *International Journal of Protective Structures*, **12**(3), pp. 347–376, 2021.
[5] Grisaro, H.Y., Packer, J.A. & Seica, M.V., Weak axis response of steel I-sections subjected to close-in detonations. *Journal of Constructional Steel Research*, **160**, pp. 189–206, 2019.
[6] Remennikov, A.M. & Uy, B., Explosive testing and modelling of square tubular steel columns for near-field detonations. *Journal of Constructional Steel Research*, **101**, pp. 290–303, 2014.
[7] Gangolu, J. & Grisaro, H.Y., Axial load ratio effect on wide-flange columns subjected to far field detonations. *Journal of Constructional Steel Research*, **223**, 109009, 2024.
[8] Nassr, A.A., Razaqpur, A.G., Tait, M.J. & Foo, S., Experimental performance of steel beams under blast loading. *Journal of Performance of Constructed Facilities*, **26**(5), pp. 600–619, 2012.
[9] Nassr, A.A., Razaqpur, A.G. Tait, M.J., Campidelli, M. & Foo, S., Strength and stability of steel beam columns under blast load. *International Journal of Impact Engineering*, **55**, pp. 34–48, 2013.
[10] Zhao, H., Wang, R., Li, Q.M., Wu, H., Hou, C.C. & An, G.Q., Experimental and numerical investigation on impact and post-impact behaviours of H-shaped steel members. *Engineering Structures*, **216**, 110750, 2020.
[11] Department of Homeland Security, *Primer to Design Safe School Projects in Case of Terrorist Attacks and School Shootings (Edition 2)*, Federal Emergency Management Agency: USA, 2012.

SECTION 3
TERMINAL BALLISTICS

MESOSCALE MODELLING OF CONCRETE SUBJECTED TO BALLISTIC IMPACT

ØYSTEIN E. K. JACOBSEN[1], MARTIN KRISTOFFERSEN[1,2], SUMITA DEY[3] & TORE BØRVIK[1,3]
[1]Structural Impact Laboratory (SIMLab), Department of Structural Engineering,
NTNU – Norwegian University of Science and Technology, Trondheim, Norway
[2]Multiconsult, Tromsø, Norway
[3]Research and Development Department, Norwegian Defence Estates Agency, Oslo, Norway

ABSTRACT

Concrete is a multiphase material commonly used in protective structures and critical infrastructure, consisting primarily of a cement matrix, aggregates and an interfacial transition zone (ITZ). The properties of the aggregates – such as volume fraction, size and shape – play a critical role in determining the overall performance and protective capabilities of such concrete structures. Modifying these aggregate properties offers the ability for enhancing the protective capacity of concrete. Given the extensive research and numerical modelling efforts on concrete subjected to various loading conditions using homogeneous models, it is worth exploring mesoscale modelling approaches where the cement matrix and aggregates are modelled separately. By doing so, we may uncover optimized configurations or designs that provide superior protective solutions. In this paper we look at three themes: (1) An aggregate placing procedure to get sufficiently high-volume fractions in the concrete and the proceeding meshing. High volume fractions are commonly difficult to obtain using create-and-drop methods, so we have implemented an aggregate compacting workflow; (2) Finite element simulations in LS-DYNA of compression tests on cubes and cylinders, and comparison to relevant experiment data; (3) Numerical simulations of ballistic impact tests to investigate the influence of aggregate properties on the ballistic perforation resistance. With information available at the design stage, we create realistic two-phase heterogeneous meshes for finite element simulations in LS-DYNA and compare the results to a homogeneous mesh and literature data. We show that some of the scatter commonly seen in concrete experiments can be attributed to variation in aggregate placement when the cumulative volume fraction remains constant. The results are promising, the models are efficient, and the modelling workflow is streamlined.

Keywords: mesoscale modelling, ballistic impact, numerical simulations, material heterogeneity.

1 INTRODUCTION

Concrete is a material frequently used in protective structures and critical infrastructure due to its versatility, availability, low price and high strength. The material is therefore constantly subject to threats of accidental or intended impact loads, such as blast or ballistic impact. A great amount of research effort has been put into numerical modelling of concrete exposed to such loads, where the material is often assumed to be homogenous [1]. On a component level, the assumption of homogeneity can give decent and accurate results, but for more localised problems like ballistic impacts, scatter in the structural behaviour cannot be effectively characterised without a heterogeneous model with a randomised aggregate structure.

In recent years, mesoscale modelling approaches have been studied frequently used in finite element simulations of concrete to study heterogeneity [1], [2]. Two of the main challenges in mesoscale modelling is compatibility between aggregate creation and meshing, as well as reaching realistically high aggregate volume fractions with reasonable computational effort. In this study, we present a simulation framework to build realistic heterogeneous meshes of a concrete material that can be used for material tests simulations and on component level simulations of ballistic perforation of concrete slabs. We compare

WIT Transactions on The Built Environment, Vol 215, © 2025 WIT Press
www.witpress.com, ISSN 1743-3509 (on-line)
doi:10.2495/SUSI250101

the results from three realistic, heterogeneous, two-phase meshes to a homogeneous mesh with concrete materials parameters and compare the results to experiment data from the open literature [3].

2 AGGREGATE CREATION AND PLACEMENT PROCEDURE

The following aggregate placement procedure has been developed to reach high aggregate volume fractions in the concrete. The aggregate placement procedure is divided in four steps: (1) creating aggregates based on material information available in the design stage, such as volume fraction and particle size distribution; (2) place aggregates inside a prismatic volume using Python; (3) compact those aggregates in a finite element simulation; (4) loop over a background mesh to see if the element centre is inside an aggregate. If so, the element is flagged as an aggregate element and assigned an aggregate material formulation.

2.1 Aggregate creation

The aggregate creation is performed in Python and follows a frequently used principle: define a sphere with a diameter D_p around a centre point, and sample i vertices on the sphere, encompassing the centre point. This sampling method produces a convex polyhedron that can be triangulated. Each triangular face (with three vertices) can be connected to the centre point to create a tetrahedron element that can be written to a mesh-file for FE analysis. By looping over the elements in a separate background FE mesh, we can check if each element centre of the regular background mesh is inside the aggregate polyhedron and assign aggregate material properties if it is. The procedure is shown in Fig. 1.

(a) (b) (c) (d) (e)

Figure 1: Aggregate creation and meshing shown for one aggregate. (a) i = 25 vertices (black) sampled on a sphere surround the red centre point; (b) Grouping three vertices to the centre point to create a tetrahedron; (c) All tetrahedrons that make up one aggregate; (d) Transferring the Python aggregate to a finite element mesh; and (e) The aggregate mesh in a surrounding matrix mesh.

2.2 Aggregate compacting procedure

Previous studies on mesoscale modelling on concrete have shown that it is difficult to reach aggregate volume fractions above 30% with create-and-place methods. Liang and Zou [4] developed a method where the aggregates were created in a volume with a larger height than the final volume that should be meshed. Then, the aggregates were compacted into the final volume in an FE simulation. The results from the FE analysis can be read by e.g., Python, and by looping over a background mesh, a final mesh can be created based on the final aggregates. The compaction procedure is shown in Fig. 2 and happens between (c) and (d) in Fig. 1.

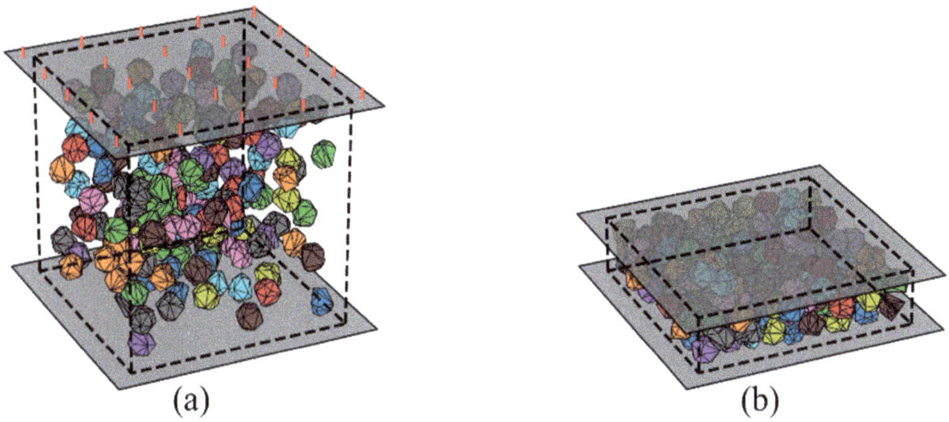

Figure 2: Aggregate compacting. (a) Aggregates in the initial state; and (b) Aggregates compacted into the final volume.

3 MATERIAL MODEL AND MATERIAL PARAMETERS

We used a modified version of the Holmquist–Johnson–Cook concrete (denoted MHJC) model for both the matrix and the aggregates in the numerical simulations of ballistic impact, and for the homogeneous concrete slab. The material model has previously been coined a good compromise between accuracy and ease-of-calibration [5] and has produced decent results in previous papers [3], [6]. The pressure-dependent deviatoric response in the model is described by eqn (1):

$$\sigma_{\mathrm{eq}}^{*} = \frac{\sigma_{\mathrm{eq}}}{f_{\mathrm{c}}} = \begin{cases} B(P^{*} + T^{*}(1-D))^{N} F(\dot{\varepsilon}_{\mathrm{eq}}^{*}) R(\theta, e) \ \text{for} \ P^{*} \geq -T^{*}(1-D) \\ 0 \ \text{for} \ P^{*} < -T^{*}(1-D) \end{cases}, \quad (1)$$

where σ_{eq}^{*} is the equivalent stress normalised with respect to the unconfined compressive strength f_{c}. B and N are pressure-hardening parameters, P^{*} and T^{*} are the hydrostatic pressure and maximum tensile hydrostatic pressure, both normalised with respect to f_{c}. D is a damage variable that accounts for both compaction and shear damage, while $F(\dot{\varepsilon}_{\mathrm{eq}}^{*})$ is a rate-dependent two-parameter term with the strain-rate sensitivity parameter C. Finally, $R(\theta, e)$ further accounts for triaxial stress states with the Lode angle θ. The model also describes and accounts for the equation-of-state through a three-part formulation. P_{lock} is part of the equation-of-state and is the hydrostatic pressure at which all the air voids in the concrete are crushed.

Material parameters in the following simulations were not directly calibrated from material tests but found using literature data and reasonable assumptions. The main idea is to have high-strength aggregates and a lower-strength matrix material. When simulating two concretes with different strengths, only the material parameters for the matrix material would have to change. Three sets of material parameters have to be found for the simulations in this study: one set of parameters for the concrete material in the homogeneous mesh – and one set of parameters for each of the matrix and aggregates in the heterogeneous meshes. The main material parameters are listed in Table 1.

4 MESOSCALE MODELLING OF BALLISTIC IMPACTS

The numerical simulations of ballistic impact herein are a proof-of-concept rather than precise recreations of the experiment results.

Table 1: Main material parameters of the MHJC model used in the simulations.

Parameter	Homogeneous model	Heterogeneous model	
	Concrete	Matrix	Aggregate
ρ (kg/m^3)	2,400	2,000	2,700
f_c (MPa)	75	70	160
B (−)	1.7	1.8	1.6
N (−)	0.7	0.7	0.7
T (MPa)	4.0	4.0	4.0
C (−)	0.02	0.02	0.02
P_{lock} (MPa)	800	400	800

4.1 Setup of the ballistic model

For the simulations, we used the base information from Kristoffersen et al. [3] and their C75 50 mm thick concrete slabs to set it up. The slab was modelled as a 400 × 400 × 50 mm prism part. The elements were modelled as 8-node solids, with side length 2 mm and 1 integration point, as well as a viscous hourglass control. The outer edges of the slab were fixed for all translations. The ogival projectile was modelled as a rigid material with mass 196 g and contact was defined as node-to-surface (projectile nodes). We set up two cases for the slab – one homogeneous slab modelled with a concrete material model, and one where the matrix and aggregates were modelled separately, but both with the MHJC material model. Three heterogeneous meshes were created with the same initial conditions, but different aggregate structure. The concrete in the experiments had a volume fraction of coarse aggregates (8–16 mm) of roughly 30%, so the same volume fraction is used for the heterogeneous meshes.

Fig. 3 shows how the aggregates were distributed in one of the heterogeneous meshes. Aggregates have only been compacted into the middle prism of the slabs with size 200 × 200 × 50 mm, as the projectile will not affect elements outside this zone.

4.2 Results from the numerical modelling

All simulations had projectile initial velocity v_i of $250\ m/s$. The projectile residual velocities v_r were found and compared to an experiment with approximately the same initial velocity. Fig. 4 shows the final state from the simulations when the projectiles have completely perforated the slabs. The projectiles in the simulations with heterogeneous meshes are rotated post perforation, and the three scabbing areas differ in shape. Both the homogeneous and heterogeneous meshes resulted in lower residual velocities compared to the experiments.

While both approaches yield reasonable results, the heterogeneous models offer the advantage of capturing scatter, allowing for a more detailed investigation. Both meshes produced scabbing areas on the backside that resembled the experiment in a qualitative manner. However, the numerical models were not able to capture any of the dust that occurred in the experiments. This is likely due to the element size – an element size of 2 mm cannot capture this effect when element erosion is used.

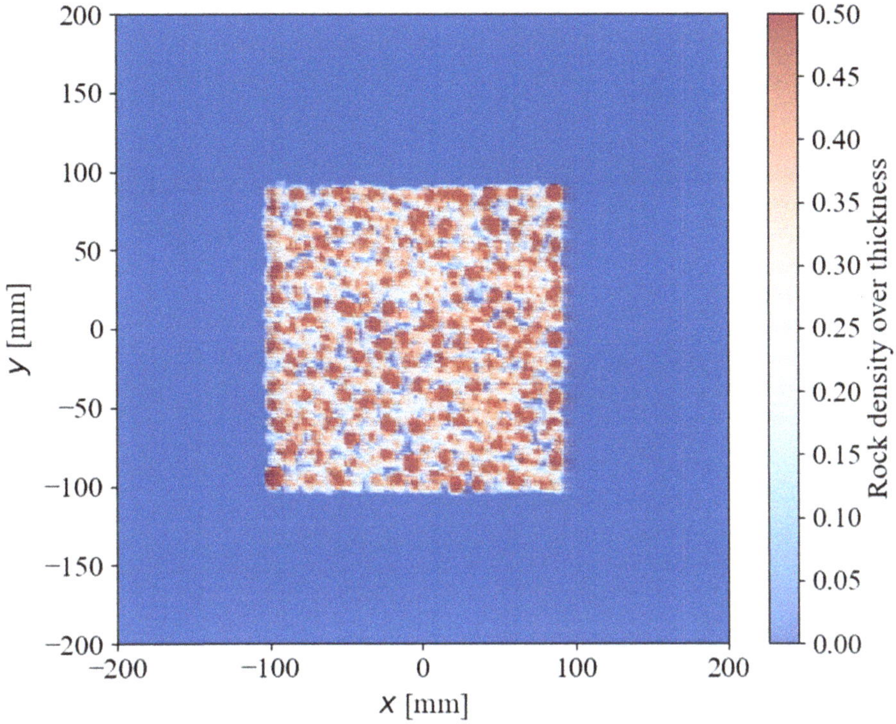

Figure 3: The heterogeneous mesh used in one of the three mesoscale simulations. Only a middle prism is modelled with aggregates, the rest of the slab is only matrix material. The projectile impacted the slab in $(x, y, z) = (0, 0, 50)$.

Table 2: Projectile velocities from experiment, homogeneous mesh and heterogeneous meshes.

Parameter	Experiment [3]	Homogeneous mesh	Heterogeneous mesh		
			No. 1	No. 2	No. 3
v_i (m/s)	253.4	250.0	250.0	250.0	250.0
v_r (m/s)	170.3	147.2	141.6	136.6	142.3

5 CONCLUSIONS AND OUTLOOK

We set up a workflow to create realistic, heterogeneous meshes of concrete material meshes using data that usually is available at the design stage. Numerical simulations of ballistic experiments served as proof-of-concept of the method, with relevant comparisons to a homogeneous mesh.

The mesoscale modelling approach can be used to simulate material tests to investigate scatter seen in experimental testing. Extensive testing would be necessary for characterisation of the matrix and aggregate material to provide accurate predictions of concrete material behaviour. The method could in principle be used on any available load case that is usually modelled using homogeneous concrete models, which would provide further details into the structural behaviour of concrete under ballistic impact.

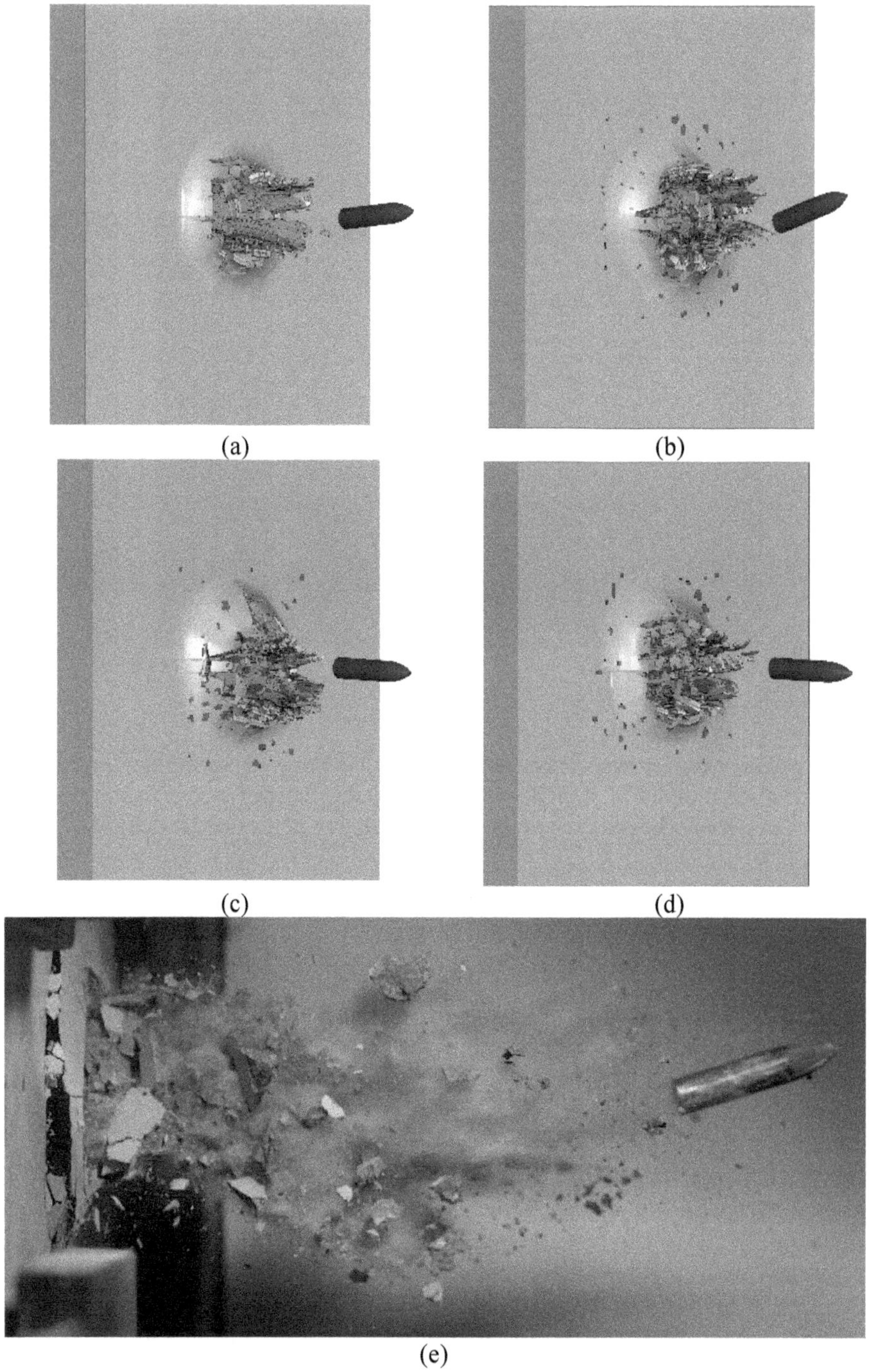

Figure 4: Pictures of the final state for (a) a homogeneous mesh; and (b), (c), (d) three heterogeneous meshes; (e) shows the projectile after perforation of the slab from a concurrent experiment [3].

REFERENCES

[1] Ren, H., Rong, Y. & Xu, X., Mesoscale investigation on failure behavior of reinforced concrete slab subjected to projectile impact. *Engineering Failure Analysis*, **127**, 2021.

[2] Yao, Z., Qin, F., Li, J., Zhang, X. & Huang, F., Experimental and 3D mesoscopic numerical simulation study of kinetic projectile penetrating into concrete. *International Journal of Impact Engineering*, **195**, 2025.

[3] Kristoffersen, M., Toreskås, O.L., Dey, S. & Børvik, T., Ballistic perforation resistance of thin concrete slabs subjected to ogive-nosed steel projectiles. *International Journal of Impact Engineering*, **156**, 2021.

[4] Liang, Y. & Zou, S., A method to rapidly build complex three-dimensional models of rock-filled concrete. *Computers and Structures*, **281**, 2023.

[5] Polanco-Loria, M., Hopperstad, O.S., Børvik, T. & Berstad, T., Numerical predictions of ballistic limits for concrete slabs using a modified version of the HJC concrete model. *International Journal of Impact Engineering*, **35**, 2008.

[6] Jacobsen, Ø.E.K., Kristoffersen, M., Rudshaug, J., Dey, S., Hauge, K.O. & Børvik, T., Fragment impact on concrete slabs: An experimental and numerical study. *International Journal of Protective Structures*, 2025.

BALLISTIC LIMIT PERFORMANCE OF BI-LAYER CERAMIC/METAL COMPOSITE ARMOUR UNDER RADIAL CONFINING PRESSURE

JIAQI DAI[1] & SUWEN CHEN[1,2]
[1]College of Civil Engineering, Tongji University, China
[2]State Key Laboratory of Disaster Reduction in Civil Engineering, Tongji University, China

ABSTRACT
The bi-layer armour composed of a front ceramic plate and a metal backplate is a typical composite protective structure. The ceramic serves to erode the projectile. Ceramics are sensitive to multi-axial stress states. To investigate the anti-penetration performance under confining pressure, a verified SPH–FEM numerical simulation model was established. The penetration experiments under higher confining pressures were conducted. Two parameters, the maximum biaxial confining pressure and the confining pressure ratio, were introduced to evaluate the prestress state of the ceramic. Finally, based on experimental and simulation results, a ballistic limit velocity calculation method and a fast computation formula considering the confining pressure were proposed. The validity of the proposed formula was verified through comparison of results from physical experiments and numerical simulations. This research provided significant insights for further research on the anti-penetration performance of ceramic/metal composite armour.
Keywords: ballistic performance, ceramic composite armour, ballistic limit velocity, energy dissipation.

1 INTRODUCTION

Ceramics are utilized on the impact face of armour protection due to the low density and high hardness. The metal backing plate attached to the ceramic front plate can inhibit the bending and cracking of ceramics under impact. The bi-layer ceramic/metal composite armour has been demonstrated to be a cost-effective protective structure.

Ceramics have a compressive strength of 10^0–10^1 GPa, while the tensile strength is only about 10% of the compressive strength. To overcome the significant difference between tensile and compressive strengths of ceramics, pre-compressive stress is introduced. Prestress ceramics can be categorized into three types based on applied prestress: hydrostatic prestress, radial prestress and axial prestress. Due to the difficulty of physical realization of hydrostatic prestress, radial and axial prestressing of ceramics has become the main approach for studying the anti-penetration performance of prestressed ceramics. Experiment and simulation on the anti-penetration performance of ceramic composite armour under confinement and prestress were performed in Sherman [1] and Holmquist and Johnson [2]. Results showed that both hydrostatic prestress and radial prestress can improve the ballistic performance of composite armour at the same level, while the enhancement of axial prestress was limited. In Farbaniec et al. [3], microstructural observations revealed the different impact failure modes between axial prestress and radial prestress ceramics, which demonstrated that radial prestress was more effective in improving impact resistance. Long rod projectile experiments on radial constrained ceramic armour were conducted in Chi et al. [4], Lundberg et al. [5] and Xu et al. [6], showing that radial prestress can extend the dwell time, thereby increasing the erosion of the projectile.

The anti-penetration performance of composite armour is evaluated through the ballistic limit velocity. This represents the initial velocity of a test projectile that just penetrates a specific type of composite armour. In research on non-prestress ceramic composite armour,

WIT Transactions on The Built Environment, Vol 215, © 2025 WIT Press
www.witpress.com, ISSN 1743-3509 (on-line)
doi:10.2495/SUSI250111

empirical formulas [7], [8] or theoretical analytical models [9]–[11] for calculating the ballistic limit velocity were developed. Based on existing researches, ballistic limit tests of prestress ceramics were carried out in Zhang et al. [12], [13], which indicated that radial prestress can significantly enhance the ballistic limit velocity of ceramic/metal composite armour. Prestress with 10^2 MPa level increased the ballistic limit velocity by approximately 30%.

In summary, there are still the following issues on the ballistic limit velocity of ceramic composite armour under radial prestress: there is no clear indicator for measuring the prestress level of different ceramics, and the theoretical calculation method for ballistic limit velocity under radial prestress is lacking. This paper presents the following research: a prestress ratio is proposed to measure the ceramic prestress level, a calculation method for the compressive, tensile, and shear strength of ceramics under radial prestress is established based on the JH-2 constitutive model, a method for calculating the ballistic limit velocity of prestress ceramic composite armour is derived based on the energy balance principle.

2 ANALYTIC MODEL BASED ON THE PRINCIPLE OF ENERGY DISSIPATION

The principle of energy dissipation is a commonly used method in theoretical research on ballistic performance. This method has the following two main advantages: (1) a clear physical meaning; and (2) avoidance of the dynamic analysis of the penetration process, directly obtaining the state parameters of the projectile at the ballistic terminal.

This section first introduces how to determine the strength of ceramics under confining pressure using the JH-2 constitutive model. Subsequently, an energy dissipation formula for the armour system is established considering the strength enhancement. Finally, the ballistic performance of composite armour is derived.

2.1 Determination of ceramic strength under confining pressure

The JH-2 constitutive model [14] is widely used to describe the mechanical behaviour of brittle geomaterials, ceramics, and glass under dynamic conditions. The strength surface is characterized by hydrostatic pressure and equivalent stress. As shown in Fig. 1, the loading curve under uniaxial loading is a straight line passing through the origin with a slope of 3. Similarly, under biaxial loading conditions, the loading curve has a slope of 1.5. The intersection points of the uniaxial and biaxial loading curves with the strength surface of the JH-2 model represent the uniaxial compressive strength of the material and the maximum radial confining pressure, respectively.

Eqn (1.1) presents the stress tensor of ceramics when reaching the normal strength $\sigma_{\mathrm{c,c/c,t}}$ in the third direction under radial confining pressure. Compression is considered positive in the equation. The equivalent stress and hydrostatic pressure for this stress state can be obtained from the stress tensor, as shown in eqn (1.2). Substituting these into the strength surface equation of the JH-2 constitutive model allows solving for the strength in the third direction. eqn (1.3) has two solutions with opposite signs, corresponding to the compressive strength and tensile strength under a confining pressure of P_{conf}. The tensor given in eqn (2) is used to solve for the shear strength $\tau_{\mathrm{c,s}}$ under confining pressure.

$$\boldsymbol{\sigma} = \begin{bmatrix} P_{\mathrm{conf}} & & \\ & P_{\mathrm{conf}} & \\ & & \sigma_{\mathrm{c,c/c,t}} \end{bmatrix}, \tag{1.1}$$

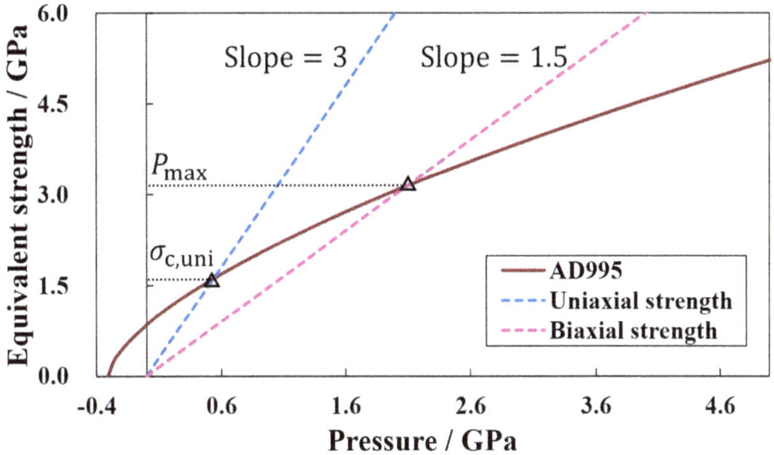

Figure 1: Determination of uniaxial and biaxial compressive strength of AD995 ceramics based on JH-2 constitutive model.

$$\begin{cases} \sigma^* = \sqrt{\left(P_{\text{conf}} - \sigma_{\text{c,c/c,t}}\right)^2} \\ P^* = \frac{1}{3}\left(2P_{\text{conf}} + \sigma_{\text{c,c/c,t}}\right) \end{cases}, \tag{1.2}$$

$$\frac{\sigma^*}{\sigma_{\text{HEL}}} = A\left(\frac{P^*+T}{P_{\text{HEL}}}\right)^N. \tag{1.3}$$

$$\boldsymbol{\sigma} = \begin{bmatrix} P_{\text{conf}} & & \tau_{\text{c,s}} \\ & P_{\text{conf}} & \\ \tau_{\text{c,s}} & & \end{bmatrix}. \tag{2}$$

- P_{conf} is the confining pressure.
- $\sigma_{\text{c,c/c,t}}$ is the normal strength of ceramic.
- A and N are JH-2 parameter.
- σ_{HEL}, P_{HEL} and T are Hugoniot strength, hydrostatic pressure at Hugoniot stress and hydrostatic tensile strength.

Fig. 2 shows the variation of the compressive strength of AD995 ceramics with the confining pressure ratio. The results indicate that the compressive strength of the ceramic increases monotonically with the confining pressure ratio, reaching up to five times the initial compressive strength. The tensile and shear strengths, however, exhibit non-monotonic behaviour as the confining pressure ratio increases. At higher confining pressure ratio, both the tensile and shear strengths decrease, which shows agreement with the phenomenon observed by Sherman [1] in ballistic experiments that more tensile cracks were detected on the backside of the ceramic. In addition, the maximum radial confining pressure corresponds to the point where the tensile strength decreases to zero. This is consistent with the physical meaning of this parameter, that is, at this confining pressure, the ceramic can no longer resist normal loads.

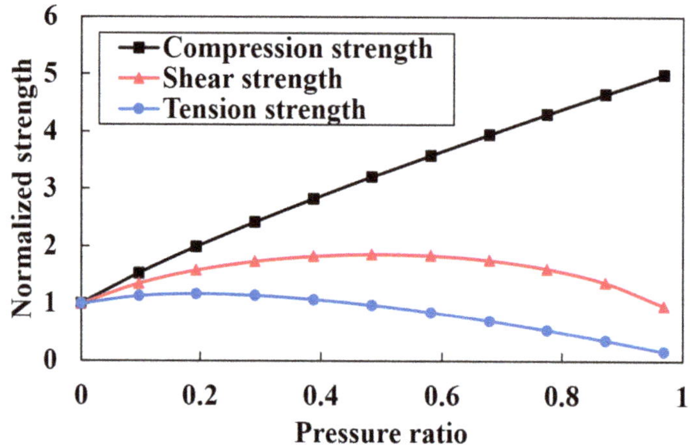

Figure 2: Variation of AD995 strength with confining pressure ratio.

2.2 Energy dissipation of composite armour system

Fig. 3 illustrates the penetration process for establishing the energy dissipation theory. During the penetration of the bi-layer composite armour, the projectile experiences mass loss and plastic deformation at the head. A crater forms on the surface of the ceramic due to crushing at first. Then, due to upward tensile cracks intersecting with the projectile head, a conical crack is formed [15].

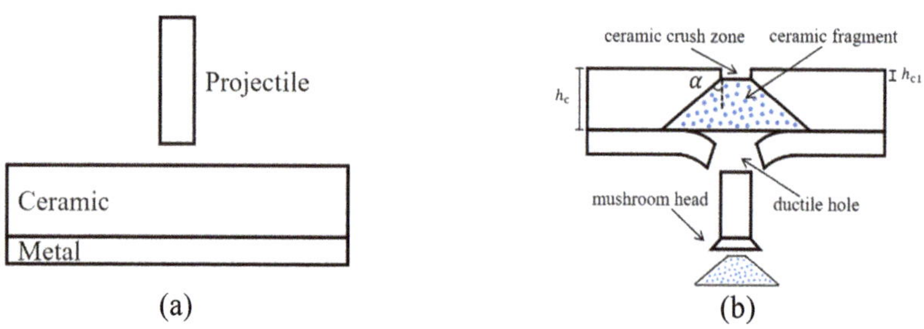

Figure 3: Schema of the penetration process. (a) Initial state; and (b) Terminal state.

2.2.1 Energy dissipation of projectile

The plastic deformation of the mushroom head has an established theoretical formula [16]. This formula can be used to calculate the length of the plastic zone and the plastic strain at each point. The plastic strain energy of the projectile can be calculated using eqn (3).

$$E_{\mathrm{p,d}} = \int_V \varepsilon_{\mathrm{p,d}} \sigma_{\mathrm{p}} dV. \tag{3}$$

- $E_{\mathrm{p,d}}$ is the plastic strain energy of the projectile.
- $\varepsilon_{\mathrm{p,d}}$ is plastic strain.
- σ_{p} is the yield strength of the projectile.

Eqn (4) serves to calculate the kinetic energy loss due to mass erosion. The key parameter $m_{\mathrm{p,e}}$ is difficult to determine using analytical methods. Eqn (5.1) determines three dimensionless variables affecting the erosion mass through dimensional analysis. Based on existing ballistic test results [8], [12], [17], the empirical expression for it is determined using symbolic regression, as shown in eqn (5.2).

$$E_{\mathrm{p,e}} = \frac{1}{2} m_{\mathrm{p,e}} v_{\mathrm{p}}^2. \tag{4}$$

$$\frac{m_{\mathrm{p,e}}}{\rho_{\mathrm{p}} d_{\mathrm{p}}^3} = f\left(\frac{\sigma_{\mathrm{c,c}}}{Y_{\mathrm{p}}}, \frac{h_{\mathrm{c}}}{d_{\mathrm{p}}}, \frac{v_{\mathrm{p}}}{\sqrt{Y_{\mathrm{p}}/\rho_{\mathrm{p}}}}\right), \tag{5.1}$$

$$\frac{m_{\mathrm{p,e}}}{\rho_{\mathrm{p}} d_{\mathrm{p}}^3} = \delta \frac{\sigma_{\mathrm{c,c}} h_{\mathrm{c}} v_{\mathrm{p}}}{d_{\mathrm{p}} \sqrt{Y_{\mathrm{p}}^3/\rho_{\mathrm{p}}}}. \tag{5.2}$$

- $E_{\mathrm{p,e}}$ is the erosion kinematic energy.
- v_{p} is the initial velocity of the projectile.
- ρ_{p} is the density of the projectile.
- d_{p} is the projectile diameter.
- Y_{p} is the flow stress of the projectile that equals to $1.7\sigma_{\mathrm{p}}$.
- h_{c} is the thickness of the ceramic front plate.

The form of eqn (5.2) is similar to the equation used in Tang and Wen [11] for studying the ballistic performance of composite armour without confining pressure. Instead of using the σ_{HEL}, the present theory uses compressive strength for dimensionless normalization to reflect the contribution of confining pressure. Fig. 4 compares the validity of the two dimensionless normalizations. The ballistic experiments in Zhang et al. [12] were conducted with a confining pressure target, while those in Sanchez-Galvez [8] and Hou et al. [17] were based on targets without confining pressure. The results show the advantage of the present theory which reveals the different erosion mass by ceramics with the same σ_{HEL} but different compressive strength.

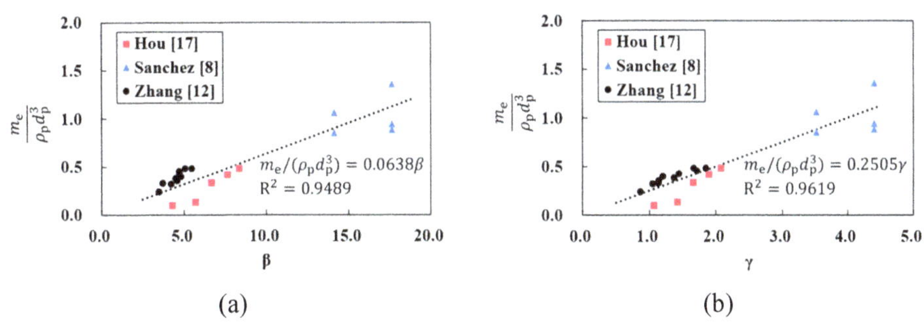

Figure 4: Normalized erosion mass of projectiles. (a) The abscissa is normalized by σ_{HEL}; and (b) The abscissa is normalized by $\sigma_{\mathrm{c,c}}$.

The energy dissipation of the projectile is given by eqn (6).

$$E_p = E_{p,e} + E_{p,d}. \tag{6}$$

2.2.2 Energy dissipation of ceramic front plate
The energy dissipation in the ceramic front plate can be divided into the crushing part and the shear part. The key parameter for the calculation is the height h_{c1} shown in Fig. 3(b). The following assumptions are adopted: (1) The conical crack is caused by the intersection of upward tensile cracks with the projectile head [15]; and (2) The propagation speed of the tensile cracks is 1/5 of the ceramic stress wave velocity [18]. Based on these assumptions, h_{c1} can be calculated using eqn (7).

$$\frac{h_{c1}}{h_c} = \frac{6u}{5u+c_c}. \tag{7}$$

- u is the penetration velocity, $u = 0.15v_p$ for steel projectile and $u = 0.4v_p$ for tungsten alloy projectile [11].
- c_c is the ceramic stress wave velocity.

The crushing energy and the shear fracture energy are given by eqn (8) and eqn (9), respectively.

$$E_{c,c} = \frac{\pi d_p^2}{4} h_{c1}\sigma_{c,c}, \tag{8}$$

$$E_{c,s} = \tau_{c,s}\pi\left(d_p + (h_c - h_{c1})\tan\alpha\right)\left(\frac{h_c - h_{c1}}{\cos\alpha}\right)\Delta_c, \tag{9.1}$$

$$\Delta_c = \gamma_c \frac{h_c - h_{c1}}{\cos\alpha}. \tag{9.2}$$

- α is the half angle of ceramic cone.
- Δ_c is the fracture displacement along the crack direction which is corresponded to the ductility coefficient γ_c and the length of the crack.

Based on existing research results [9], [10], [19], the half-cone angle α under impact loading is 53°. The method for determining γ_c is given in Section 4.1.
Hence the total energy dissipated by the ceramic can be written as eqn (10).

$$E_c = E_{c,c} + E_{c,s}. \tag{10}$$

2.2.3 Energy dissipation of metal back plate
The failure modes of the thin metal backing plate in composite armour can be classified into ductile hole failure and punching failure. This is primarily determined by the velocity of the projectile and the fractured ceramic cone. According to the ballistic experiments in Zhang et al. [12], within the velocity range studied in present paper, the failure mode of the backing plate is ductile hole failure. The energy dissipation formula can be given by the equation proposed by Woodward and Cimpoeru [20] as shown in eqn (11).

$$E_m = \frac{\pi d_p \sigma_m h_m (r_p + \pi h_m)}{4}. \tag{11}$$

- σ_m is the tensile strength of the back plate.
- h_m is the thickness of the back plate.

In summary, the ballistic performance of the composite armour considering the effect of confining pressure can be expressed by eqn (12).

$$v_{\mathrm{res}} = \sqrt{\frac{m_{\mathrm{p}} v_{\mathrm{p}}^2 - 2(E_{\mathrm{p}} + E_{\mathrm{c}} + E_{\mathrm{m}})}{m_{\mathrm{p,res}} + m_{\mathrm{c,res}}}}, \quad (12.1)$$

$$m_{\mathrm{p,res}} = m_{\mathrm{p}} - m_{\mathrm{p,e}}, \quad (12.2)$$

$$m_{\mathrm{c,res}} = \frac{1}{3}\rho_{\mathrm{c}}\pi(h_{\mathrm{c}} - h_{\mathrm{c1}})\left(r_{\mathrm{top}}^2 + r_{\mathrm{top}} r_{\mathrm{bot}} + r_{\mathrm{bot}}^2\right), \quad (12.3)$$

$$r_{\mathrm{top}} = r_{\mathrm{p}}, r_{\mathrm{bot}} = r_{\mathrm{p}} + \tan\alpha\,(h_{\mathrm{c}} - h_{\mathrm{c1}}). \quad (12.4)$$

- v_{res} is the residual velocity.
- $m_{\mathrm{c,res}}$ and $m_{\mathrm{p,res}}$ are the mass of the ceramic cone and the residual projectile respectively.

3 RESULTS AND DISCUSSIONS

3.1 Comparison between test and analytical model

Each determined confining pressure state corresponds to a specific ductility coefficient γ_{c}. According to the theory in Section 2, a ballistic performance prediction curve can be obtained. Based on the ballistic performance tests in Zhang [12], the data sets for three different confining pressures (0, 125 MPa, 274 MPa) are predicted through the present theory. The ductility coefficient for each curve is calculated by the principle of minimizing squared residuals.

Fig. 5 shows the optimal ballistic performance prediction curve and the corresponding ductility coefficient γ_{c}. Fig. 6 indicates that for AD995 ceramics, when the confining pressure ratio is less than 10%, the ductility coefficient exhibits a highly linear relationship with the confining pressure ratio. Table 1 presents a comparison between the prediction for the ballistic limit velocity in this paper and test data, demonstrating that the theory proposed in this paper also shows high accuracy in predicting the ballistic limit velocity.

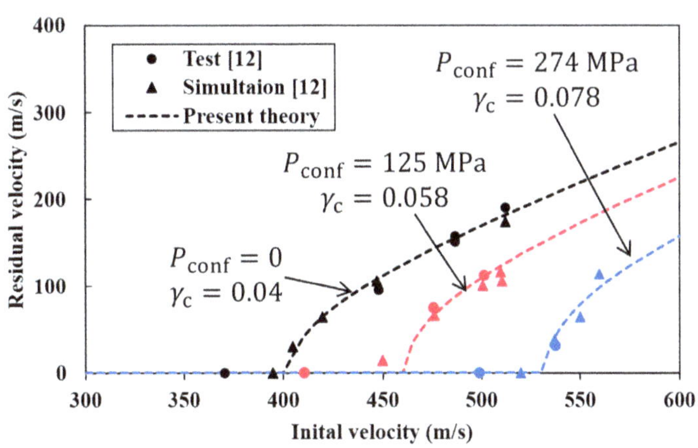

Figure 5: Prediction of ballistic performance with experiments and simulations.

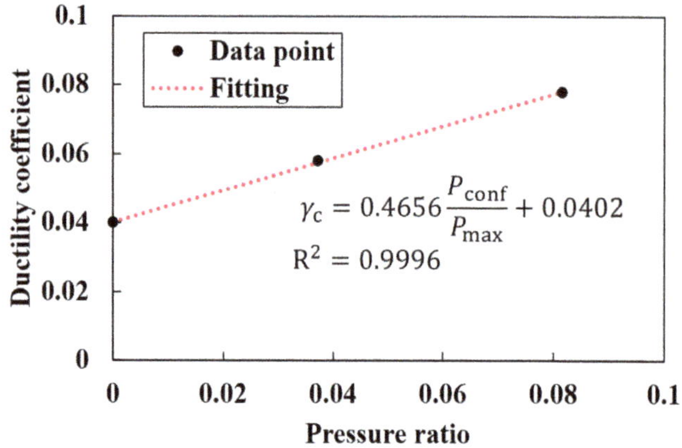

Figure 6: Fitting curve of ductility coefficient and confining pressure ratio.

Table 1: Ballistic limit velocity under confining pressure test.

P_{conf} (MPa)	h_c (mm)	h_m (mm)	Test BLV ($m \cdot s^{-1}$) [12]	Cal BLV ($m \cdot s^{-1}$)	Error (%)
0	6	3	398	400	1.8
125	6	3	440	465	5.7
274	6	3	499–537	530	−1.3 to 6.2

3.2 Contribution of strength and ductility in analytic model

This paper introduces two improvements compared to existing theories: the enhancement of ceramic strength and ductility under confining pressure. To verify the significance of improvements, contribution coefficient analyses for both strength enhancement and ductility enhancement are conducted.

Eqn (13) and eqn (14) are defined for contribution coefficient of strength and ductility, respectively.

$$\delta_D|_{P_{conf}} = \left(\frac{1}{n} \sum_{i=1}^{n} \frac{1}{|\gamma_{c,0} - \gamma_{c,i}|} \frac{\int_0^{v_{p,max}} |v_{res,0} - v_{res,i}| dv}{\int_0^{v_{p,max}} v_{res,0} dv} \right) \Bigg|_{P_{conf}}, \tag{13}$$

$$\delta_S|_{P_{conf}} = \left(\frac{\int_0^{v_{p,max}} \left| v_{res,P_{conf}} - v_{res,P_{conf}=0} \right| dv}{\int_0^{v_{p,max}} v_{res,P_{conf}} dv} \right) \Bigg|_{P_{conf}}. \tag{14}$$

The contributions of strength and ductility enhancement are measured by the ratio of the area difference enclosed by the curves and the absolute value of the parameter changes. The range of the contribution coefficient is $[0, +\infty)$. A larger contribution coefficient indicates the higher significance in the ballistic performance calculation.

Fig. 7 and Table 2 show the ductility and strength contribution coefficients for different target configurations. The results indicate that the contribution coefficients increase with the rise of the pressure ratio. This suggests that the two correction parameters proposed in this paper are essential for the ballistic performance calculation of composite armour under confining pressure. Furthermore, these two coefficients are numerically close, revealing that the importance of these two corrections for the theoretical calculation is consistent. Removing either parameter would lead to deviations for the results.

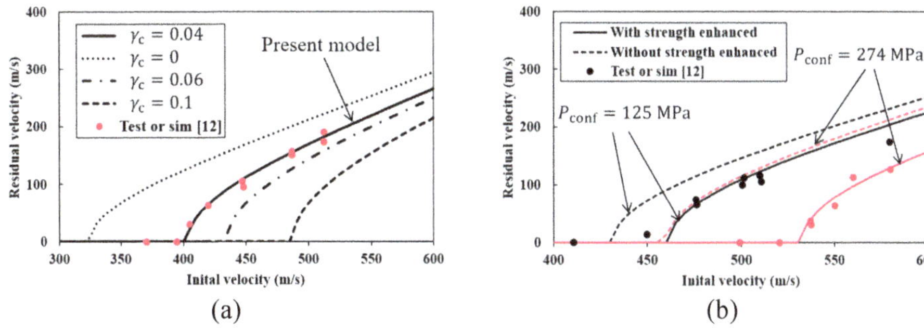

Figure 7: Contribution of ductility and strength. (a) Results with different ductile coefficient γ_c for $P_{conf} = 0$; and (b) Results with and without strength enhanced.

Table 2: Contribution coefficient of ductility and strength.

Contribution coefficient	P_{conf} (MPa)	Pressure ratio (%)	δ_D or δ_S
Ductility	0	0	0.106
	125	3.72	0.197
	274	8.15	0.355
Strength	125	3.72	0.101
	274	8.15	0.293

3.3 Applicability of the analytic model

To verify the applicability of the theory proposed in this paper, this section selects ballistic experiments involving different projectiles penetrating various configurations of bi-layer ceramic/metal composite armour. The tensile strengths σ_m of 2024-T3 and GFRP are 600 MPa [21] and 1000 MPa [22], respectively. The yield strength σ_p of the 4340 steel projectile is 785 MPa [12], and the key JH-2 parameters for the ceramics are provided in Table 3.

Table 3: JH-2 parameters of AD995, AD95 and AD85.

Ceramic	ρ_c (kg·m^{-3})	σ_{HEL} (GPa)	P_{HEL} (GPa)	A	N
AD995 [12]	3890	6.54	3.51	0.88	0.64
AD95 [23]	3750	6.0	3.2	0.889	0.764
AD85 [24]	3480	2.0	1.46	0.93	0.6

Table 4 demonstrates the applicability of the theory proposed in this paper. Additionally, for conditions with higher projectile velocities, the predicted residual velocity is lower. This is mainly due to two reasons: (1) Under high-speed penetration, the failure mode of the backing metal transitions from ductile hole failure to adiabatic shear failure that dissipates less energy during the penetration process; and (2) The fracture ceramic will not form a complete ceramic cone moving together with the projectile, meaning the actual ceramic mass $m_{c,res}$ will be lower than the predicted.

Table 4: Ballistic performance prediction accuracy under different armour configurations.

Armour configuration	h_c (mm)	h_m (mm)	v_p (m·s^{-1})	$v_{res,exp}$ (m·s^{-1})	$v_{res,cal}$ (m·s^{-1})	Error (%)
Projectile: 4340 steel $\phi 7.62 \times 20$ Target: AD995+2024T3 [12]	6	3	447	96	107	11.9
	6	3	486	152	154	1.4
	6	3	512	191	181	−5.0
	6	3	476	75	68	−9.6
	6	3	501	112	111	−0.6
Projectile: 4340 steel $\phi 7.56 \times 30.5$ Target: AD95+2024T3 [23]	6.1	6	650	341	330	−3.3
	6.12	6	655	351	335	−4.7
	6.14	6.03	830	559	515	−7.8
	6.1	5.87	712	430	400	−7.0
	8.96	6	775	370	359	−3.0
	8.98	5.97	845	463	419	−9.5
	6.14	6.88	860	580	527	−9.2
Projectile: 4340 steel $\phi 12.7 \times 45$ Target: AD85+GFRP [22]	4	5	893	832	722	−13.2
	6	5	880	800	678	−15.2
	9	5	898	693	636	−8.2
	9	8	880	658	575	−12.7
	9	10	882	621	535	−13.9
	18	5	895	425	402	−5.4
	18	8	888	329	359	9.1
	18	10	876	299	311	3.9

4 CONCLUSIONS

This paper proposes a ballistic performance calculation method that considers the enhancing effect of confining pressure on ceramics based on the energy dissipation principle. The theoretical prediction results are in good agreement with various ballistic experiments. The main conclusions are as follows:

1. The present theory can predict the ballistic performance of targets with a confining pressure ratio below 10% (including no confining pressure). When the initial speed is below 850 m/s, the prediction error is within 10% compared to the experiments. For projectile speed exceeding 850 m/s, the prediction error remains within 15%.
2. The two theoretical improvements introduced which are the ductility coefficient of ceramic and the enhancement of ceramic strength due to confining pressure, both contribute equally to the increment of the theory accuracy. The degree of improvement increases with the rise of confining pressure ratio.

ACKNOWLEDGEMENTS

This research was supported by the National Natural Science Foundation of China (Grant No. 52278526) and the Natural Science Foundation of Shanghai (Grant No. 22dz1202900).

REFERENCES

[1] Sherman, D., Impact failure mechanisms in alumina tiles on finite thickness support and the effect of confinement. *International Journal of Impact Engineering*, **24**(3), pp. 313–328, 2000.

[2] Holmquist, T.J. & Johnson, G.R., Modeling prestressed ceramic and its effect on ballistic performance. *International Journal of Impact Engineering*, **31**(2), pp. 113–127, 2005.

[3] Farbaniec, L. et al., Damage evolution of hot-pressed boron carbide under confined dynamic compression. *International Journal of Impact Engineering*, **99**, pp. 75–84, 2017.

[4] Chi, R. et al., Pre-stress effect on confined ceramic armor ballistic performance. *International Journal of Impact Engineering*, **84**, pp. 159–170, 2015.

[5] Lundberg, P., Renström, R. & Andersson, O., Influence of confining prestress on the transition from interface defeat to penetration in ceramic targets. *Defence Technology*, **12**(3), pp. 263–271, 2016.

[6] Xu, W. et al., Effects of dynamic prestress on silicon carbide ceramic against long rod impact. *Journal of Physics: Conference Series*, **2478**(7), 2023.

[7] Florence, A.L., Interaction of Projectiles and Composite Armour Part II. Report, Defense Technical Information Center, USA, 1969.

[8] Sanchez-Galvez, V., Impact behaviour of lightweight armours. *WIT Transactions on The Built Environment*, vol. 25, WIT Press: Southampton and Boston, pp. 91–101, 1970.

[9] Woodward, R.L., A simple one-dimensional approach to modelling ceramic composite armour defeat. *International Journal of Impact Engineering*, **9**(4), pp. 455–474, 1990.

[10] Feli, S., Aalami-Aaleagha, M.E. & Ahmadi, Z., A new analytical model of normal penetration of projectiles into the light-weight ceramic-metal targets. *International Journal of Impact Engineering*, **37**(5), pp. 561–567, 2010.

[11] Tang, R.T. & Wen, H.M., Predicting the perforation of ceramic-faced light armors subjected to projectile impact. *International Journal of Impact Engineering*, **102**, pp. 55–61, 2017.

[12] Zhang, R. et al., Influence of prestress on ballistic performance of bi-layer ceramic composite armors: Experiments and simulations. *Composite Structures*, **227**, 2019.

[13] Zhang, R., Han, B. & Lu, T.J., Confinement effects on compressive and ballistic performance of ceramics: A review. *International Materials Reviews*, **66**(5), pp. 287–312, 2020.

[14] Johnson, G.R. & Holmquist, T.J., Response of boron carbide subjected to large strains, high strain rates, and high pressures. *Journal of Applied Physics*, **85**(12), pp. 8060–8073, 1999.

[15] Zaera, R. & Sanchez-Galvez, V., Analytical modelling of normal and oblique ballistic impact on ceramic/metal lightweight armours. *International Journal of Impact Engineering*, **21**(3), pp. 133–148, 1998.

[16] Taylor, G., The use of flat-ended projectiles for determining dynamic yield stress I: Theoretical considerations. *Proceedings of the Royal Society of London Series A Mathematical and Physical Sciences*, **194**(1038) pp. 289–299, 1948.

[17] Hou, H.L., Zhu, X. & Li, W., Investigation on bullet proof mechanism of light ceramic/steel composite armor. *Acta Armamentarii*, **34**(1), pp. 105–114, 2013. (In Chinese.)

[18] Reijer, P.C.D., Impact on ceramic faced armour. Doctoral thesis, TU Delft, The Netherlands, 1991.

[19] Wilkins M.L., Fourth Progress Report of Light Armor Program. Technical Report, U.S. Department of Energy Office of Scientific and Technical Information, USA, 1969.
[20] Woodward, R.L. & Cimpoeru, S.J., A study of the perforation of aluminium laminate targets. *International Journal of Impact Engineering*, **21**(3), pp. 117–131, 1998.
[21] Abotula, S. & Chalivendra, V.B., An experimental and numerical investigation of the static and dynamic constitutive behaviour of aluminium alloys. *The Journal of Strain Analysis for Engineering Design*, **45**(8), pp. 555–565, 2010.
[22] Hetherington, J.G. & Rajagopalan, B.P., An investigation into the energy absorbed during ballistic perforation of composite armours. *International Journal of Impact Engineering*, **11**(1), pp. 33–40, 1991.
[23] Serjouei, A. et al., Experimental validation of BLV model on bi-layer ceramic-metal armor. *International Journal of Impact Engineering*, **77**, pp. 30–41, 2015.
[24] Ma, G. & Wang, F., Modeling the confinement and prestress effects on dynamic failure of ceramic deep beam. *International Journal of Protective Structures*, **4**(2), pp. 189–207, 2013.

NUMERICAL MODELLING OF ADDITIVELY MANUFACTURED AND WROUGHT MARAGING STEEL TARGETS UNDER BALLISTIC IMPACT

MAISIE EDWARDS-MOWFORTH[1,2], MIGUEL COSTAS[3], MARTIN KRISTOFFERSEN[3,4], TORE BØRVIK[3] & FILIPE TEIXEIRA-DIAS[1]
[1]Institute for Infrastructure and Environment, University of Edinburgh, UK
[2]IMPETUS Afea AS, Flekkefjord, Norway
[3]Structural Impact Laboratory (SIMLab), Norwegian University of Science and Technology, Norway
[4]Multiconsult AS, Tromsø, Norway

ABSTRACT

The advantages of on-demand production and customised or optimised parts make additive manufacturing (AM) an exciting prospect for optimised armour systems. High-strength maraging steel is a martensitic steel, widely used in aerospace and tooling industries. The combination of high-strength and the flexibility of AM gives maraging steel significant potential for impact protection applications. However, the enhanced strength of maraging steel, achieved through heat-treatment, simultaneously causes reduced fracture elongation and impact toughness. Therefore, the improved ballistic perforation resistance comes at a compromise of reduced ductility, with the effects of AM fabrication potentially exacerbating the brittle behaviour. This study predicts the ballistic performance of heat-treated AM maraging steel targets, in comparison to wrought targets of the same material, through simulations validated by experiment data. Numerical models calibrated from material characterisation tests are established in the IMPETUS Afea Solver. The ballistic limit velocity was estimated conservatively within the experimentally determined value for both AM and wrought maraging steel with good agreement. Comparisons are drawn concerning how well the numerical failure techniques implemented, node splitting and element erosion, were able to capture the fragmentation behaviour of the brittle heat-treated, high-strength targets.

Keywords: ballistic impact, maraging steel, additive manufacture, numerical modelling, fragmentation.

1 INTRODUCTION

As additive manufacturing (AM) brings the advantages of on-demand production and customised or optimised parts to the defence industry, important questions arise regarding the structural integrity of safety-critical AM components in armour systems [1]–[4]. Laser powder-bed fusion (LPBF), whereby metallic powder is selectively fused by an energy source to build a part, is one promising AM technology where the uncertainty in resulting material properties has hindered its application in protective structures. The nature of layer-by-layer fabrication, with large thermal gradients in the substrate, can induce discontinuities in the microstructure leading to anisotropy [5], [6]. Printing defects such as porosity, lack of fusion between melt layers and microsegregation, frequently reported, can lead to tendencies for brittle fracture and fragmentation during an impact event [7].

Excessive fragmentation of high-strength AM maraging steel, produced by LPBF, under ballistic impact has recently been evidenced in experimental campaigns [7], [8]. Maraging steel is a martensitic steel, widely used in aerospace and tooling industries, with a low carbon composition that makes it suited to AM through LPBF techniques [6]. Ultra high-strength is achieved after heat-treatment through the development of intermetallic precipitates. The combination of high-strength and the flexibility of AM gives maraging steel significant potential for impact protection applications. AM maraging steel in its as-printed state has been found to exhibit similar ballistic performance to wrought maraging steel of the same

WIT Transactions on The Built Environment, Vol 215, © 2025 WIT Press
www.witpress.com, ISSN 1743-3509 (on-line)
doi:10.2495/SUSI250121

composition [8]. However, the precipitates responsible for enhancement of strength in heat-treated material simultaneously cause reduced fracture elongation and impact toughness [5]. Therefore, a compromise must be sought between improved ballistic perforation resistance and reduced ductility, which may be further exacerbated by AM fabrication defects, as shown by Costas et al. [7].

Numerical models form an important part of the research and design process into new materials in armour systems, especially where the possibility arises for optimisation of geometries and materials. Achieving accurate finite element (FE) simulations of ballistic impact on high-strength AM maraging steel poses two key challenges: firstly, including the effects of AM processing appropriately in material models and, secondly, capturing the fragmentation. A commonly used technique to model failure in explicit simulations, element erosion or deletion, may lead to unrealistic results for large amounts of fragmentation due to the loss of mass in the model. Success has been found with an alternative method in place of element erosion in several studies [9], [10], namely, node splitting: a technique whereby cracks are propagated by separating elements rather than removing elements, which is implemented in the IMPETUS Solver [11].

This study evaluates the performance of numerical models using various failure techniques to predict the ballistic performance of AM maraging steel targets in comparison to wrought targets of the same material. Material models calibrated from material characterisation tests are established in the IMPETUS Solver. Node splitting technology is used to simulate material failure in both the AM maraging steel targets and the APM2 (armour piercing) projectile, in comparison to element erosion. The results of bullet core residual velocity obtained from the simulations are compared with experimental ballistic impact tests, as well as a qualitative analysis of the perforation processes. Comparisons are drawn concerning how well the numerical techniques implemented were able to capture the fragmentation behaviour of the brittle, high-strength targets.

2 EXPERIMENTAL BACKGROUND

The numerical work in this study draws on experimental tests carried out by the authors in a proof-of-concept examination of the ballistic performance of AM and wrought maraging steel. Material characterisation tests, including quasi-static and dynamic tensile tests, and ballistic impact tests were carried out on AM and wrought maraging steel, both before and after an industry standard heat-treatment procedure, using experimental facilities at the Norwegian University of Science and Technology (NTNU). A summary of the main results from the ballistic impact tests, relevant for comparison to numerical models in this study, follows. For a full description of the AM processing, experimental set-ups and results, the reader is referred to Edwards-Mowforth et al. [8].

The experimental ballistic impact tests were conducted in a 16 m^3 chamber by firing APM2 bullets at 2 × 5 mm heat-treated cylindrical target plates, depicted in Fig. 1, using a smooth-bored Mauser rifle. As shown in Fig. 2, fragmentation was present in the target during impact for both the AM and wrought maraging steel, but much more significantly for the former. The extensive fragmentation in the AM material may be attributed to anisotropy and lack of fusion defects induced by the AM processing. In addition, the AM material post heat-treatment was capable of breaking the hard steel core of the armour piercing bullet in each test, whereas this occurred to a lesser extent for the wrought maraging steel, despite both exhibiting similar ballistic limit velocities. Fragmentation in both the target and projectile plays a large role in the perforation mechanisms and therefore requires consideration in numerical models of the impact.

www.witpress.com, ISSN 1743-3509 (on-line)

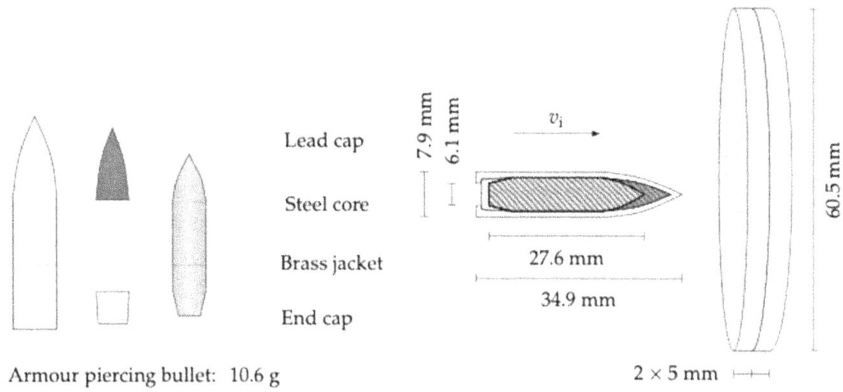

Figure 1: Experimental ballistic impact test set-up. Components of the APM2 bullet (left) and section view of the projectile with 2 × 5 mm heat-treated target plates (right).

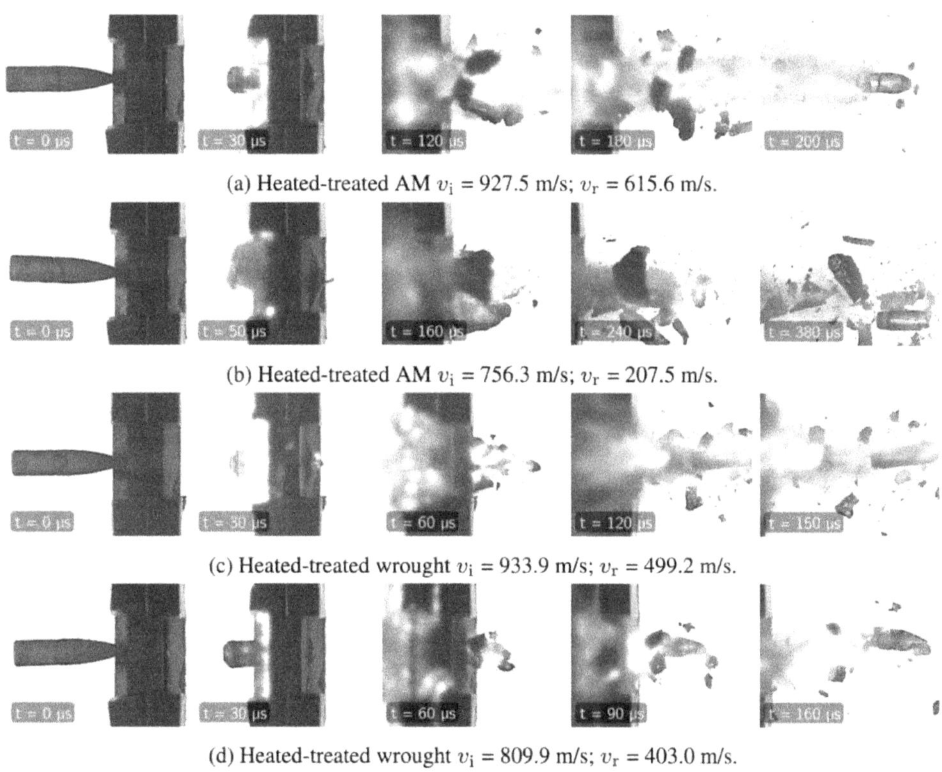

Figure 2: Sequential images of ballistic impact experiments capture with a high-speed camera. Two different impact velocity examples are shown for AM, (a) and (b), and wrought, (c) and (d), heat-treated maraging steel target plates.

3 NUMERICAL SIMULATIONS

3.1 Material modelling

The modified Johnson–Cook (MJC) model is a widely used phenomenological relation for metallic materials under large deformations, high strain rates and high temperatures [12]. Having proved reliable in numerous studies on ballistic impact and AM materials [13]–[15], MJC is also utilised in this study to model the maraging steel target plates and components of the APM2 projectile. The same models were chosen for both target materials to evaluate how well the AM material behaviour could be captured using standard modelling techniques. Empirical parameters to populate the MJC model were calibrated through inverse modelling of the quasi-static tensile tests taken from 0° and 90° to the build direction in the case of the AM maraging steel, and from quasi-static axial compression tests for the APM2 steel bullet core. The AM maraging steel exhibited little anisotropy in work hardening, therefore an average best fit between the calibration at 0° and 90° to the build direction was used [8].

The Cockcroft–Latham (CL) criterion [17] was used to include damage in the model for relative simplicity, requiring only one tensile test to calibrate the CL failure parameter, W_C. The CL parameters were extracted from simulations of quasi-static tensile tests using the calibrated work hardening parameters for each material variation, identifying the critical element at the point of failure. The final value of W_c was averaged from all tests for each material variation. A similar procedure was used for the AMP2 bullet core from simulation of a Brazilian tensile test, using the work hardening calibration obtained from the axial compression test calibration. Crack propagation in the node splitting formulation is defined by the fracture energy parameter, G, taken as a typical value of 5,000 kJ/m^2 for steel [18]. The calibrated work hardening and failure parameters used are reported in Table 1. The typical values for the component materials of the APM2 bullet and maraging steel listed in Fig. 3 were assumed from the literature [8], [13], [16].

Table 1: MJC work hardening parameters for the AM and wrought heat-treated maraging steel, calibrated from quasi-static tensile tests.

	Yield stress	Strain hardening				Strain rate	Failure
	A (MPa)	$Q1$ (MPa)	$C1$	$Q2$ (MPa)	$C2$	C	W_c (MPa)
AM	1,678.7	−73.8	75.9	354.0	200.8	0.0077	170
Wrought	1,730.0	178.7	4.4	120.1	200.0	0.0077	1,274.5

3.2 Numerical set-up

Simulations of the experimental ballistic impact tests were carried out in the explicit FE code IMPETUS Solver. The Ø60.5 mm maraging steel target plates were discretised into elements of approximately 0.1 mm within a Ø15 mm impact zone region, and around 50 elements over the 5 mm target plate thickness, as shown in Fig. 3. A sufficiently small element size consistent with the mesh used for material model calibrations was ensured, as per a computational cell approach [19], while maintaining computational efficiency. Simulations were carried out using both a structured mesh, generated within IMPETUS, and an unstructured mesh, using the open source Gmsh software [20]. As shown in Fig. 3, the resulting unstructured mesh has non-uniform element sizes and shapes. Irregular element

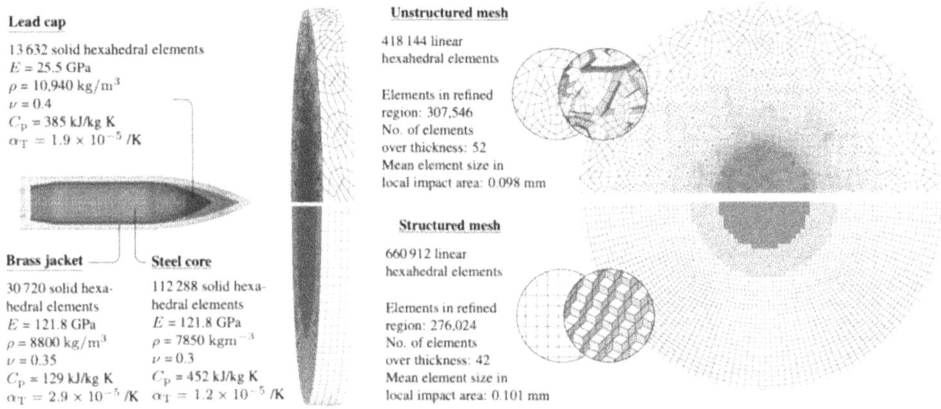

Figure 3: Ballistic impact simulation set-up detailing the mesh specifications for each component, as well as thermoelastic and adiabatic heating constants. The APM2 bullet mesh is displayed to the left, and the target plate to the right shows a half-cut plate of the unstructured and structured mesh types.

sizes were intended to minimise any influence of mesh uniformity on crack patterns of the plate, and to introduce stochastic variation, given that the CL criterion is mesh dependent. The brass jacket, lead cap and bullet core of the APM2 projectile were discretised with regular meshes of 30,720, 13,632 and 71,824 elements, respectively, as also depicted in Fig. 3.

The bullet model was assigned initial velocities to replicate the experimental tests. A frictionless, surface-to-surface, penalty stiffness-based contact algorithm was employed to define interaction between all parts including self-contact, as a conservative assumption. Boundary conditions were representative of the clamp set-up in experimental tests. All components were discretised by solid linear hexahedral elements with selectively reduced integration. Node splitting with the crack plane orthogonal to the direction of maximum principal stress was used as the penalty for reaching the failure criterion and was compared to repeated simulations using element erosion.

3.3 Simulation results

The initial velocity, v_{i}, and residual velocity, v_{r}, of the projectile were extracted from the simulations and fitted to the Recht–Ipson model [21] to estimate the ballistic limit velocity, v_{bl} for each material configuration. The Recht–Ipson model is given as

$$v_{\mathrm{r}} = a(v_{\mathrm{i}}^{p} - v_{\mathrm{bl}}^{p})^{1/p} \tag{1}$$

where a and p are constants fixed at 1 and 2, respectively, for the purpose of clear comparison. The resulting ballistic limit curves and v_{bl} values for AM and wrought heat-treated maraging steel are shown in Fig. 4 and listed Table 2, respectively.

Using an unstructured mesh with node splitting achieved a v_{bl} value within 2% and 3% of the experimental values for AM and wrought targets, respectively, as shown in Fig. 4. Table 2 reports these v_{bl} values alongside the corresponding results for the structured mesh

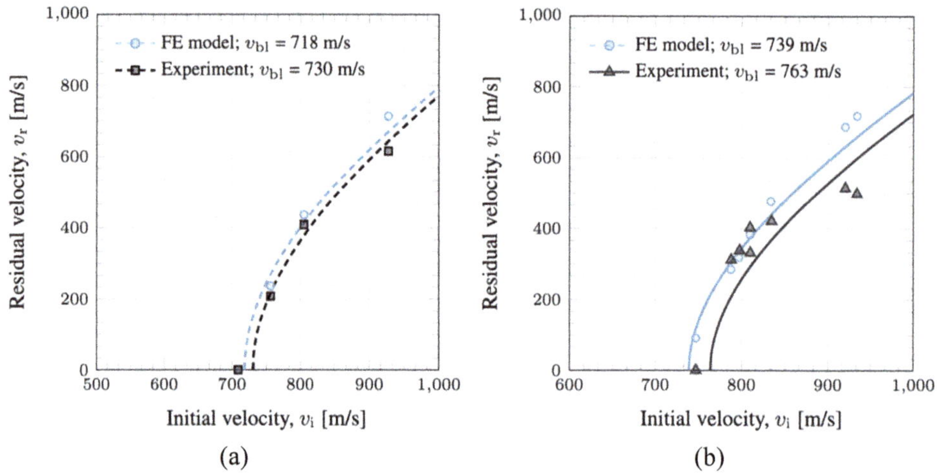

Figure 4: Experimental ballistic impact curves in comparison to simulation predictions using node splitting and the unstructured mesh for (a) AM and (b) wrought heat-treated maraging steel.

Table 2: Ballistic limit velocity for different failure modelling approaches and structured and unstructured meshes for AM and wrought heat-treated maraging steel.

	Node splitting Unstructured mesh v_{bl} (m/s)	Node splitting Structured mesh v_{bl} (m/s)	Element erosion Unstructured mesh v_{bl} (m/s)	Experiment v_{bl} (m/s)
AM	718.2	703.4	636.5	730.1
Wrought	739.1	728.5	820.6	763.3

and element erosion. Using node splitting with a structured mesh achieved a slightly larger, 4% and 5%, difference from the experimental v_{bl} values for AM and wrought targets, respectively. The percentage difference increased to approximately 13% and 11%, respectively, using element erosion. The increase in deviation from the experimentally determined value for the latter may be attributed to the shortfalls of the element erosion technique to capture target fragmentation, found to be significant for the heat-treated maraging steel, especially so for the AM material.

Differences in the capability of node splitting and element erosion to capture target plate fragmentation are documented qualitatively in Fig. 5, which shows the projectile-exit face of the target plates post-perforation. Photos of the perforated experimental test target are also shown in Fig. 5 alongside the corresponding simulation results. For the AM example, a large, irregular shaped perforation channel is shown both for the experiment and from the node splitting simulation with both mesh types, whereas the hole created by element erosion is smaller, evidence of less fragmentation and cracking than that observed in the experiment. For the wrought example, all simulation variations capture a small, circular shaped perforation channel, while only the node splitting simulations capture the radial cracks visible in the experiment. The simulated cracks, however, do not reach the edge of the target plate as they are seen to do in the experiment. This may be partially attributed to the gradual

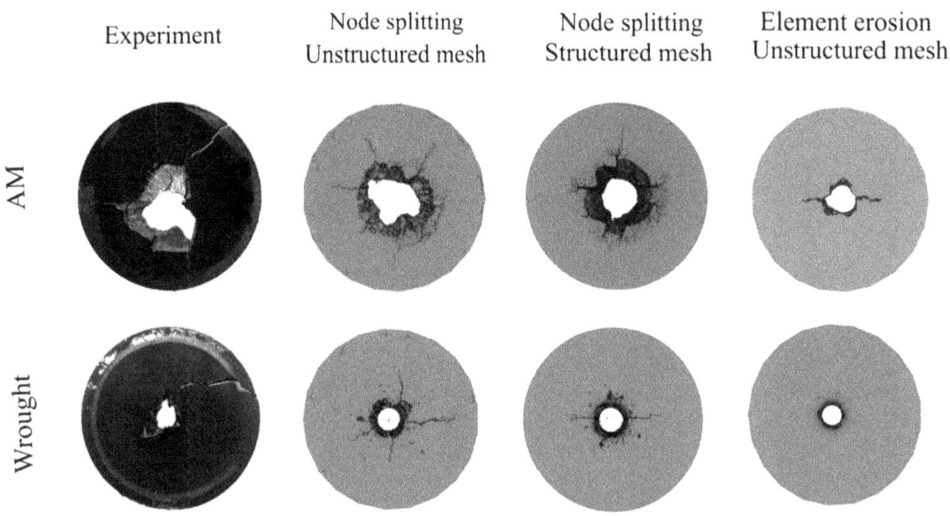

Figure 5: Projectile-exit face of the heat-treated AM (v_i = 756.5 m/s) and wrought (v_i = 933.9) maraging steel target plates impacted by armour piercing bullets: Experiment compared to simulations using different failure techniques and meshes.

increase in element size outside the impact zone prematurely arresting crack growth. Cracks simulated by the node splitting algorithm appear to follow the straight mesh-lines in the structured mesh, whereas the crack patterns appear more natural for the unstructured mesh.

Snapshots throughout the simulated perforation process are displayed in Fig. 6, which further provide evidence of differences between the node splitting and element erosion technique. For both the AM and wrought examples (Figs 6(b) and 6(d)) element erosion appears to generate fewer, smaller fragments compared to the node splitting simulations (Figs 6(a) and 6(c)), resulting in a higher residual velocity of the bullet core after perforation. The simulation with element erosion for the wrought material was the only example in which the hard steel bullet core fragmented, however this did not occur for the corresponding velocity in the experiment, as shown in Fig. 2(c).

Several differences between simulation and experiment in the projectile response were identified in all cases, including with node splitting. The brass jacket was removed during perforation in the experiments on the AM material, but remained in place in the simulations, which were also unable to capture breakages in the bullet core. However, the lack of fragmentation in the bullet core did not induce significantly or hinder accurate ballistic limit velocity predictions. This could indicate that target plate fragmentation is more influential to residual velocity than bullet core fragmentation. Nevertheless, accurate modelling of bullet core fracture requires further experimental tests and validation to improve the failure model.

4 CONCLUSIONS

- Commonly used constitutive models and failure criteria, along with node splitting in the IMPETUS Solver, were able to accurately predict the ballistic limit velocity of heat-treated AM and wrought maraging steel target plates impacted by APM2 projectiles.

(a) Node-splitting, heat-treated AM, $v_i = 756.5$ m/s; $v_r = 264.5$ m/s.

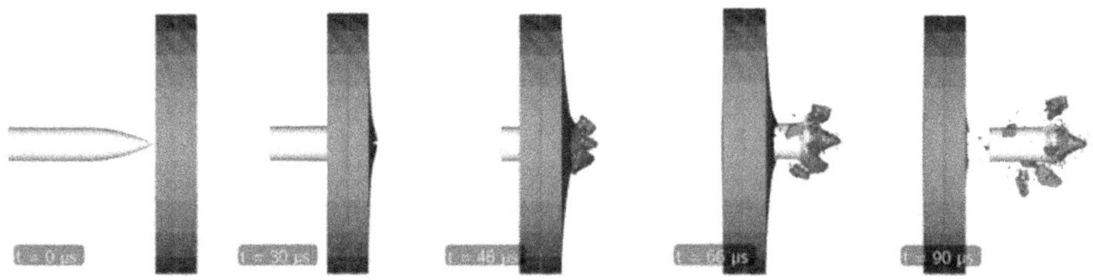

(b) Element erosion, heat-treated AM, $v_i = 756.5$ m/s; $v_r = 468.2$ m/s.

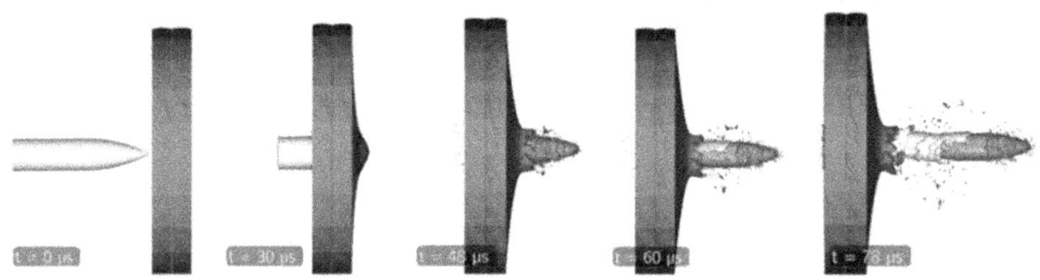

(c) Node-splitting, heat-treated wrought $v_i = 933.9$ m/s; $v_r = 716.5$ m/s.

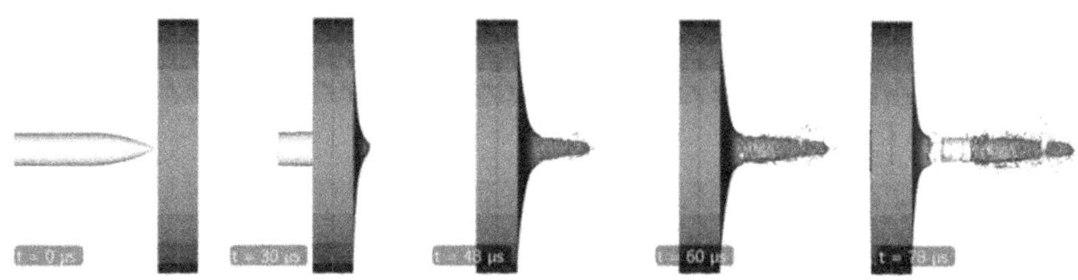

(d) Element erosion, heat-treated wrought $v_i = 933.9$ m/s; $v_r = 684.5$ m/s.

Figure 6: Sequential images throughout simulations of 2×5 mm heat-treated AM, (a) and (b), and wrought, (c) and (d), maraging steel target plates impacted by armour piercing bullets, comparing the node splitting and element erosion failure techniques.

- The node splitting algorithm reproduced large, irregular fragments and radial cracks in target plates for the AM and wrought material, respectively, however, to a lesser extent than the severe fragmentation observed in experimental tests, especially for the AM material. The inclusion of AM-induced defects in the material model to capture fragmentation more accurately is left for further studies.

- Damage in the hard steel APM2 bullet core was included by use of the Cockcroft–Latham failure criterion, which was unable to capture fracturing of the bullet core as seen in experiments despite the accurate ballistic limit velocities achieved.
- Both structured and unstructured meshes paired with node splitting predicted ballistic limit velocities to within 4–5% of experiment. The unstructured mesh produced more natural crack patterns, whereas the structured mesh guided cracks along mesh grid lines.
- While node splitting gave ballistic limit velocity predictions within 2–3% of the experiment, element erosion yielded overestimations of roughly 11–13%. This discrepancy corresponds to an under-prediction of target plate fragmentation and cracking in the element erosion simulations.

REFERENCES

[1] Medvedev, A.E., Maconachie, T., Leary, M., Qian, M. & Brandt, M., Perspectives on additive manufacturing for dynamic impact applications. *Materials and Design*, **221**, 110963, 2022.

[2] Stavropoulos, P., *Metal Additive Manufacturing Technology Applications in Defense Organizations*, Springer International Publishing, pp. 1–35, 2021.

[3] Colorado, H.A., Cardenas, C.A., Gutierrez-Velazquez, E.I., Escobedo, J.P. & Monteiro, S.N., Additive manufacturing in armor and military applications: research, materials, processing technologies, perspectives, and challenges. *Journal of Materials Research and Technology,* **27**, pp. 3900–3913, 2023.

[4] Ngo, T., Kashani, A., Imbalzano, G., Nguyen, K. & Hui, D., Additive manufacturing (3D printing): A review of materials, methods, applications and challenges. *Composites Part B: Engineering*, **143**, pp. 172–196, 2018.

[5] Dehgahi, S., Alaghmandfard, R., Tallon, J., Odeshi, A. & Mohammadi, M., Microstructural evolution and high strain rate compressive behavior of as-built and heat-treated additively manufactured maraging steels. *Materials Science and Engineering: A*, **815**, 141183, 2021.

[6] Mooney, B., Kourousis, K. & Raghavendra, R., Plastic anisotropy of additively manufactured maraging steel: Influence of the build orientation and heat treatments. *Additive Manufacturing*, **25**, pp. 19–31, 2019.

[7] Costas, M., Edwards-Mowforth, M., Kristoffersen, M., Teixeira-Dias, F., Brøtan, V., Paulsen, C. & Børvik, T., Ballistic impact resistance of additive manufactured high-strength maraging steel: An experimental study. *International Journal of Protective Structures*, **12**(4), pp. 577–603, 2021.

[8] Edwards-Mowforth, M., Costas, M., Kristoffersen, M., Teixeira-Dias, F. & Børvik, T., On the ballistic perforation resistance of additively manufactured and wrought maraging steel: Experiments and numerical models. *International Journal of Impact Engineering*, **201**(8), 105271, 2025.

[9] Olovsson, L., Limido, J., Lacome, J.L., Hanssen, A.G. & Petit, J., Modeling fragmentation with new high order finite element technology and node splitting. *EPJ Web of Conferences*, **94**, 04050, 2015.

[10] Rakvåg, K., Børvik, T. & Hopperstad, O., A numerical study on the deformation and fracture modes of steel projectiles during taylor bar impact tests. *International Journal of Solids and Structures*, **51**(3), pp. 808–821, 2014.

[11] IMPETUS Afea, Home. https://www.impetus.no. Accessed on: Mar. 2025.

[12] Børvik, T., Hopperstad, O., Berstad, T. & Langseth, M., A computational model of viscoplasticity and ductile damage for impact and penetration. *European Journal of Mechanics – A/Solids*, **20**(5), pp. 685–712, 2001.

[13] Børvik, T., Dey, S. & Clausen, A., Perforation resistance of five different high-strength steel plates subjected to small-arms projectiles. *International Journal of Impact Engineering*, **36**(7), pp. 948–964, 2009.
[14] Dey, S., Børvik, T., Hopperstad, O. & Langseth, M., On the influence of fracture criterion in projectile impact of steel plates. *Computational Materials Science*, **38**(1), pp. 176–191, 2006.
[15] Kristoffersen, M., Costas, M., Koenis, T., Brøtan, V., Paulsen, C. & Børvik, T., On the ballistic perforation resistance of additive manufactured AlSi10Mg aluminium plates. *International Journal of Impact Engineering,* **137**, 103476, 2020.
[16] Bracq, A., Brest, J., de Sampaio, J., Moitrier, F. & Demarty, Y., Characterization and modelling of the mechanical behaviour of metal rings: Application to a brass bullet jacket. *Forces in Mechanics*, **4**, 100030, 2021.
[17] Cockcroft, M. & Latham, D., Ductility and the workability of metals. *Journal of the Institute of Metals*, **96**, pp. 33–39, 1968.
[18] Defense Technical Information Center, High Strain Rate Material Behavior. Technical Report ADA165214, Defense Technical Information Center, 1985.
[19] Ruggieri, C., Panontin, T.L. & Dodds, R.H., Numerical modeling of ductile crack growth in 3-D using computational cells. *International Journal of Fracture*, **82**, pp. 67–95, 1996.
[20] Geuzaine, C. & Remacle, J.F., Gmsh: A 3-D finite element mesh generator with built-in pre-and post-processing facilities. *International Journal for Numerical Methods in Engineering*, **79**(11), pp. 1309–1331, 2009.
[21] Recht, R. & Ipson, T., Ballistic perforation dynamics. *International Journal of Applied Mechanics (Trans ASME)*, **30**, pp. 384–390, 1963.

SECTION 4
EXPERIMENTS AND NUMERICAL MODELLING

DEVELOPMENT AND TESTING OF A COMBINED HEAT, AXIAL LOAD AND LOW IMPACT TESTING RIG

LAURENCE G. CLOUGH & ZANA KURDO
School of Civil Engineering and Surveying, University of Portsmouth, UK

ABSTRACT

The effect of combined static, impact and thermal loads on structures is a relatively little-known area of research with large potential applicability to sectors including defence and energy. In order to study the effect of these combined loads on structural elements, it is important to conduct controlled laboratory testing in order to validate computational analyses. There are currently few or no testing apparatus which are able to repetitively test structural elements to combined heat, axial and impact loads at a low cost. This paper details the development of a bespoke combined low-impact testing rig capable of testing 500 mm long, 100 mm wide structural elements. The rig can apply combined load sets including a thermal load (up to 400°C) an axial load (up to 150 kN) and an impact load from a 7.5 kg drop weight. For the developmental testing 500 mm long, square and rectangular concrete sections were tested as part of a wider study into the performance of concrete containing hybrid polypropylene (PPE) and steel fibres at high temperatures. Recorded results from the study include external and internal concrete temperatures via thermo-couples, longitudinal strains and vertical deflections. Conclusions regarding the suitability of the bespoke testing rig to impart the combined load sets on structural columns are presented along with discussions on the effects of the hybrid fibres in concrete columns. The study found that the rig is able to impart the combined load sets effectively and accurately enabling multiple repeated testing on elements at a low cost. The presence of PPE fibres allowed the concrete to reach higher temperatures without spalling and improved its overall resistance.
Keywords: impact, thermal, axial load, concrete, polypropylene fibres.

1 INTRODUCTION

A bespoke experimental rig was designed and developed for the testing of combined thermal, axial and low velocity impact loads on concrete structural elements. Sixty trials were undertaken on a range of concrete column types under different heating regimes. The tested concrete columns were 500 mm in length, containing a mix of high strength steel fibres (SF), standard reinforcing bars, Polypropylene (PPE) fibres and plain concrete. Tests were conducted to observe the effect of the addition of PPE fibres on the resistance of concrete columns subject to combined thermal, compression and low velocity impact loads. Conclusions regarding the suitability of the developed rig are made along with the effect of the inclusion of PPE fibres within the concrete mixes. Further development of the rig is proposed including the testing of different structural elements, improved thermal load application, static loading configurations and impactor head types.

2 POLYPROPELENE FIBRES IN CONCRETE

High strength steel fibres (SF) are used in concrete applications to increase load bearing capacity and durability. Research into SF use for exceptional, non-standard service loads, e.g. blast and impact loads [1] has shown an increased tensile resistance and ductility. SF can also be used in thin structural slabs and walls which would require low standard covers to normal reinforcement. If plain or standard reinforced concrete (RC) is subject to high temperatures in a rapid heating regime there is a high probability that explosive spalling will occur. This spalling is due to trapped and expanding water vapour pressures within the concrete matrix [2]. After spalling, the concrete compressive capacity is compromised and

WIT Transactions on The Built Environment, Vol 215, © 2025 WIT Press
www.witpress.com, ISSN 1743-3509 (on-line)
doi:10.2495/SUSI250131

the reinforcing bars are exposed enabling quicker heat transfer to the rebar and subsequent degradation of yield strength and Young's modulus. The addition of PPE fibres in concrete has been previously researched to show an improved resistance to explosive spalling at high temperatures [3], [4]. The PPE fibres do not add any additional tensile or compressive capacity to the concrete. At high temperatures the PPE fibres melt at 160°C to170°C, leaving a network of fine channels through the concrete matrix. These channels act as routes for trapped water vapour to escape thus lowering the probability of explosive spalling at high temperatures. With no explosive spalling the concrete retains its structural integrity and the reinforcement remains unexposed from direct thermal radiation. The use of PPE fibres in combination with normally reinforced or SF reinforced concrete increases structural resistance to heat loads from fire and other thermal loads, this consequently maintains the structural resistance to any subsequent impact or blast loads. Prior to the combined thermal, compression and impact testing on the concrete columns documented in this paper, a series of material characterisation trials were undertaken to determine the optimum PPE fibre dosage [5]. The optimum dosage determined was 2.5%, this was subsequently adopted within the concrete columns for the combined load testing.

3 RIG DEVELOPMENT

A bespoke rig was developed at the University of Portsmouth (UoP) Civil Engineering laboratories to impart a combination of thermal, compression and impact load on to structural concrete columns. The rig is a low-cost alternative to full scale field testing, with the ability to test consecutive multiple samples with varying characteristics in controlled conditions. The following describes the rig development detailing the separate load application methods.

3.1 Compressive load

A critical goal of the column testing was to recreate in-service conditions as closely as practical in a controlled laboratory. Therefore, the structural columns should be subject to stresses akin to a standard building axial load. Options were investigated included the use of an in-house hydraulic rig, but it was deemed impractical due to the high risk of damage to the rig instrumentation from the thermal load. A manually operated floor standing hydraulic press, capable of applying 200 kN (Fig. 1(a)), was adopted. The press was adjusted and laid flat, enabling test samples to be positioned horizontally, which allowed a drop rig to be positioned vertically above the centre of any test sample. A 500 kN capacity load cell (Fig. 1(b)) was positioned between the end of the concrete column and the press plate to record the applied compression, ensuring all tests were subject to the same initial load (150 kN).

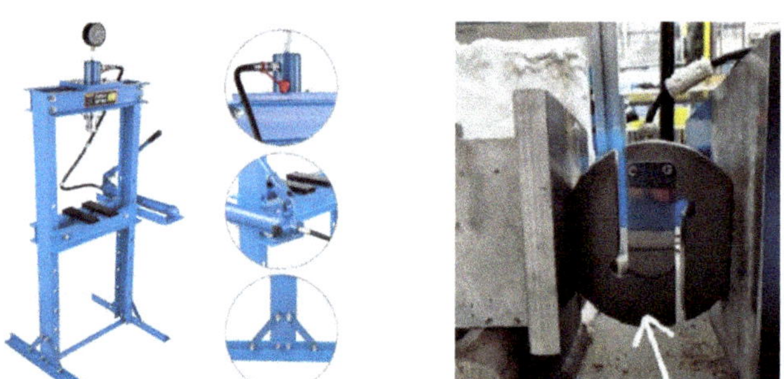

(a) (b)

Figure 1: (a) Hydraulic press [6]; and (b) Load cell.

3.2 Heat load

A main research requirement was that the heat load applied to the structural columns should be akin to a fire load and that it should cause a significant, observable thermo-mechanical response in the concrete columns. From prior research, the PPE fibres would melt at 160°C to 170°C, therefore this was set as the minimum required temperature with a maximum achievable temperature of 400°C. This maximum temperature was set due to the health and safety requirements within the laboratory and the capacity of the heating system. The use of an in-house rapid heating chamber furnace (1200°C maximum) was not appropriate for the trials there was no ability to apply the compression and impact loads to the column samples within the chamber. The option of heating the samples within the furnace, then subsequently removing and testing them with applied compression and impact loads in a separate rig was deemed unpractical and unsafe. Therefore, ceramic heating elements were used to heat the sides of the columns to the required temperatures. Two 1000 W full quartz ceramic heating elements [7] were fixed inside aluminium boxes with protective wire meshes. These boxes were placed within a thermally insulated aluminium channel positioned to face the columns in the centre, with a 20 mm gap between the sides of the columns and the heating boxes (Fig. 2). The central 250 mm of each column were heated via the ceramic heaters. The ceramic heaters are able to reach surface temperature of 770°C, however, due to the air gap and safety considerations the maximum achievable concrete surface temperature was set at 400°C.

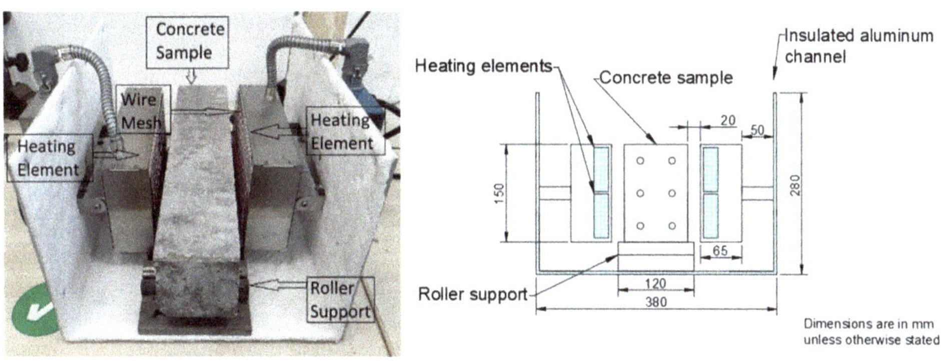

Figure 2: Thermal load application.

3.3 Impact load and combined rig

The main requirements for the impact load included a lateral application over the central section of the columns, the practical lifting of the impactor, the imparting an intense load over a very short duration and no sustained damage to the impactor during tests. A 720 mm long, 41 mm diameter, high-strength steel rod was manufactured for this purpose. Galvanised steel tubes positioned above the columns guided the 7.5 kg steel rod to impact the concrete columns at their midpoint. The columns were supported by roller supports at 450 mm centres, thus the vertical impact load imparted three-point bending on to the columns. The steel tubes were machined with guide holes and pins to allow the rod to be dropped from up to 2 m height (imparting 147 J), for these trials a consistent height of 1 m (73.5 J) was selected. The rod was manually lifted between tests via a pulley system using strong wire connected to the top (Fig. 3(a)).

(a) (b)

Figure 3: (a) Impactor rod; and (b) Combined rig.

The three separate loading mechanisms were configured to work together in a combined rig (Fig. 3(b)). The rig was fixed via brackets to the strong floor of the laboratory. A lightweight aluminium frame held the impactor guide tube and rod in place over the columns.

4 EXPERIMENTAL METHODOLOGY

After the development and calibration of the rig, the trials were undertaken. Sixty concrete columns were tested with a variety of fibre/reinforcement configurations under a range of thermal loads. The applied compressive load and impact load were constant across the tests. The following section details the concrete column mix design and experimental procedure.

4.1 Concrete column preparation and mix designs

Table 1 details the concrete column sections tested; six square, and six rectangular column types were tested with a mix of 10 mm reinforcement bars, plain concrete, PPE fibres, steel fibres and two hybrid mixes: PPE + rebar and PPE + steel fibre. Five tests were conducted per mix and the mean values subsequently calculated and presented. The concrete columns were cast based on a high strength concrete (HSC) mix design as shown in Table 2. This mix design had been previously determined from a material characterisation study [5], to find the optimum dosage of PPE fibres using compressive tests on 100 mm × 100 mm concrete cubes. The concrete columns were cast in 600 mm length moulds (with 50 mm sacrificial polystyrene at each end). The concrete was placed in two layers to allow compaction and placing of thermocouples at centre of columns. Bespoke steel bar holders which also acted as shear links were specially manufactured for the RC columns (Fig. 4). These bar holders ensured the bars were placed at consistent 50 mm centres with 20 mm concrete cover. Columns were set for 24 h under a damp hessian sheets then water cured for 28 days in accordance with BS EN 12390-2:2019 [8].

Table 1: Concrete column samples.

Column sizes	Naming convention	PPE fibre content (kg/m^3)	SF content (%)	No. rebar	No. specimens
100×100×500	SQ_CTRL	0	0	0	5
100×100×500	SQ_CTRL_PP	2.5	0	0	5
100×100×500	SQ_RC	0	0	4	5
100×100×500	SQ_RC_PP	2.5	0	4	5
100×100×500	SQ_SF	0	2	0	5
100×100×500	SQ_SF_PP	2.5	2	0	5
150×100×500	RECT_CTRL	0	0	0	5
150×100×500	RECT_CTRL_PP	2.5	0	0	5
150×100×500	RECT_RC	0	0	6	5
150×100×500	REC_RC_PP	2.5	0	6	5
150×100×500	REC_SF	0	2	0	5
150×100×500	REC_SF_PP	2.5	2	0	5

Table 2: Concrete mix designs.

Constituent material	Plain HSC	HSC and PPE	HSC, PPE and SF
Cement (kg/m^3)	415	415	415
Water (kg/m^3)	139	139	139
Superplasticiser	13.75	13.75	13.75
Steel fibre (kg/m^3)	–	–	157
Silica fumes (kg/m^3)	45	45	45
Polypropylene (%)	–	2.5	2.5
Silica sand (kg/m^3)	432	432	432
Sharp sand (kg/m^3)	439	439	439
Fine aggregate 10 mm (kg/m^3)	488	488	488
Coarse aggregate 20 mm (kg/m^3)	561	561	561

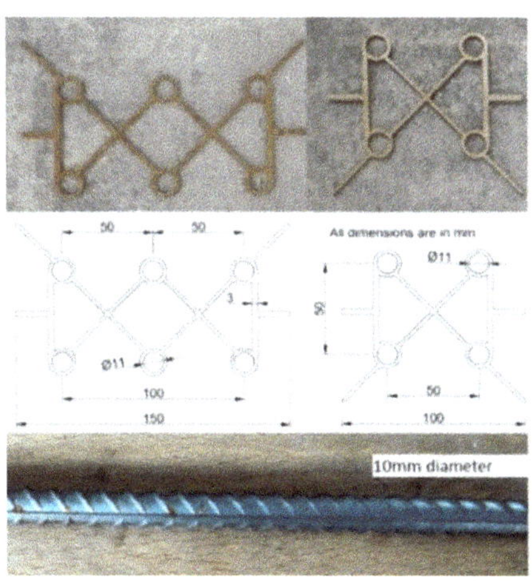

Figure 4: Shear links / reinforcement bar holders.

4.2 Instrumentation

The instrumentation used for the concrete column tests is detailed below:

- Thermocouples: K-type, max. four per test: 1× each side facing heaters, one top, one cast in concrete.
- Strain gauges: 150/120 LY41, 150 mm length HBK strain gauges, fixed with heat resistant adhesive to top face of column (longitudinally).
- Vertical displacement gauge: Novatech DC2.5 type linear variable differential transducer (LVDT), fitted with long rod to reach top face of concrete column.
- Load cell: 500 kN capacity.
- Temperature control: Proportional integral derivative (PID) controller connected to external thermocouples.
- Data acquisition system: DT80 dataTaker acquisition system, connected to laptop via USB.

4.3 Testing procedure

The following flow chart (Fig. 5) details the testing procedure for each concrete column.

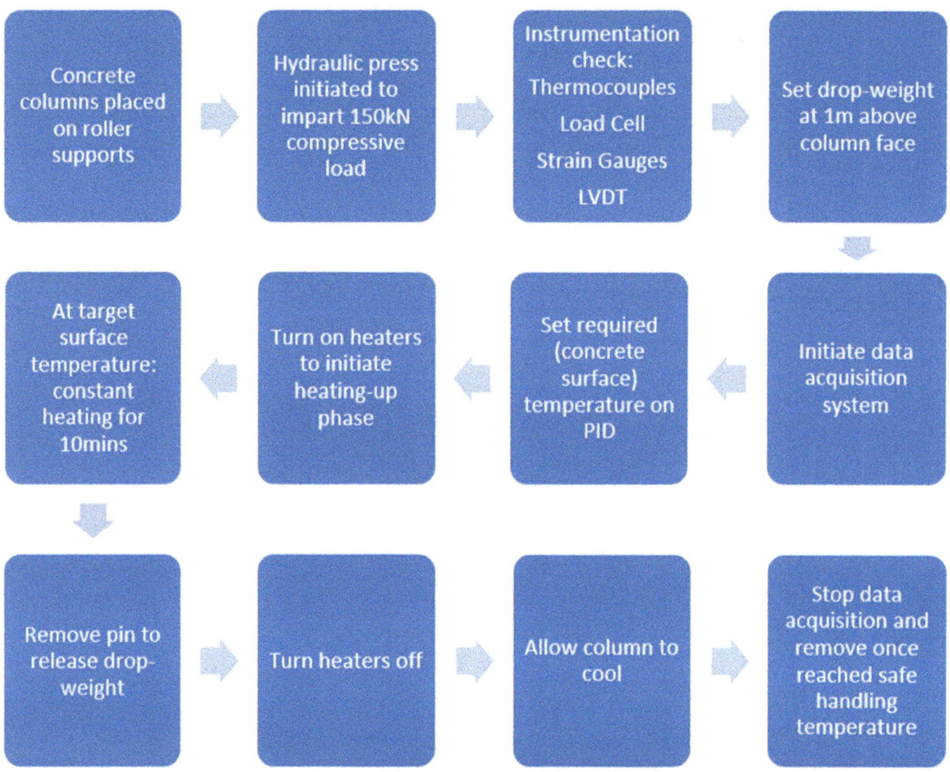

Figure 5: Experimental testing procedure.

5 EXPERIMENTAL RESULTS

For each test the columns were heated at a rate of 27°C/min (due to lab conditions), until the target temperatures were reached. All tests were set to reach 400°C maximum on the concrete surface; once the target was achieved, the temperature was maintained for 10 min to allow further heat soak into the concrete, increasing the internal temperature further. The following details results from the trials. The majority of columns were tested to full temperature and subsequent impact, however several columns failed prior to reaching the target temperature due to the explosive spalling under the thermal and compressive loads alone.

5.1 Thermal results

Fig. 6 shows the surface temperature vs. time for the rectangular columns with different concrete mixes; RECT_CTRL_PP, RECT_RC_PP and RECT_SF_PP. The different column samples had an identical heating rate, constant temperature phase and end time. This pattern was repeated on all column tests. Internal temperatures were recorded for the rectangular concrete columns via thermocouples set in the centre of the section during pouring of the wet concrete. Fig. 7 shows the recorded internal temperatures for the three rectangular sections. There is an initial delay, followed by a temperature increase rate of approximately 11°C/min. At approximately 14 min (400°C external temperature), the applied heat is constant. Fig. 7 shows that the internal temperature continues to rise after 14 min and also rises further after the heaters stop at approximately 24 min. A slight deviation is observed at 14 min which correlates to the application of the impact loads. The CTRL_PP shows a slightly higher rate and maximum internal temperature (228°C compared to 215°C), this could be due to a combination of early crack development and slight thermocouple misalignment.

Fig. 8 shows the temperature differential calculated from the surface and internal temperatures for the square columns. At the end of the heating phase the average temperature differential was 178°C, equating to 3.56°C/mm across the sections. The temperature differentials increased rapidly during the first 5 to 10 min, reaching a peak then followed by a slower drop. The columns with the lowest temperature differential increase (therefore

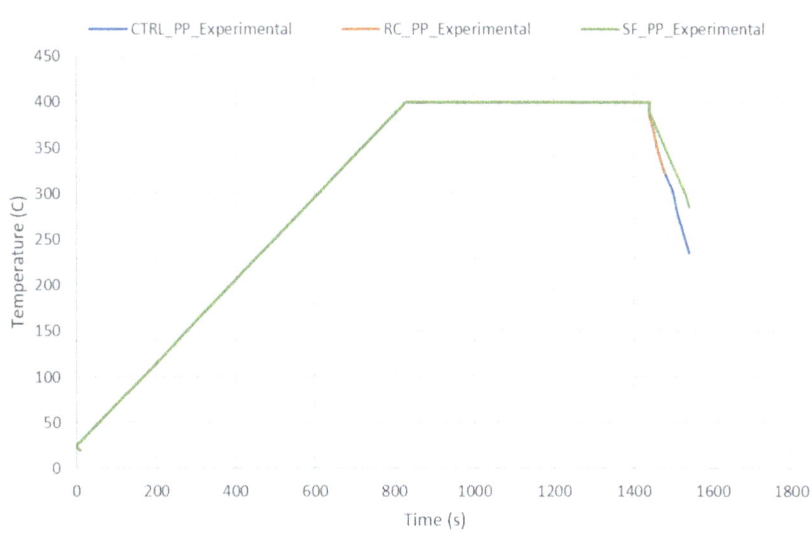

Figure 6: Surface temperature of rectangular columns.

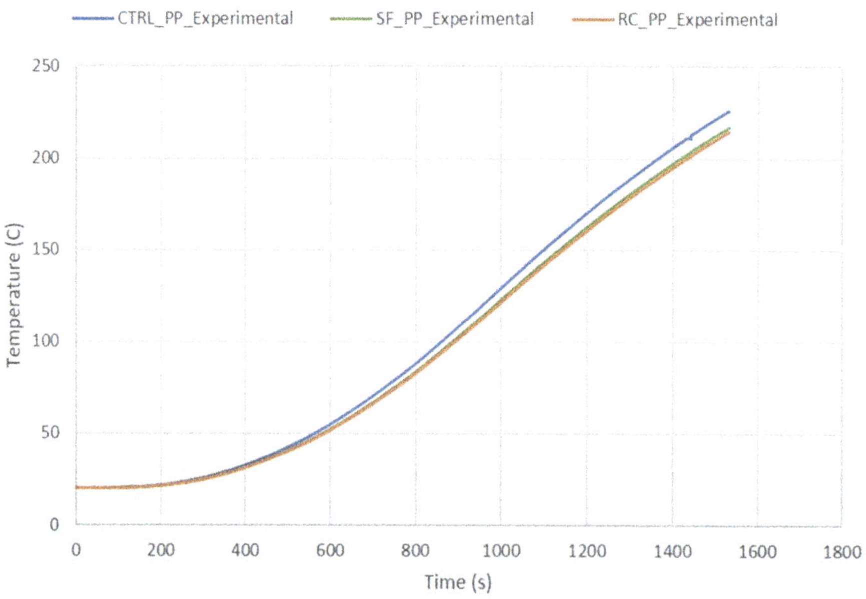

Figure 7: Internal temperature of rectangular columns.

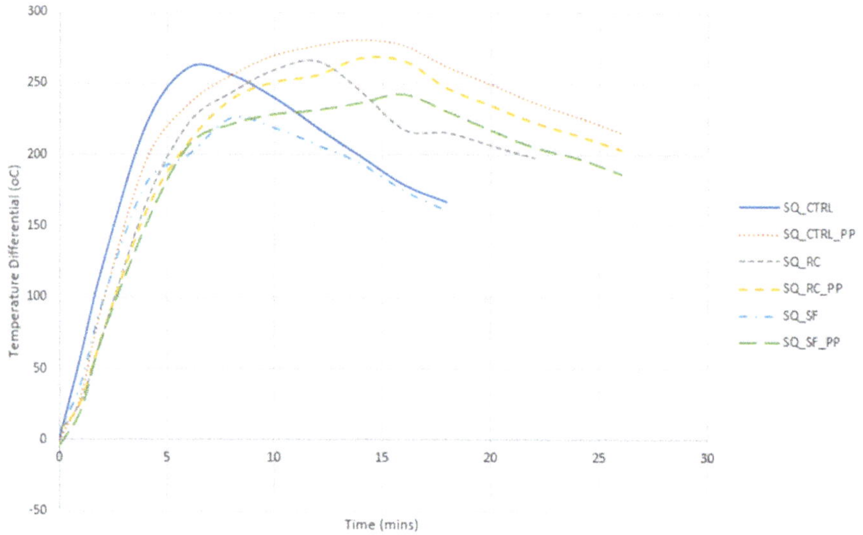

Figure 8: Surface to internal temperature differential.

higher internal temperatures) contained either steel fibre, or reinforcement bars. Some notable phenomena include the sudden drop in differential at approximately 175°C to 225°C; this could be due to surface cracks, PPE fibres melting and the change of state of any embedded water.

5.2 Concrete cracks and visual observations

During heating a network of fine cracks were observed at approximately 300°C on the thermally exposed column faces, with evidence of moisture migration to surface. At 350°C several control samples with no PPE exhibited explosive spalling on the thermally exposed faces. For safety and equipment damage reasons the tests were subsequently stopped (heaters turned off and no impact load applied) if explosive spalling occurred prior to reaching 400°C. No explosive spalling was observed in the concrete samples containing PPE fibres. At 400°C a reddish colour was observed on the surface of the concrete columns. Fig. 9 shows the CTRL column heated to 350°C (post-test) with no impact load; the explosively spalled region is clearly noted. After testing, deep cracks were also observed on samples heated to 400°C.

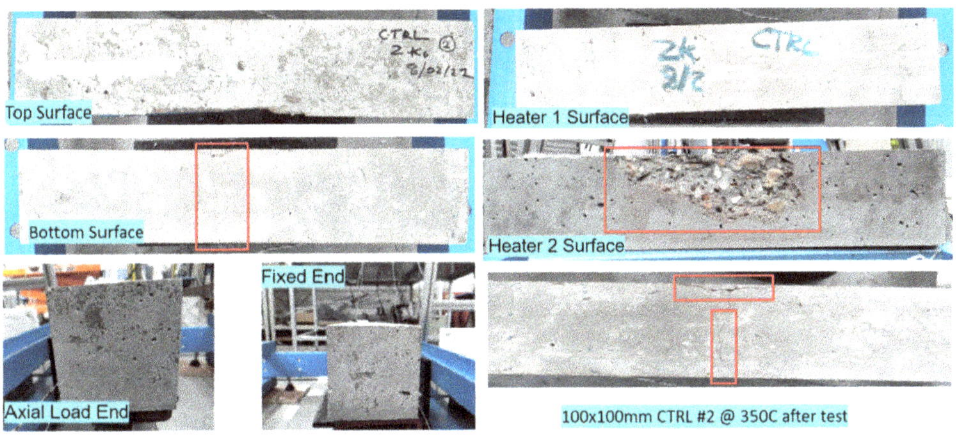

Figure 9: CTRL square column post-test.

5.3 Vertical deflections

The deflections during the trials were measured with an LVDT, placed adjacent to the midspan of the beam (avoiding the central impact area). Table 3 shows the recorded peak deflections for the different column sections heated from 250°C to 400°C. Most columns exhibited an increase in vertical deflection under impact from 250°C to 400°C, with some exceptions. For the square sections the columns with reinforcement bars and steel fibres exhibited lower deflections compared to the plain columns; CTRL = 13.9 mm, RC = 9.7 mm and SF = 11.1 mm. There was also a small decrease in the vertical deflection between the columns without and with PPE; 13.9 mm to 13.2 mm (CTRL), 9.7 mm to 9.3 mm (RC), 11.1 mm to 10.8 mm (SF). Similar correlations between columns without and with PPE are noted with the rectangular columns. The recorded vertical displacement during three tests are shown in Fig. 10 and Fig. 11. Fig. 10 shows the vertical deflections of three different square columns (control + PP, RC + PP and SF + PP) during the heat loading phase. Up to 900 s, the deflections of the three different samples are relatively linear and consistent. After which the control column deflection increases more rapidly than the RC and PP columns. Fig. 11 shows the deflections of the same samples, during the subsequent impact load phase (at

Table 3: Column peak deflections.

Column naming convention	Peak deflection (mm)			
	250°C	300°C	350°C	400°C
SQ_CTRL	–	13.5	–	13.9
SQ_CTRL_PP	–	–	–	13.2
SQ_RC	10.6	–	–	9.7
SQ_RC_PP	–	9.9	–	9.3
SQ_SF	–	9.45	–	11.1
SQ_SF_PP	–	8.25	–	10.8
RECT_CTRL	–	9.2	–	10.04
RECT_CTRL_PP	9.5	–	–	9.95
RECT_RC	–	–	–	9.37
RECT_RC_PP	9.39	–	9.27	8.8
RECT_SF_PP	9.04	9.10 and 7.59	–	9.17

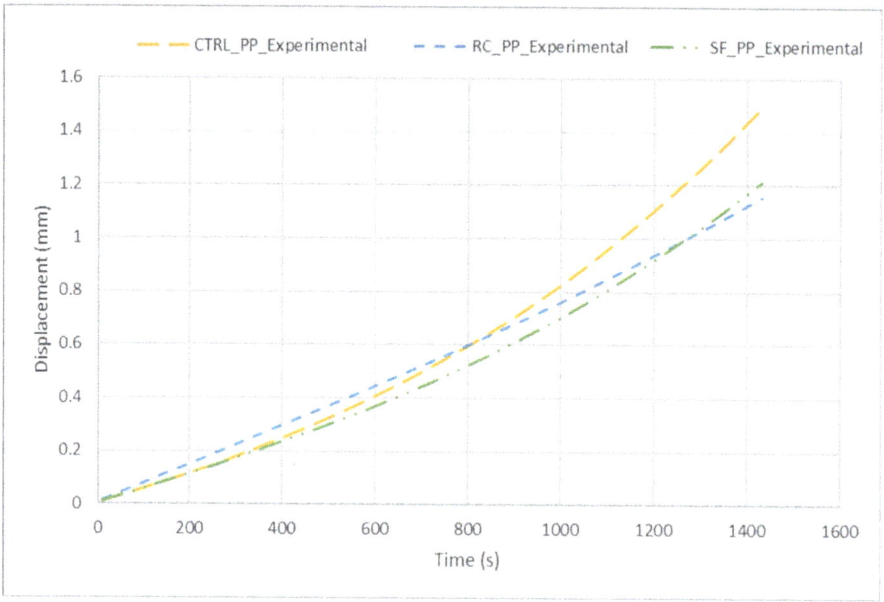

Figure 10: Vertical deflections during thermal load.

1,440 m/s). All samples exhibit a sudden increase in deflection with peak deflections of 13.2 mm (control + PP), 9.3 mm (RC+PP) and 10.8 mm (SF + PP). The inclusion of RC and SF improves resistance to the impact load, with the standard RC exhibiting higher resistance than SF. The post-impact residual deflections are all approximately 4 mm.

6 FURTHER RIG DEVELOPMENTS AND APPLICATIONS

For these experiments the compression rig was configured to apply a compressive axial load to the column sections. Modifications to the positioning of the columns or the compression

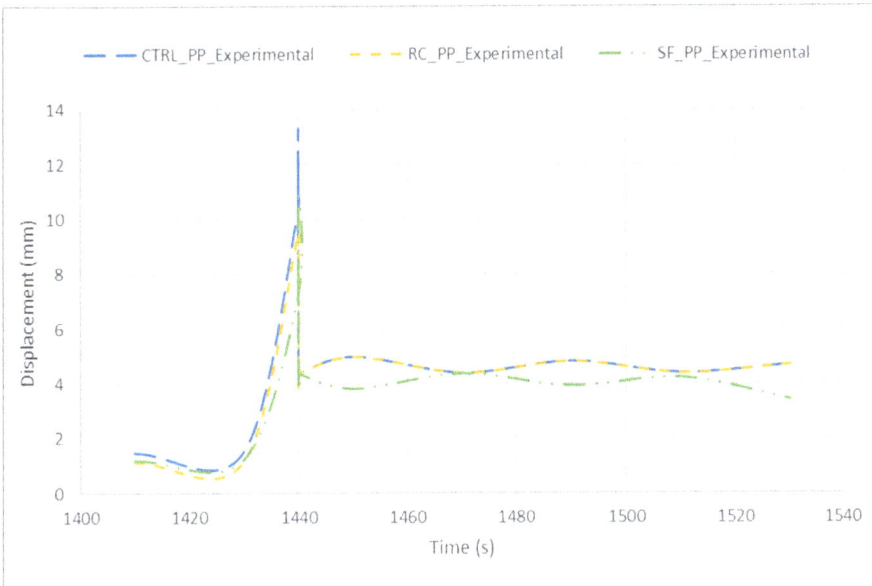

Figure 11: Vertical deflections under impact load.

rig will enable the application of a non-concentric axial load to the columns. This eccentric axial load will develop bending stresses within the structural element, allowing the combination of bending stresses, thermal and impact load to be researched. This combination of load sets would be more akin to beams and structural frames. Initial experimental testing has begun on rectangular, square and circular hollow steel sections. Further investigation could also look into the response of thin gauge steel sections, different materials and section types used for secondary structural applications including cladding and curtain wall systems, such as aluminium channels. This work will be validated with detailed computational studies using non-linear finite element analysis. Alterations to the impact rig include dropping the impact rod from a higher height to achieve a higher imparted kinetic energy. The use of interchangeable impact heads types such as conical, pointed and triangular could be adopted. In order to increase the applied temperature to the columns, the rig could be developed with wrap around insulated panels to maintain high temperatures and reduced the heat loss via radiation and convection. Further ceramic tiles along the full length (500 mm) of the column could be adopted.

7 CONCLUSIONS

This paper documents the development of a combined impact, thermal and compression rig for testing short structural columns. A series of concrete columns were tested in the rig with PPE, SF, RC and plain concrete. All columns with PPE were able to reach the maximum set temperature of 400°C without explosive spalling, whereas several column sections without PPE exhibited explosive spalling before reaching 400°C. The vertical deflection of all column types (CTRL, RC and SF) containing additional PPE were lower than the non-PPE equivalents, demonstrating that the use of PPE in the concrete mix improves the resistance at higher temperatures and enables stability against explosive spalling. The columns subject to higher temperatures also exhibited the higher deflections from the subsequent impact.

The testing rig was successful in producing high thermal loads in combination with compressive axial loads (akin to a column in service), and impact loads. The results from the concrete column trials have demonstrated that the rig can replicate extreme loading events on short structural elements, repeated at a low cost with a reasonable level of accuracy. The current limitations including limited maximum temperature (400°C), low impact energy and applied compressive stresses will be overcome with further rig developments.

ACKNOWLEDGEMENT

The authors acknowledge the support of staff at the UoP Civil Engineering laboratories.

REFERENCES

[1] Schleyer, G.K., Barnett, S.J., Millard, S.G. & Wight, G., Modelling the response of UHPFRC panels to explosive loading. *WIT Transactions on the Built Environment*, vol. 113, WIT Press: Southampton and Boston, pp. 173–184, 2010.

[2] Ukala, D.C., Effect of heat intensity and duration on the compressive strength of concrete. *Applied Sciences and Environmental Management*, **23**(9), p. 1637, 2019.

[3] Hager, I. & Mróz, K., Role of polypropylene fibres in concrete spalling risk mitigation in fire and test methods of fibres effectiveness evaluation. *Materials*, **12**(23), p. 3869, 2019.

[4] Kalifa, P., Chéné, G. & Gallé, C., High-temperature behaviour of HPC with polypropylene fibres from spalling to microstructure. *Cement and Concrete Research*, **31**, pp. 1487–1499, 2001.

[5] Kurdo, Z. & Clough, L., Thermo-mechanical behaviour of high strength concrete containing polypropylene under thermal loading. *Proceedings of the Eleventh EFEE World Conference on Explosives and Blasting*, pp. 411–422, 2022.

[6] Vevor, Vevor 20 tons/2000 kg hydraulic shop press floor press with pump and manometer. https://eur.vevor.com/hydraulic-shop-press-c_10088/workshop-garage-hydraulic-press-20-t-bench-separate-hydraulic-pump-with-gauge-p_010979778585. Accessed on: 17 Mar. 2025.

[7] Ceramicx, Full quartz infrared heating element (FQE). https://www.ceramicx.com/products/quartz-elements/standard-quartz-elements/full-quartz-element/. Accessed on: 17 Mar. 2025.

[8] British Standards Institution, BS EN 12390-2:2019: Testing hardened concrete – Making and curing specimens for strength tests. BSI: London, 2019.

A COMPRESSION-TO-TENSION STRESS CONVERSION JIG SET AND A DUMBBELL-SHAPED SPECIMEN FOR ARTIFICIALLY INDUCING NECKING DURING A SPLIT HOPKINSON PRESSURE BAR TEST

HYUNHO SHIN[1], GYUNGGEUN SHIN[1], JAEHYUCK LEE[2],
HYUNSUK CHO[2], JUNMOO LEE[1] & KYONG YOP RHEE[2]
[1]Mechanics of Materials and Design Laboratory, Department of Materials Engineering, Gangneung-Wonju National University, Republic of Korea
[2]Mechanics of Materials and Design Laboratory, Department of Mechanical Engineering, Kyunghee University, Republic of Korea

ABSTRACT

A jig set was designed to convert compressive stress into tensile stress in a specialized dumbbell-shaped tensile specimen during a split Hopkinson pressure bar (SHPB) test. This setup intended to artificially induce necking during the SHPB test. The functionality of the aforementioned setup, made of 4340 steel, has been investigated using explicit finite element analysis. Two types of dumbbell-shaped specimens were examined: one with a uniform diameter in the tension-receiving section and the other with a reduced central diameter compared to the diameters at its ends. In the case of the uniform diameter specimen, no significant necking occurred until the first pulse passed through it. Conversely, necking occurred in the specimen with the reduced central diameter during the first pulse duration, resulting in higher strain rates, strains and temperatures in the neck region, confirming the effectiveness of the proposed CTC structure. Therefore, if the specimen's constitutive parameters are inversely identified by tracking the bar signals derived from the necking specimen, these parameters are expected to accurately predict the material's deformation behaviour up to higher strain rates, strains and temperatures, in contrast to the parameter set determined from the specimen that did not experience necking.

Keywords: constitutive model calibration, inverse parameter identification, Considère criterion.

1 INTRODUCTION

Two representative variants of split Hopkinson bars (SHB), specifically the split Hopkinson pressure bar (SHPB) [1]–[14] and the split Hopkinson tension bar (SHTB) [15]–[22], are utilized to measure the stress–strain and strain rate–strain curves of various materials at high strain rates, typically ranging from 100 to 10,000 s^{-1}. These curves, along with those obtained at quasistatic strain rates [23], [24], are crucial for calibrating a strain-rate-dependent constitutive model [25]–[27], which is essential for simulating the dynamic deformation behaviour of materials and their structures [28], [29].

The constitutive models have conventionally been calibrated using nonlinear curve fitting of various stress–strain curves measured across a range of strain rates and temperatures. This process presents challenges due to experimental uncertainties and associated costs particularly at high temperatures. Alternatively, calibration can be accomplished by replicating the measured bar signals through repeated simulations until the simulated signals closely align with the measured counterparts, thereby enabling the inverse identification of constitutive parameters from the consistency of the measured and simulated bar signals [30]–[33]. While the simulation-based inverse identification method for the constitutive model is considered both accurate and cost-effective [32], [33], it requires precise calibration of the bar properties, including elastic modulus, Poisson's ratio, and density, before they are input

WIT Transactions on The Built Environment, Vol 215, © 2025 WIT Press
www.witpress.com, ISSN 1743-3509 (on-line)
doi:10.2495/SUSI250141

into the simulation. This requirement can be fulfilled by utilizing either the one-dimensional (1D) method [7], [8] or the three-dimensional (3D) method [9], [10].

According to the rationale outlined in Shin et al. [32] and Cho et al. [33], the combination of a compression-to-tension converting (CTC) specimen structure, the SHPB test, and the inverse parameter identification method allows for the benefits of both the SHPB and the SHTB tests to be achieved while mitigating their respective drawbacks. Three key advantages, among others, are as follows: First, the equivalent stress–equivalent plastic strain curve at a quasi-static strain rate, can be obtained suitably even for materials that exhibit necking through standard tensile tests [23], [24]. This curve is crucial for quantifying strain rate dependency by comparing it with the results from SHB tests. Second, the nature of SHPB signals [9], [10], [33], which can be simulated, can be exploited, while SHTB signals are rarely simulated due to the complexities involved in the pulse generation process. Third, the plastically deforming region in the CTC structure is located away from the contact surfaces between the CTC structure and the bars, thereby minimizing the influence of friction on the measured properties [33], [34], particularly the determined constitutive parameters.

Based on the rationale outlined above, we recently calibrated a constitutive model for high-strength steel using a CTC specimen structure, as illustrated in Fig. 1 [32], [33], referred to as the Type I CTC structure. This method involved conducting repeated simulations of the experimentally measured bar signals derived from the CTC specimen structure until the simulated signals closely matched their experimental counterparts.

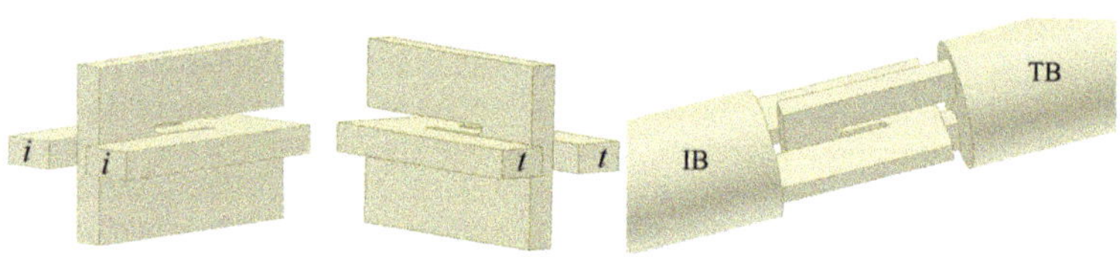

Figure 1: Type I compression-to-tension (CTC) structure. *i* represents the contact surface to the incident bar (IB) and *t* denotes that to the transmitted bar (TB).

If necking occurs in a CTC structure during the SHPB test, the strain rate, strain, and temperature in the necked region increase significantly [32], [33]. Consequently, if the constitutive parameters are identified by inversely tracking the bar signals derived from the necking CTC structure, these identified parameters will accurately simulate the deformation behaviour of the materials and structures up to higher strain rates, strains, and temperatures. Therefore, achieving necking in a CTC structure is essential.

According to the Considère criterion [35], necking occurs when the strain hardening slope ($d\sigma_t/d\varepsilon_t$) equals or falls below the flow stress (σ_t): $d\sigma_t/d\varepsilon_t \leq \sigma_t$, where σ_t represents the true stress and ε_t signifies the true strain. This equation exclusively considers the properties of a homogeneous material with ideal geometry. However, in the presence of material inhomogeneity or geometric imperfections, these factors can act as local stress concentration sources, leading to necking even if the Considère criterion has not yet been satisfied.

For a homogeneous material with ideal geometry, necking is less likely to occur in materials that exhibit significant strain-hardening behaviour, such as AISI 4340 steel and oxygen-free high-conductivity copper. However, if a geometric flaw is artificially introduced into a CTC structure, it can promote necking in these materials. Introducing a geometric imperfection into the central column, which experiences tensile stress in the Type I CTC structure (see Fig. 1), poses challenges. This requirement limits the applicability of this structure for materials with a high strain-hardening slope, as it is challenging to induce necking during the SHPB test, particularly during the first stress pulse duration.

In the context presented above, we introduce an alternative CTC specimen structure, referred to as the Type II CTC structure, as illustrated in Fig. 2. The most distinctive feature of this structure, compared to Type I, is its ability to artificially introduce a geometric imperfection in the tension-receiving section of the dumbbell-shaped specimen by reducing the central diameter relative to the diameters at its ends. This modification is expected to facilitate artificial necking during SHPB tests, significantly enhancing strain, strain rate, and temperature. The objective of this paper is to numerically verify the capability of inducing necking in the Type II CTC structure during the first pulse duration in SHPB tests, with increased strain rates, strains, and temperatures.

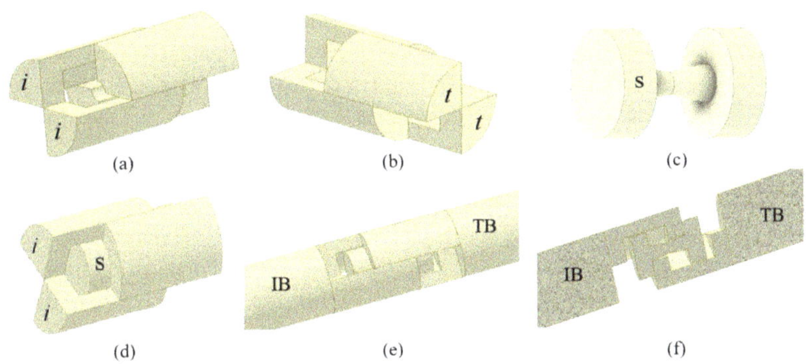

Figure 2: Type II CTC structure. Views of the jig from: (a) the incident bar; and (b) the transmitted bar. (c) Dumbbell-shaped specimen; (d) Assembly of the CTC structure; (e) CTC structure sandwiched between the bars; and (f) Cross-sectional view of the CTC structure and the bars. The symbols *i, t,* IB, and TB have the same meanings as those in Fig. 1.

2 NUMERICAL ANALYSIS

A general-purpose finite element (FE) software, ABAQUS Explicit [36], was employed as the solver. The lengths of the striker, incident bar, and transmitted bar were 300 mm, 2,000 mm, and 944 mm, respectively, while the diameter of the formers was 19.05 mm. One quarter of the geometry was modelled with appropriate boundary conditions, taking into account the symmetry of the model. The FE models for two dumbbell-shaped specimens are illustrated in Fig. 3: one with $d = d_c = 3.5$ mm, and the other with $d = 3.5$ mm and $d_c = 3.0$ mm. Here, d represents the end diameter of the tension-receiving section and d_c denotes its central diameter. The length of the CTC assembly is 30 mm.

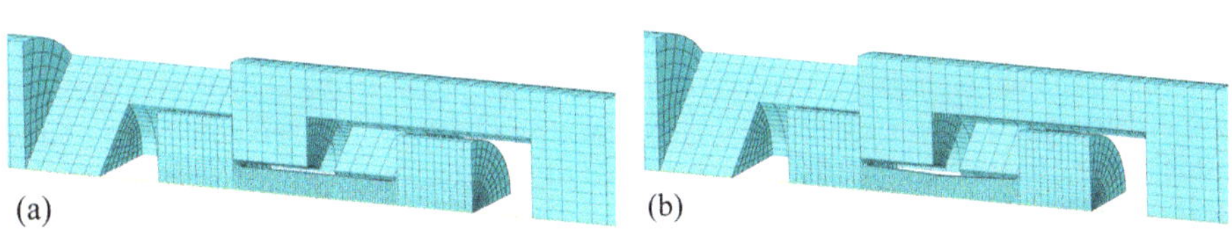

Figure 3: FE models of Type II CTC structures: (a) with $d = d_c = 3.5$ mm; and (b) with $d = 3.5$ mm and $d_c = 3.0$ mm, for the dumbbell-shaped specimen.

Eight-node linear brick elements (C3D8R) were employed to discretize the model space. The mesh sizes of the bar and the portion receiving tension were established through separately conducted mesh size sensitivity test in our previous studies [9], [33]. The initial velocity of the striker, which was initially separated by 0.1 mm from the incident bar, was 15 m/s. A master-slave contact algorithm was applied to all contacting surfaces. A Coulomb friction coefficient of 0.1 was applied to both the surfaces in contact with the CTC structure and the internal surfaces of the CTC structure [33].

Maraging steel was considered for the striker and incident bar, while AA 6061-T6 for the transmitted bar. The density, elastic modulus, and Poisson's ratio of Maraging steel are 8,058.9 kg/m^3, 169.99 GPa and 0.3356, respectively. These values were determined by referencing the authors' calibration results [7]–[10]. The corresponding properties for AA 6061-T6 are 2,700 kg/m^3, 68.9 GPa and 0.33, respectively [37].

AISI 4340 steel was selected as the material for the CTC structure, which comprises the jig and the dumbbell-shaped specimen. Its properties are as follows: a density of 7,830 kg/m^3, an elastic modulus of 200 GPa, a Poisson's ratio of 0.29 and a specific heat capacity of 477 J/kgK. The plastic deformation behaviour is described using the Johnson–Cook model [26]:

$$\sigma = [A + B\varepsilon^n][1 + C \ln \dot{\varepsilon}^*][1 - T^{*m}] \tag{1}$$

where σ represents the equivalent stress, ε denotes the equivalent plastic strain, $\dot{\varepsilon}$ indicates the rate of equivalent plastic strain, and T stands for the current temperature; $\dot{\varepsilon}_0$ is the reference strain rate and T_{ref} is the reference temperature; $\dot{\varepsilon}^* = \dot{\varepsilon}/\dot{\varepsilon}_0$ and $T^* = (T_m - T)/(T_m - T_{ref})$; A, B, n, C, m, and T_m are material constants. The inelastic heat fraction (IHF) was set to one as the configuration variable based on the rationale presented in our previous study [33]. The material parameters used for the simulation are [26]: $A = 792$ MPa, $B = 510$ MPa, $n = 0.26$, $C = 0.014$, $m = 1.03$, and T_m=1,793 K (T_{ref}= 298 K, $\dot{\varepsilon}_0$=1 s^{-1}, and IHF = 1).

3 RESULTS AND DISCUSSION

3.1 Bar signals

The strain signals monitored on the surfaces of the incident and transmitted bars, positioned 1,000 mm and 100 mm from the CTC specimen structure, are illustrated in Fig. 4. This figure presents data for two types of dumbbell-shaped specimens: one with $d = d_c = 3.5$ mm and the other with $d = 3.5$ mm and $d_c = 3.0$ mm. While the incident bar signals exhibit relatively minor differences between the two specimen types, the transmitted bar signals demonstrate significant variation. Notably, the magnitudes of all pulses in the incident bar exceed 1,000 μstrain, whereas those in the transmitted bar are approximately 600 μstrain or higher. Since these magnitudes are routinely measured using a metal foil strain gauge in an SHPB experiment, the strain signals derived from the proposed CTC structure can be utilized for the inverse parameter identification of AISI 4340 material. Cho et al. [33] illustrates the sensitivity of the bar signals, particularly the transmitted pulse, to the constitutive parameters of a CTC structure (Type I). This sensitivity was also observed in a separate trial conducted for this study (results not shown). Such sensitivity forms the basis for inverse parameter identification by tracking the bar signals.

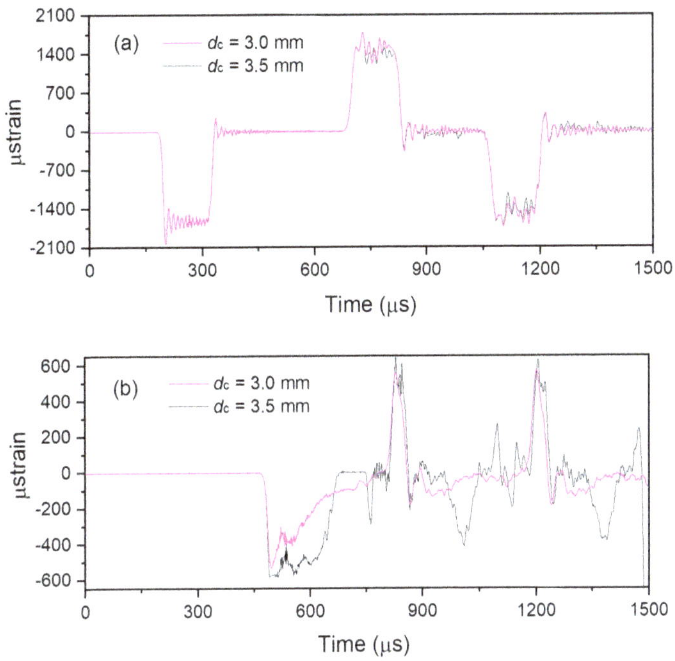

Figure 4: Strain signals on the (a) incident; and (b) transmitted bars during the numerical experiments for the two types of specimens: $d = d_c = 3.5$ mm and $d = 3.5$ mm with $d_c = 3.0$ mm.

3.2 Deformed shapes

The distorted shapes of the two types of CTC structures, along with their equivalent stress contours at 525 μs, are illustrated in Fig. 5. At this point, the dumbbell-shaped specimen with a uniform diameter in the tension-receiving section (Fig. 5(a)) shows no significant necking due to the highly strain-hardening nature of AISI 4340 steel. In contrast, the specimen with a reduced central diameter (Fig. 5(b)) exhibits notable necking. This observation confirms the effectiveness of the proposed CTC structure, which is designed to artificially induce necking during the SHPB test, particularly during the first pulse duration.

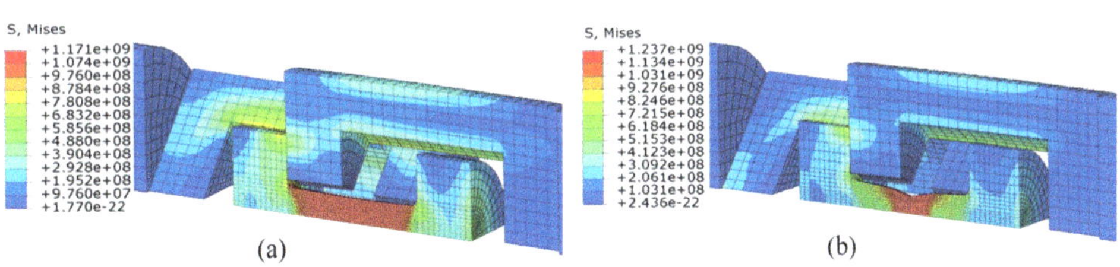

Figure 5: Deformed shapes of the two types of specimens at 525 μs: (a) $d = d_c = 3.5$ mm; and (b) $d = 3.5$ mm with $d_c = 3.0$ mm. The numbers in the legend are expressed in pascals (Pa).

3.3 Strain, strain rate and temperature

The time history of equivalent plastic strain, its time derivative (strain rate), and temperature were also extracted from the central portion of two types of dumbbell-shaped specimens. The results are illustrated in Fig. 6. The specimen exhibiting no apparent necking ($d = d_c$ = 3.5 mm) demonstrates a maximum strain rate of approximately 4,100 s^{-1}, a strain of 0.33, and a temperature of 395 K. In contrast, the specimen exhibiting apparent necking (d = 3.5 mm; d_c = 3.0 mm) experienced a maximum strain rate of approximately 15,000 s^{-1}, a strain of 1.55, and a temperature of 784 K. This observation confirms the effectiveness of the proposed Type II CTC structure. If the constitutive parameters are identified inversely by tracking the bar signals derived from the proposed jig and the dumbbell-shaped specimen with a reduced central diameter, the determined parameters are reliable at least up to the specified strain rate, strain, and temperature. This limit is significantly higher than what can be achieved when identifying parameters using signals derived from a specimen that does not exhibit necking. Our future studies will involve inducing fractures in the neck during the first pulse duration to calibrate a fracture model that is dependent on stress state, strain rate and temperature.

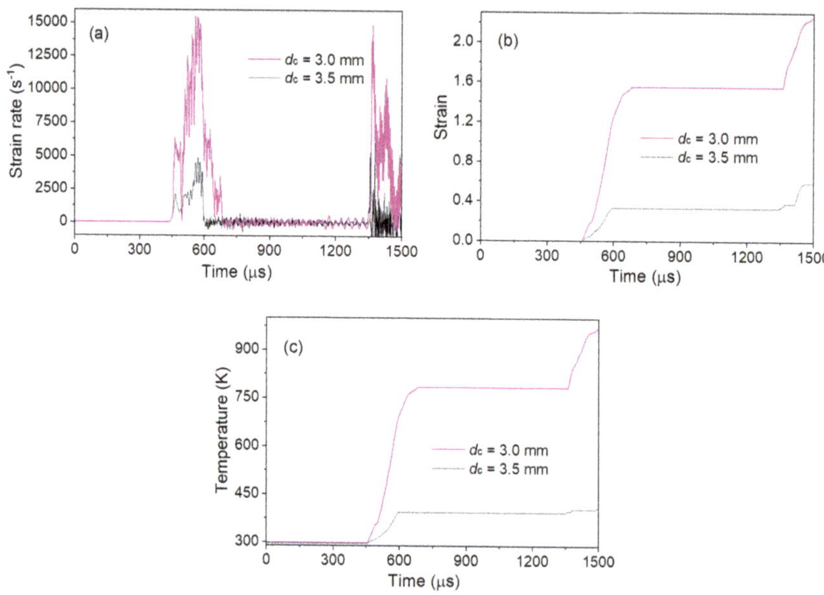

Figure 6: (a) Strain rate; (b) Strain; and (c) Temperature of the dumbbell-shaped specimens: $d = d_c$ = 3.5 mm and d = 3.5 mm with d_c = 3.0 mm.

4 CONCLUSION

A jig set and a specialized dumbbell-shaped specimen were designed for compression-to-tension conversion (CTC) during a split Hopkinson pressure bar (SHPB) test, referred to as the Type II CTC structure. The most distinctive feature of this structure, compared to the Type I CTC structure, is its ability to artificially induce necking. The capabilities of the two types of dumbbell-shaped specimens were investigated using explicit finite element analysis: one with a uniform diameter in the tension-receiving section and the other with a reduced central diameter compared to the diameters at its ends. AISI 4340 steel was selected as the material for both the jig set and the specimen.

When the uniform diameter specimen was subjected to the SHPB test, no significant necking occurred until the first stress pulse passed through it. The specimen experienced a strain rate of approximately 4,100 s^{-1}, a strain of 0.33, and a temperature of 395 K. In contrast, necking was observed in the specimen with the reduced central diameter during the duration of the first pulse. This specimen experienced a strain rate of approximately 15,000 s^{-1}, a strain of 1.55, and a temperature of 784 K. This outcome confirms the effectiveness of the proposed Type II CTC structure. Therefore, it can be concluded that if the specimen's constitutive parameters are inversely identified by tracking the bar signals derived from the necking specimen through repeated simulations, these parameters are expected to accurately predict the material's deformation behaviour up to higher strain rates, strains, and temperatures.

ACKNOWLEDGEMENT

This study was financially supported by the Near-Ideal Split Hopkinson Pressure Bar (NI-SHPB) Program, funded by Dyntech Inc., located in Gangneung, South Korea.

REFERENCES

[1] Kolsky, H., An investigation of the mechanical properties of materials at very high rates of loading. *Proceedings of the Physical Society of London, Section B*, **62**(11), pp. 676–700, 1949. https://doi.org/10.1088/0370-1301/62/11/302.

[2] Zhao, H. & Gary, G., A new method for the separation of waves: Application to the SHPB technique for an unlimited duration of measurement. *Journal of the Mechanics and Physics of Solids*, **45**(7), pp. 1185–1202, 1997. https://doi.org/10.1016/S0022-5096(96)00117-2.

[3] Shin, H. & Kim, J.-B., Evolution of specimen strain rate in split Hopkinson bar test. *Proceedings of the Institution of Mechanical Engineers, Part C: Journal of Mechanical Engineering Science*, **233**(13), pp. 4667–4687, 2019. https://doi.org/10.1177/0954406218813386.

[4] Shin, H. & Kim, D., One-dimensional analyses of striker impact on bar with different general impedance. *Proceedings of the Institution of Mechanical Engineers, Part C: Journal of Mechanical Engineering Science*, **234**(2), pp. 589–608, 2020. https://doi.org/10.1177/0954406219877210.

[5] Shin, H. & Kim, J.-B., Understanding the anomalously long duration time of the transmitted pulse from a soft specimen in a Kolsky bar experiment. *International Journal of Precision Engineering and Manufacturing*, **17**(2), pp. 203–208, 2016. https://doi.org/10.1007/s12541-016-0026-8.

[6] Shin, H., Pochhammer–Chree equation solver for dispersion correction of elastic waves in a (split) Hopkinson bar. *Proceedings of the Institution of Mechanical Engineers, Part C: Journal of Mechanical Engineering Science*, **236**(1), pp. 80–87, 2021. https://doi.org/10.1177/0954406220980509.

[7] Shin, H., Sound speed and Poisson's ratio calibration of (split) Hopkinson bar via iterative dispersion correction of elastic wave. *ASME Journal of Applied Mechanics*, **89**(6), 061007, 2022. https://doi.org/10.1115/1.4054107.

[8] Shin, H., Manual for calibrating sound speed and Poisson's ratio of (split) Hopkinson bar via dispersion correction using Excel and Matlab templates. *Data*, **7**(5), p. 55, 2022. https://doi.org/10.3390/data7050055.

[9] Shin, H., Calibration of bar properties, measured strain, and impact velocity in a bar impact test. *Mechanics of Advanced Materials and Structures*, **31**(24), pp. 6044–6056, 2023. https://doi.org/10.1080/15376494.2023.2224797.

[10] Shin, H., Manual of GUI program governing ABAQUS simulations of bar impact test for calibrating bar properties, measured strain, and impact velocity. *Data*, **8**(3), p. 54, 2023. https://doi.org/10.3390/data8030054.

[11] Shin, H., Lee, H., Kim, J.-B. & Yoo, Y.-H., A numerical verification of the reliability of a split Hopkinson pressure bar with a total bar length of 3 m and a diameter of 1 inch. *Applied Mechanics and Materials*, **799–800**, pp. 681–684, 2015. https://doi.org/10.4028/www.scientific.net/AMM.799-800.681.

[12] Kim, D. & Shin, H., Minimum required distance of strain gauge from specimen for measuring transmitted signal in split Hopkinson pressure bar test. *MATEC Web of Conference*, **308**, 04005, 2020. https://doi.org/10.1051/matecconf/202030804005.

[13] Verleysen, P. & Degrieck, J., Optimized calculation of high strain rate stress–strain curves. *WIT Transactions on the Built Environment*, vol. 98, WIT Press: Southampton and Boston, pp. 369–377, 2008. https://doi.org/10.2495/SU080361.

[14] Verleysen, P., Degrieck, J., Van Slycken, J., De Cooman B. & Samek, L., Experimental investigation of the influence of the specimen geometry on Hopkinson tensile test results. *WIT Transactions on the Built Environment*, vol. 73, WIT Press: Southampton and Boston, pp. 497–505, 2004. https://doi.org/10.2495/SU040461.

[15] Gerlach, P., Kettenbeil, C. & Petrinic, N., A new split Hopkinson tensile bar design. *International Journal of Impact Engineering*, **50**, pp. 63–67, 2012. https://doi.org/10.1016/j.ijimpeng.2012.08.004.

[16] Haugou, G., Leconte, N. & Morvan, H., Design of a pre-stretched tension Hopkinson bar device: Configuration, tail corrections, and numerical validation. *International Journal of Impact Engineering*, **97**, pp. 89–101, 2016. https://doi.org/10.1016/j.ijimpeng.2016.05.018.

[17] Shin, H., Lee, J.-H., Kim, J.-B. & Sohn, S.-I., Design guidelines for the striker and transfer flange of a split Hopkinson tension bar and the origin of spurious waves. *Proceedings of the Institution of Mechanical Engineers, Part C: Journal of Mechanical Engineering Science*, **234**(1), pp. 137–151, 2020. https://doi.org/10.1177/0954406219869984.

[18] Shin, H., Kim, S. & Kim, J.-B., Stress transfer mechanism of flange in split Hopkinson tension bar. *Applied Sciences*, **10**(21), p. 7601, 2020. https://doi.org/10.3390/app10217601.

[19] Nguyen, K.-H., Kim, H.-C., Shin, H., Yoo, Y.-H. & Kim, J.-B., Numerical investigation into the stress wave transmitting characteristics of threads in the split Hopkinson tensile bar test. *International Journal of Impact Engineering*, **109**, pp. 253–263, 2017. https://doi.org/10.1016/j.ijimpeng.2017.07.004.

[20] Kim, Y.-B. & Kim, J. Redesign of split-Hopkinson tensile bar to eliminate spurious wave and verification of strain signal using FEM. *Advanced Mechanical Engineering*, **15**(10), pp. 1–11, 2023. https://doi.org/10.1177/16878132231199843.

[21] Jakkula, P., Cohen, A., Lukić, B., Levi-Hevroni, D., Rack, A., Ganzenmüller, G. & Hiermaier, S., Split Hopkinson tension bar and universal testing machine for high-speed X-ray imaging of materials under tension. *Instruments*, **6**(3), pp. 38, 2022. https://doi.org/10.3390/instruments6030038.

[22] Ruiz-Ripoll, M.L., De Pedraza, V.R. & Roller, C., Influence of the projectile shape on the dynamic tensile characterization of concrete using a split Hopkinson bar. *WIT Transactions on the Built Environment*, vol. 209, WIT Press: Southampton and Boston, pp. 155–166, 2022. https://doi.org/10.2495/HPSU220141.

[23] Shin, H., Kim, S., Choi, M.K. & Ju, Y., Extraction of equivalent stress versus equivalent plastic strain curve of necking material in tensile test without assuming constitutive model. *ASME Journal of Engineering Materials and Technology,* **146**(1), 021037, 2024. https://doi.org/10.1115/1.4064372.

[24] Lee, J., Cho, H., Shin, H., Rhee, K.Y., Kim, W., Lee, H.H., Kim, H.S., Seo, S. & Lee, J., Determination of the equivalent stress–equivalent plastic strain curve of ECAP-processed copper. *Transactions of the Korean Society of Mechanical Engineers, A*, **48**(8), pp. 553–563, 2024. https://doi.org/10.3795/KSME-A.2024.48.8.553.

[25] Shin, H., Ju, Y., Choi, M.-K. & Ha, D.-H., Flow stress description characteristics of some constitutive models at wide strain rates and temperatures. *Technologies*, **10**(2), p. 52, 2022. https://doi.org/10.3390/technologies10020052.

[26] Johnson, G.R. & Cook, W.H., A constitutive model and data for metals subjected to large strains, high strain rates and high temperatures. *Proceedings of the 7th International Symposium on Ballistics*, The Hague, Netherlands, 19–21 Apr., pp. 541–547, 1983.

[27] Shin, H. & Kim, J.-B., A phenomenological constitutive equation to describe various flow stress behavior of materials in wide strain rate and temperature regimes. *ASME Journal of Engineering Materials and Technology,* **132**(2), 021009, 2010. https://doi.org/10.1115/1.4000225.

[28] Shin, H., Lee, H.-J., Yoo, Y.-H. & Lee., W., A determination procedure for element elimination criterion in finite element analysis of high-strain-rate impact/penetration phenomena. *JSME International Journal*, **47**(1), pp. 35–41, 2004. https://doi.org/10.1299/jsmea.47.35.

[29] Kim, J.-B. & Shin, H., NET3D calculation on the depth of penetration by a long-rod penetrator with varying obliquity. *Applied Mechanics and Materials*, **619**, pp. 58–62, 2014. https://doi.org/10.4028/www.scientific.net/AMM.619.58.

[30] Jang, T.J., Kim, J.-B. & Shin, H., Identification of plastic constitutive Johnson–Cook model parameters by optimization-based inverse method. *Journal Computational Design and Engineering*, **8**(4), pp. 1082–1097, 2021. https://doi.org/10.1093/jcde/qwab033.

[31] Lee, J.H., Shin, H., Kim, J.-B., Kim, J.Y., Park, S.T., Kim, G.L. & Oh, K.W., Determination of the flow stress–strain curve of aluminum alloy and tantalum using the compressive load–displacement curve of a hat-type specimen. *ASME Journal of Applied Mechanics*, **86**(3), 031012, 2019. https://doi.org/10.1115/1.4042138.

[32] Shin, H., Cho, H., Lee, J., Lee, J., Shin, G., Rhee, K.Y., Choi, M.K. & Ju, Y., Three types of compression-to-tension stress conversion structures for split Hopkinson pressure bar test for constitutive model calibration in tensile mode. *Abstracts Book of the 14th International Conference on Mechanical and Physical Behaviour of Materials Under Dynamic Loading (DYMAT 2024)*, ETH, Zurich, Switzerland, 9–13 Sep., pp. 254–258, 2024.

[33] Cho, H., Shin, H., Lee, J., Lee, J., Shin, G., Rhee, K.Y., Choi, M.K. & Ju, Y., Constitutive model calibration of high-strength steel in tensile mode using a compression-to-tension stress-converting specimen structure in split Hopkinson pressure bar tests. *Mechanics of Advanced Materials and Structures*, 2025. https://doi.org/10.1080/15376494.2025.2457125.

[34] Shin, H., Lee, J.-H., Kim, J.-B., Seo, S.-J., Lee, J.-G., Lee, J.-O., Yoon, T.-S. & Jung, C.-S., Numerical verification of the Schroeder–Webster surface types and friction compensation models for a metallic specimen in axisymmetric compression test. *ASME Journal of Tribology*, **141**(10), 101401, 2019. https://doi.org/10.1115/1.4044131.
[35] Considère, A., Mémoire sur l'emploi du fer et de l'acier dans les constructions. *Annales des Ponts et Chaussées*, **9**, pp. 574–775, 1885.
[36] Smith, M., ABAQUS/Standard User's Manual, version 6.9, Dassault Systèmes Simulia Corp: Providence, RI, USA, 2009.
[37] MatWeb, Material Property Data. https://matweb.com.

FIELD TESTING AND NUMERICAL MODELLING OF BRIDGE COLUMN TO CONTACT EXPLOSION

HRVOJE DRAGANIĆ, GORAN GAZIĆ, MARIO JELEČ & IVAN RADIĆ
Faculty of Civil Engineering and Architecture Osijek,
Josip Juraj Strossmayer University of Osijek, Croatia

ABSTRACT

Columns are among the most vulnerable elements of bridges due to their exposure to ongoing traffic beneath the bridge and the fact that their collapse often leads to the failure of the bridge superstructure. Contact explosions can compromise their integrity by damaging the concrete protective layer, weakening the core and severing reinforcing bars, ultimately leading to loss of stability. Such explosions may result from wartime efforts to sever critical transportation routes or controlled demolitions for infrastructure development. This study aims to evaluate the structural response of a bridge column subjected to a contact explosion through experimental testing and numerical modelling. A field test was conducted using a 1:3 scale column, designed based on an actual bridge column, with a spherical contact charge moulded from plastic explosive due to its chemical stability and handling properties. A steel support structure was implemented to replicate realistic end support conditions. The contact explosion caused concrete cratering and reinforcement bar bending but did not fully penetrate the column. Following experimental testing, a numerical simulation was conducted using Ansys Autodyn hydrocode software to replicate the field test. By varying erosion parameters, contact explosion damage was accurately simulated. Model validation was performed by comparing the experimentally measured damage area with the corresponding numerical results, ensuring consistency in size, shape, and deformation characteristics. The findings suggest that while significant localised damage occurs, the column retains partial structural integrity, highlighting the role of reinforcement in mitigating complete failure.

Keywords: bridge column, collapse, contact charge, experimental testing, numerical modelling.

1 INTRODUCTION

During the last decades, terrorist activities have emphasised the critical need for a better understanding of structural behaviour under both accidental and intentional blast loads. As key elements of transportation infrastructure, bridges are particularly vulnerable to explosion hazards. Bridge columns are crucial bridge components due to their exposure to ongoing traffic beneath the bridge and their vital role in supporting the superstructure, making their failure a direct threat to overall bridge stability.

The effect of blast loads on structures generally depends on whether the explosion is areal (distributed over a surface) or contact (occurring directly against a structure). Areal explosions typically induce a widespread structural response, whereas contact explosions cause localised damage, often resulting in cratering or full-thickness penetration of an element. The extent and nature of the damage are influenced by various factors, including the explosive's weight and shape, the thickness of the structural component, the ratio and arrangement of reinforcement bars, and the concrete grade used in reinforced concrete (RC) structures. In the case of RC bridge columns, contact explosions can severely compromise their integrity by damaging the protective concrete layer, weakening the core, and rupturing reinforcing bars, ultimately leading to structural instability. These explosions may occur during wartime efforts to disrupt key transportation routes or as controlled demolitions for infrastructure development.

Several studies experimentally investigated the contact explosions on RC elements (slabs [1]–[4], beams [5]–[7] or columns [8]–[13]), examining both un-retrofitted (plain) and

WIT Transactions on The Built Environment, Vol 215, © 2025 WIT Press
www.witpress.com, ISSN 1743-3509 (on-line)
doi:10.2495/SUSI250151

retrofitted elements. Most prior research focused on evaluating the efficacy of different retrofitting materials in enhancing the blast-resistance capabilities of such elements [14]–[16]. Previous studies classified the local damage modes into cratering, spalling, and breaching. However, the main damage modes of RC columns under close-in/contact explosion were also characterised by side cover concrete spalling, which has not been reported in relevant studies for RC slabs/walls. This phenomenon indicates that the propagation of stress waves is more complex and not one-dimensional in the cross-section of the column, which is significantly different from the one-dimensional propagation in the slab/wall elements.

Previous studies also highlight the complexity of understanding the damage modes of RC columns under contact explosion loads. Information on damage criteria or threshold curves remains scarce, and experimental research alone is prohibitively expensive [17]. Therefore, developing a reliable numerical model based on experimental data is essential for simulating damage modes and establishing a foundation for RC column design against contact explosions. Furthermore, measuring blast wave parameters in contact explosions is challenging, and the response of structural elements is not yet well understood. This has prompted a growing interest in numerical modelling to simulate such scenarios.

Numerical simulations are widely used to predict the response of RC structures to blast and impact loads, including concrete fracture and spalling. One common approach is the erosion model, which removes elements that meet predefined failure criteria, preventing excessive distortion of Lagrangian meshes. A review in Luccioni et al. [18] examines various erosion criteria and limits used to simulate concrete under blast loads, discussing their influence on damage predictions and their dependence on material properties, mesh size, and scaled distance. An alternative method, the Arbitrary Lagrangian–Eulerian (ALE)–FEM–smoothed particle hydrodynamics (SPH) coupling approach, unlike erosion models, preserves elements and has demonstrated strong agreement with experimental results [19], making it a promising technique for simulating concrete fragmentation under blast conditions. Several previous investigations conducted a numerical analysis of RC columns under contact explosion loads [20]–[22]. However, current numerical studies primarily focus on specific failure mechanisms, with models related to particular research objects, limiting their applicability to different cases and parameter analysis (even with minor variations in section size or shape). Some research conclusions remain unclear and require further validation [16]. Improving the accuracy of numerical models for RC columns under contact explosions and effectively verifying them against experimental data remain critical challenges.

RC bridge columns are critical structural elements, as their failure can lead to the collapse of the entire bridge. Evaluating their damage after an explosion is essential for assessing structural safety and collapse risk. This study investigates the response of a bridge column subjected to a contact explosion through experimental and numerical analyses. A field test was conducted on a 1:3 scale bridge column designed as an indented rectangular, as typically used in Croatian highways. A spherical plastic explosive charge generated the blast, while a steel support structure ensured realistic boundary conditions. The explosion caused concrete cratering and reinforcement bar deformation but did not fully penetrate the column. A high-fidelity numerical simulation was performed using Ansys Autodyn hydrocode software to analyse damage mechanisms further. Refining erosion parameters leads to the simulation more accurately replicating the observed damage, including the explosion-affected area's size, shape, and deformation. The results indicate that the column retains partial structural integrity while severe localised damage occurs, highlighting the reinforcement's crucial role in preventing catastrophic failure.

2 EXPERIMENTAL INVESTIGATION

The experiment's main objective is to evaluate the blast load resistance of the selected bridge column to contact explosion and provide a basis for validating and calibrating the developed numerical model.

2.1 Column geometry

The column geometry was selected based on a database containing the design documentation of all overpasses on Croatian roads and highways. A column with the most common geometry was selected and reproduced on a scale of 1:3. The scaling of the column was conducted concerning the theory of model similarity [23]. The scaled geometry of the selected column is presented in Fig. 1.

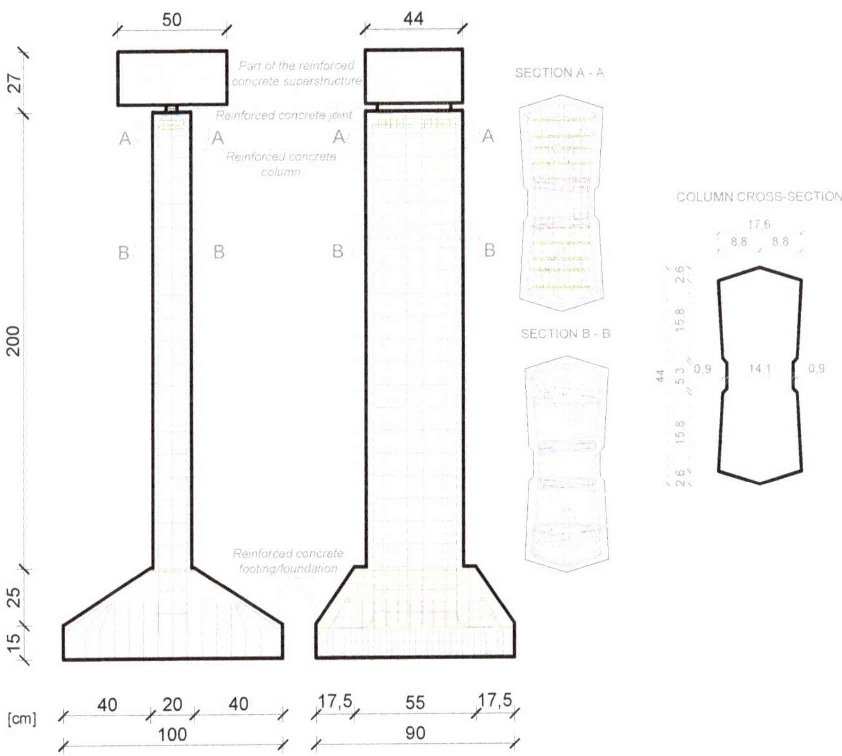

Figure 1: Tested column geometry.

2.2 Test setup

A steel reaction frame was designed to simulate real-life exploitation conditions. The reaction frame consisted of two parts, namely, a reinforced concrete foundation slab and a steel grid structure. The foundation slab was designed to accommodate and prevent horizontal displacements of the tested column footing. A steel grid structure consists of a part fixed to a foundation slab and an upper part designed as a lever. A fixed part provides boundary conditions of a column head beam and simulates the longitudinal stiffness of a bridge deck by preventing horizontal displacements and rotations. A lever part serves to introduce axial force into the column. The axial force was determined to simulate the weight of a bridge deck. A 250 g plastic explosive marked as PEP500 was used in the experiment. This type of plastic explosive is a mixture of 50% pentrite and 50% trinitrotoluene with a TNT equivalency of 1.3. The detonation was initiated using a lead-azide detonator capsule. The explosive was placed at a column surface at 40 cm from the column footing. The detonation capsule was placed horizontally, i.e. perpendicular to the column surface. The experiment was conducted in a corporation with the Croatian Army Forces. Fig. 2 shows the experimental test setup.

Figure 2: Test setup.

Free-field pressures were measured using pencil probes arranged in a semicircle during the test. Six pencil probes were used with a pressure capacity of 6895 kPa at temperatures from −73°C to +135°C, a resolution of 0.069 kPa/mV and a resonance frequency ≥ 400 kHz. Four probes were placed at 10 m and two at 2 m. All probes were placed at 1 m from the ground. The axial force was introduced by employing the Yale hydraulic press to the value of 130 kN. The axial force variation was measured by a force transducer having a capacity of 500 kN. The sensor disposition is presented in Fig. 3.

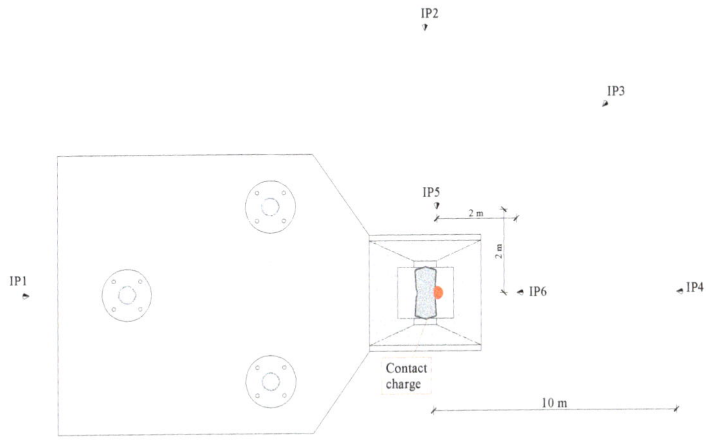

Figure 3: Sensor disposition.

2.3 Free-field pressure measurements

Obtained pressure–time history measurements are presented in Fig. 4(a) for pencil probes placed at 10 m and Fig. 4(b) for pencil probes placed at 2 m. The maximum free-field pressures are compared with the Kingery and Bulmash analytical predictions using eqns (1) and (2). The measured maximum pressures and analytical predictions are presented and compared in Tables 1 and 2.

$$U = K_0 + K_1 \times \log Z, \tag{1}$$

$$Y = 10^{(C_0 + C_1 \times U + C_2 \times U^2 + \cdots + C_N \times U^N)}, \tag{2}$$

where Z is the scaled distance, K and C are constants and Y is an incident or reflected pressure. The constants K and C change depending on the parameter Y.

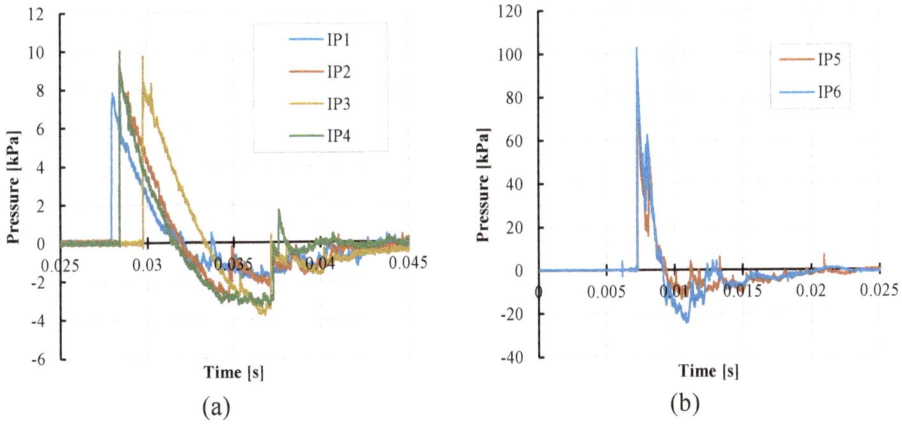

Figure 4: Free-field pressures at the distance of (a) 10 m; and (b) 2 m.

Table 1: Measured incident pressures (distance 10 m).

Range 10 m	IP1	IP2	IP3	IP4	Mean	St. dev.	Cov.	KB_IP	Rel. diff.
Maximum (kPa)	7.81	9.36	9.76	10.04	9.24	0.99	10.77	8.95	3.25
Positive phase (ms)	3.65	3.54	3.69	3.04	3.48	0.30	8.58	3.53	1.42

Table 2: Measured incident pressures (distance 2 m).

Range 2 m	IP5	IP6	Mean	St. dev.	Cov.	KB_IP	Rel. diff.
Maximum (kPa)	69.74	102.58	86.16	23.22	26.95	121.44	29.05
Positive phase (ms)	2.06	1.91	1.98	0.11	5.56	1.81	9.61

The Kingery and Bulmash analytical predictions agree with the measured values. The relative differences in the case of maximum free-field pressures measured at the distance of 10 m are negligible and equal to 3.25%. Somewhat higher, 29.05% relative differences are obtained for measurements at the distance of 2 m. Nevertheless, differences between measured and predicted positive phase duration are negligible for both measurement distances.

2.4 Damage description

Damage of the tested column is presented in Fig. 5. According to CSA-S850-12 [24], damage can be classified as hazardous failure. The observed damage is severe and localised around the explosive charge placement. The protective concrete cover was completely blown off from all sides of the column cross-section. Noticeable flying debris was found at a distance of up to 15 m. Columns' concrete core was reduced in the cross-sectional area due to spalling and fractured at a height of approximately 20 cm.

Figure 5: Damage to the tested column.

Nevertheless, a significant part of the column concrete core remained within, presumably due to rebar distance preventing dislodgement of the concrete. It is assumed that the dense layout of reinforcement bars caused a confinement effect, constraining the concrete core and preventing total perforation of the column. This assumption can be justified because the axial force reduction was negligible; an approximate loss of 10% was observed, as presented in Fig. 6.

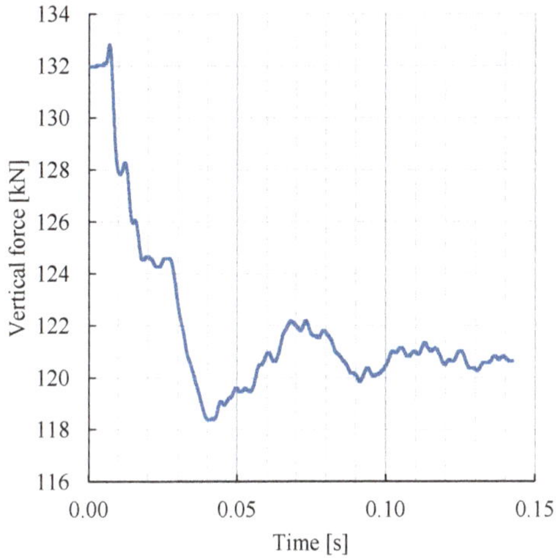

Figure 6: Axial force variation through time.

3 NUMERICAL MODELLING

3.1 Modelling approach

Contact explosion was modelled using Ansys Autodyn hydrocode software [25] specialised for simulating fluid–structure interaction. As Section 2 described, the column was supported by a steel support structure simulating fixed boundary conditions at the column head. The numerical model was simplified by directly implementing fixed boundary conditions to the column head, omitting the steel support structure numerical model. Further simplification was implemented by using symmetry. Only one-half of the column was modelled with a symmetry plane positioned in the column's longitudinal axis. The concrete part of the column was modelled as a solid element, while reinforcement was modelled as a beam element connected with a contact element, enabling their mutual deformation. Fig. 7 shows the column numerical model in Ansys Autodyn with details regarding boundary conditions and element types.

3.2 Material models

Concrete behaviour was modelled using the RHT material model [26]. The RHT (Riedel–Hiermaier–Thoma) concrete material model is widely used in numerical simulations to describe the dynamic behaviour of concrete under high strain rates and complex loading conditions, such as impacts and explosions. It incorporates pressure-dependent strength, strain rate effects, damage accumulation, and failure criteria. Due to its ability to capture

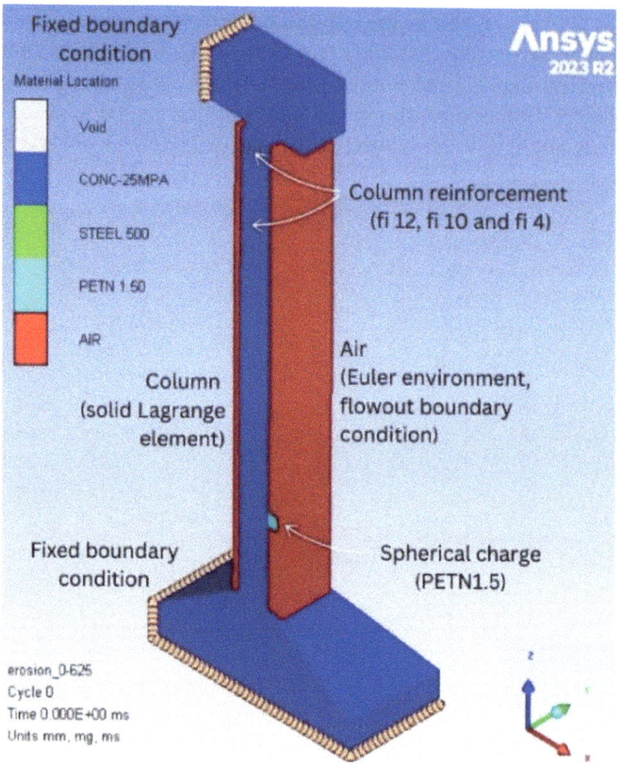

Figure 7: Contact explosion numerical model.

nonlinear stress–strain responses, progressive damage, and brittle failure modes, the RHT model is particularly suitable for simulating protective structures, impact analyses, and blast-resistant design scenarios.

Reinforcement material was modelled using the Johnson–Cook material model [27]. The Johnson–Cook (J–C) steel material model is commonly employed for simulating the behaviour of metals, particularly steel, under large strains, high strain rates, and elevated temperatures. It captures strain hardening, strain rate sensitivity, and thermal softening effects, making it especially suitable for impact, penetration, machining, and crash analyses. Due to its simplicity and robustness, the J–C model is widely used in explicit finite element analyses of dynamic problems involving metals.

The explosive was modelled as a spherical charge using the PETN1.5 material model. It incorporates the Jones–Wilkins–Lee (JWL) [28] equation of state to represent the explosive's pressure–volume–energy relationship during detonation. Due to its accuracy in capturing detonation pressure, energy release, and expansion of explosive products, PETN1.5 is particularly suitable for modelling high-explosive charges in structural response analyses, impact simulations, and blast-resistant design.

Column and charge were placed into an air environment modelled as an ideal gas [29]. The air material model as an ideal gas represents air behaviour based on the ideal gas equation of state, relating pressure, volume, and temperature through simple thermodynamic relations. It is commonly used in numerical simulations involving aerodynamic flows, blast wave propagation, shock interactions, and fluid–structure interaction problems. Due to its simplicity, computational efficiency, and reasonable accuracy for most engineering scenarios involving low to moderate pressures and temperatures, the ideal gas model for air is widely employed in structural, aerodynamic, and blast analyses.

3.3 Model verification

One of the parameters governing the damage behaviour activated was erosion. The geometric strain erosion parameter (instantaneous) defines a numerical erosion criterion used in explicit finite element simulations to remove excessively distorted elements once they reach a specified strain threshold. This parameter helps maintain numerical stability and computational efficiency by preventing unrealistic deformation and element inversion, which is particularly useful in high-strain scenarios such as impacts, penetration and blast simulations.

The default value of the erosion parameter is set at 0.5. The numerical study incorporated variations of erosion parameters to validate the model against experimental data. Erosion was varied in a step of 0.125 considering the following values: 0.375, 0.5, 0.625, 0.75, 0.875 and 1.0. For each erosion value, the figure of simulated damage is given in Tables 3 and 4. The most suitable erosion value was identified based on the visual comparison and damage area analysis. Calculated relative differences between the experimental and numerical values are given in Table 5.

4 DISCUSSION

Based on a comparison between the damaged areas observed in numerical simulations and those recorded during experimental field tests, it can be concluded that the most accurate correlation is achieved when a geometric strain erosion criterion of 0.5 is applied in the simulation (Tables 3, 4 and 5). This value corresponds to the default value provided by the RHT material model. Further analysis of the column's cross-sectional damage reveals that the numerical model predicts more extensive concrete damage than observed in the field.

Table 3: Column damage based on erosion criterion variation (views at 2 ms).

Column damage		Geometric strain erosion criterion					
		0.375	0.500	0.625	0.750	0.875	1.000
Front face		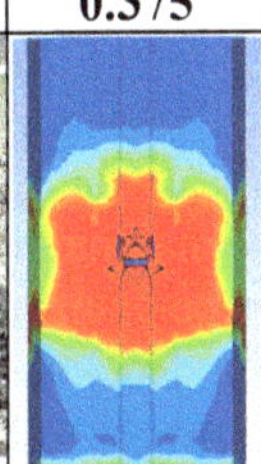	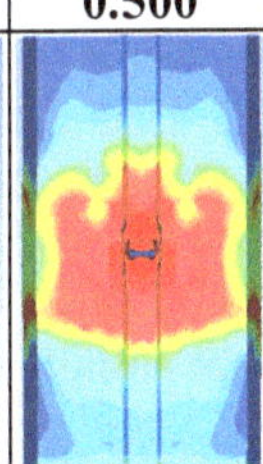	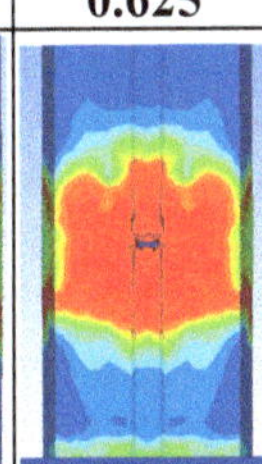	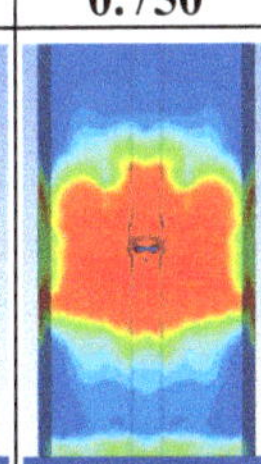	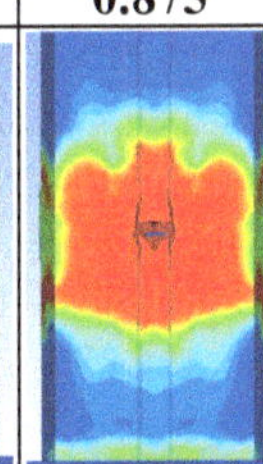	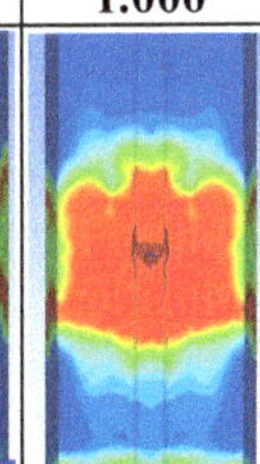
Back face		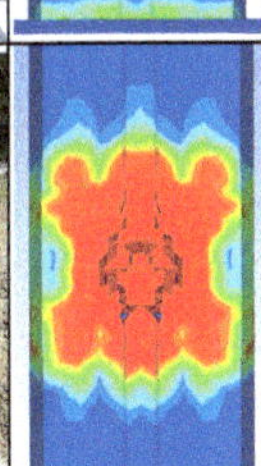	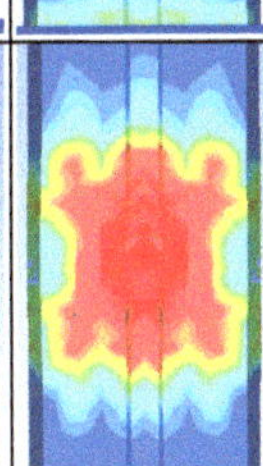	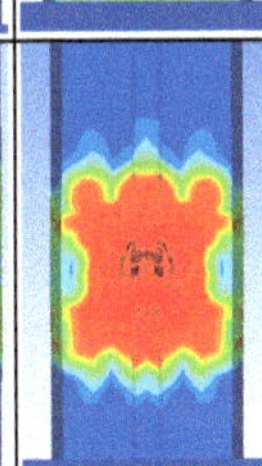	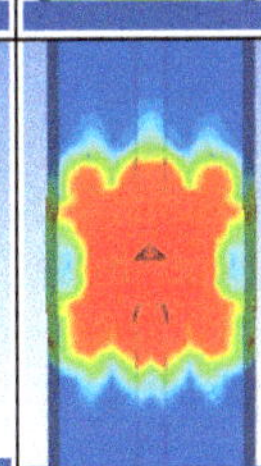	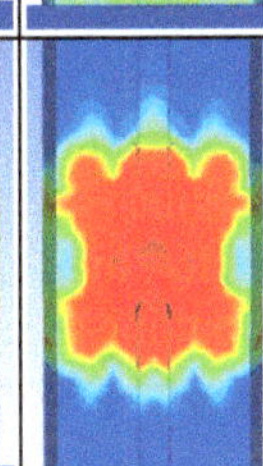	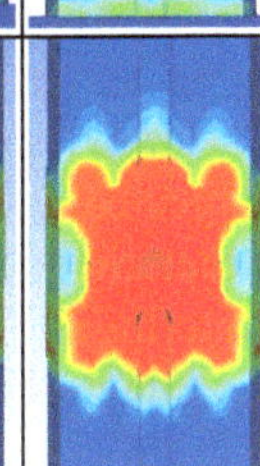

Table 4: Column damage based on erosion criteria variation (cross-section at 2 ms).

Column damage		**Geometric strain erosion criterion**					
		0.375	**0.500**	**0.625**	**0.750**	**0.875**	**1.000**
Front face		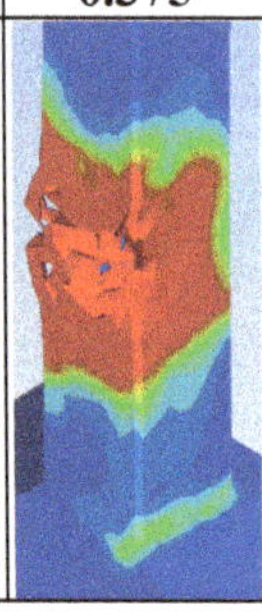	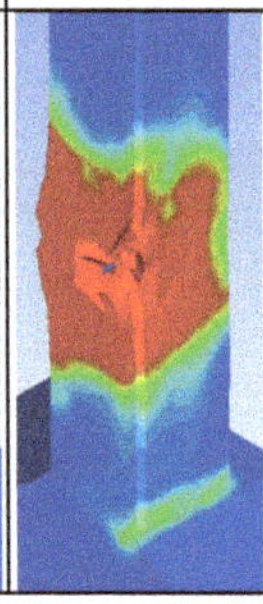	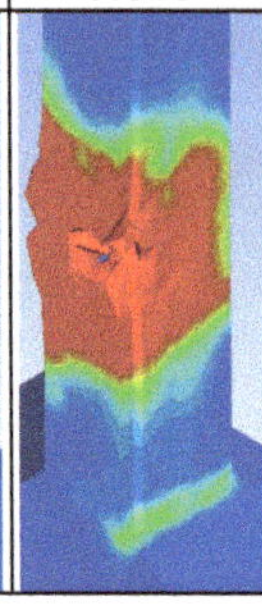	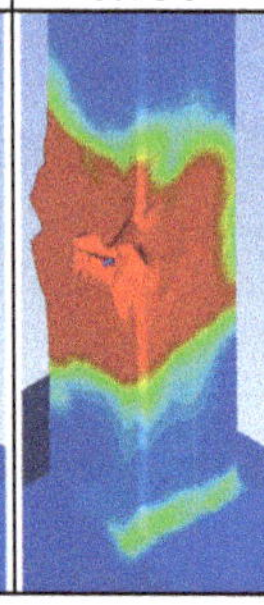	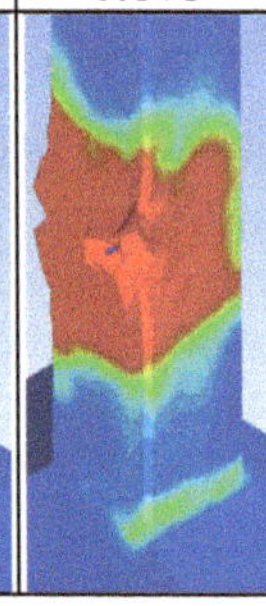	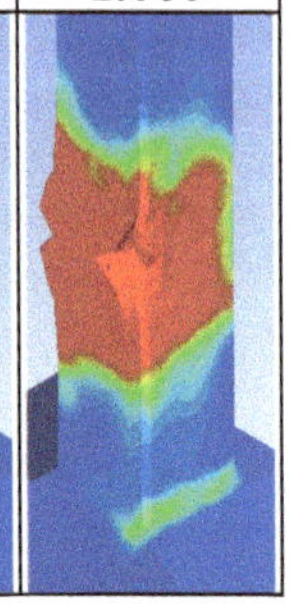

Table 5: Damage area for field test and numerical simulations.

Erosion	Front	Exp	RD (%)	Back	Exp	RD (%)
0.375	1550	1499	3.4	1885	2233	15.6
0.500	1507	1499	0.5	1940	2233	13.1
0.625	1558	1499	3.9	1889	2233	15.4
0.750	1562	1499	4.2	1694	2233	24.1
0.875	1660	1499	10.7	1768	2233	20.8
1.000	1634	1499	9.0	1828	2233	18.1

RD, relative difference.

Table 4 shows that the explosive charge used in the experimental test did not fully perforate the column. However, a significant portion of the reinforced concrete cover was removed near the detonation point. A thin layer of concrete from the core remained intact within the reinforcement 'cage', suggesting that while surface damage was severe, the inner core retained some structural integrity.

In contrast, the numerical simulation produced a damage pattern visually similar to the experimental results using the geometric strain erosion criterion. However, a closer examination of the simulated column cross-section indicates that the blast wave penetrates deeper into the structure, suggesting localised perforation at the point of charge contact. The simulation predicts a concentrated zone of damage around the detonation point, yet the erosion of elements is limited and highly localised, regardless of the erosion threshold applied. This discrepancy between visual surface damage and internal material erosion in the simulation is highlighted in Tables 3 and 4, which show that while damage is concentrated near the charge, the elements in the simulation are not entirely eroded beyond this localised area. The damage area trend based on applied erosion value can be seen in Fig. 8.

5 CONCLUSION

The field test results and corresponding numerical analyses demonstrate that bridge columns subjected to contact explosions can sustain severe localised damage but retain partial structural integrity. In the experimental investigation, a 1:3 scale column incurred significant cratering and spalling around the charge placement, with complete loss of concrete cover and core fracturing; however, the dense reinforcement layout helped to confine the concrete and prevent complete perforation. Measured free-field pressures agreed well with the Kingery and Bulmash analytical predictions, further validating the experimental methodology.

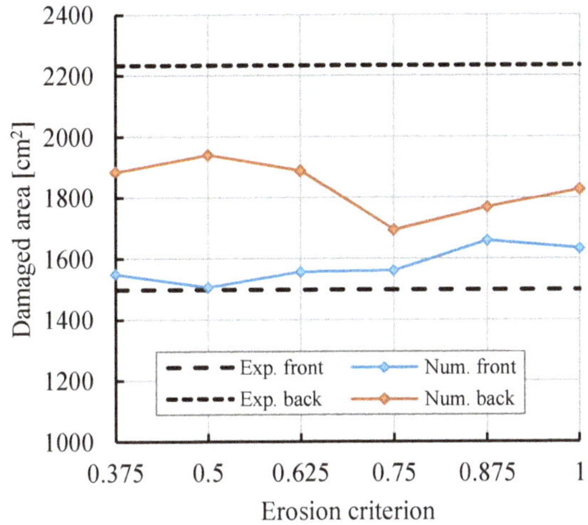

Figure 8: Damaged area for field test and numerical simulations.

In the numerical study, simulations using the RHT material model for concrete and the J–C model for steel reinforcement effectively replicated many aspects of the observed damage. A geometric strain erosion criterion of 0.5 yielded the best correlation between experimental and numerical results, capturing the extent of localised damage and the remaining core integrity. Nonetheless, the simulation predicted a slightly deeper penetration of the blast wave than the physical specimen, highlighting the need for continued refinement of numerical parameters and material modelling approaches.

Overall, the findings underscore the critical influence of reinforcement in mitigating catastrophic failure from contact explosions and confirm that properly calibrated numerical models can be valuable tools for assessing blast resistance. The ability of the column to maintain partial load-bearing capacity despite major local damage emphasises the importance of robust detailing and accurate computational methods when designing and evaluating bridge columns for blast resistance.

ACKNOWLEDGEMENTS

The research presented in this paper has been partly supported by the Croatian Science Foundation (HRZZ) under the project UIP-2017-05-7041 'Blast Load Capacity of Highway Bridge Columns', and support for this research is gratefully acknowledged. The authors would like to thank all individuals who provided help during the research.

REFERENCES

[1] Dua, A. & Braimah, A., Assessment of reinforced concrete slab response to contact explosion effects. *Journal of Performance of Constructed Facilities*, **34**(4), 04020061, 2020.

[2] Wang, W. et al., Study on antiexplosion performance of reinforced concrete slabs strengthened with POZD coated steel plate under contact explosion. *Engineering Failure Analysis*, **140**, 106589, 2022.

[3] Yang, G. et al., Experimental and numerical study of damage characteristics of RC slabs subjected to air and underwater contact explosions. *Marine Structures*, **66**, pp. 242–257, 2019.

[4] Cai, R. et al., Size effect on reinforced concrete slabs under direct contact explosion. *Engineering Structures*, **252**, 113656, 2022.
[5] Zhu, W. et al., Blast resistant performance and damage mechanism of steel reinforced concrete beams under contact explosion. *Engineering Structures*, **315**, 118472, 2024.
[6] Yang, C. et al., Damage of full-scale reinforced concrete beams under contact explosion. *International Journal of Impact Engineering*, **163**, 104180, 2022.
[7] Jun-Hua, H. et al., Dynamic response and damage assessment of a prestressed T-shaped simply supported girder bridge under contact explosion. *Structures*, **48**, pp. 40–52, 2023.
[8] Williamson, E.B., Blast-resistant highway bridges: Design and detailing guidelines. *Transportation Research Board*, **645**, 2010.
[9] Wu, K.-C., Li, B. & Tsai, K.-C., The effects of explosive mass ratio on residual compressive capacity of contact blast damaged composite columns. *Journal of Constructional Steel Research*, **67**(4), pp. 602–612, 2011.
[10] Fujikake, K. & Aemlaor, P., Damage of reinforced concrete columns under demolition blasting. *Engineering Structures*, **55**, pp. 116–125, 2013.
[11] Roller, C. et al., Residual load capacity of exposed and hardened concrete columns under explosion loads. *Engineering Structures*, **55**, pp. 66–72, 2013.
[12] Yuan, S. et al., A study of RC bridge columns under contact explosion. *International Journal of Impact Engineering*, **109**, pp. 378–390, 2017.
[13] Li, M. et al., Experimental and numerical study on damage mechanism of CFDST bridge columns subjected to contact explosion. *Engineering Structures*, **159**, pp. 265–276, 2018.
[14] Draganić, H., Gazić, G. & Varevac, D., Experimental investigation of design and retrofit methods for blast load mitigation: A state-of-the-art review. *Engineering Structures*, **190**, pp. 189–209, 2019.
[15] Lukić, S. & Draganić, H., Blast loaded columns: State of the art review. *Applied Sciences*, **11**, 2021. https://doi.org/10.3390/app11177980.
[16] Yuan, S. et al., A review on close-in blast performance of RC bridge columns. *Journal of Traffic and Transportation Engineering (English Edition)*, **10**(4), pp. 675–696, 2023.
[17] Zhang, C., Gholipour, G. & Mousavi, A.A., Blast loads induced responses of RC structural members: State-of-the-art review. *Composites Part B: Engineering*, **195**, 108066, 2020.
[18] Luccioni, B.M., Aráoz, G.F. & Labanda, N.A., Defining erosion limit for concrete. *International Journal of Protective Structures*, **4**(3), pp. 315–340, 2013.
[19] Wang, Z. et al., Numerical prediction of blast fragmentation of reinforced concrete slab using ALE–FEM–SPH coupling method. *Finite Elements in Analysis and Design*, **220**, 103948, 2023.
[20] Xiao, Y. et al., Concrete spalling behavior and damage evaluation of concrete members with different cross-sectional properties under contact explosion. *International Journal of Impact Engineering*, **181**, 104753, 2023.
[21] Mejía, N. et al., Damage assessment of RC columns under the combined effects of contact explosion and axial loads by experimental and numerical investigations. *Engineering Structures*, **254**, 113776, 2022.
[22] Liu, L. et al., Experimental and numerical study of CFRP protective RC piers under contact explosion. *Composite Structures*, **234**, 111658, 2020.
[23] Harris, H. & Sabnis, G.M., *Structural Modeling and Experimental Techniques*, 2nd ed., CRC Press, 1999.

[24] CSA Group, S850-12 Design and Assessment of Buildings Subjected to Blast Loads. Mississauga, 2012.
[25] Ansys Autodyn, *Theory Manual Revision 4.3*, Century Dynamics: Concord, CA, USA, 2005.
[26] Thoma, K., Riedel, W. & Hiermaier, S., Mesomechanical modeling of concrete shock response experiments and linking to macromechanics by numerical analysis. *Proceedings of European Conference on Computational Mechanics*, 1999.
[27] Johnson, G.R. & Cook, W.H., Fracture characteristics of three metals subjected to various strains, strain rates, temperatures and pressures. *Engineering Fracture Mechanics*, **21**(1), pp. 31–48, 1985.
[28] Lee, E., Hornig, H. & Kury, J., Adiabatic expansion of high explosive detonation products. University of California Radiation Laboratory at Livermore, Livermore, CA, USA, 1968.
[29] Rogers, G.F.C. & Mayhew, Y.R., *Thermodynamic and Transport Properties of Fluids*, John Wiley and Sons, 1995.

EXPERIMENTAL AND NUMERICAL INVESTIGATION OF POLYMERIC GYROID TPMS STRUCTURES UNDER CRUSHING LOAD

MATS OVERBECK[1,2], PIETRO BALLARIN[3], TIM LUPLOW[2] & SEBASTIAN HEIMBS[1,2]
[1]Cluster of Excellence SE^{2}A – Sustainable and Energy-Efficient Aviation,
Technische Universität Braunschweig, Germany
[2]Institute of Aircraft Design and Lightweight Structures (IFL), Technische Universität Braunschweig, Germany
[3]Department of Aerospace Science and Technology, Politecnico di Milano, Italy

ABSTRACT
Triply periodic minimal surface (TPMS) structures offer a unique set of characteristics that make them particularly attractive for aerospace applications. As open-cell structures with efficient specific strength and stiffness properties, they enable the design of lightweight and effective sandwich core materials for applications such as laminar flow control via boundary layer suction on the leading edge of an aircraft wing. In such components, bird strike often represents the critical load case for structural design, a precise understanding of the mechanical behaviour of TPMS structures is essential for their effective application in these contexts. Therefore, this study presents a comprehensive modelling methodology for gyroid structures integrating experimental and numerical methods to investigate both quasi-static and dynamic crushing behaviour. Initially, quasi-static coupon experiments are conducted on polymeric materials tailored for this application. Digital image correlation techniques are employed to analyse the stress states. Subsequently, dynamic tests are performed to assess the performance of TPMS structures under high loading rates. A finite element model is developed to simulate the mechanical response of the TPMS structures. This model is then validated against experimental data. The results demonstrate strong numerical–experimental agreement, with material parameters calibrated from coupon tests enabling an accurate representation of the stress–strain response and acceptable results were obtained for the first part of the dynamic crushing simulation of gyroid structures.
Keywords: TPMS, gyroid, compression, crushing, finite element modelling.

1 INTRODUCTION

Triply periodic minimal surface (TPMS) structures represent a viable solution to design and build aerospace structural elements, whose major requirement is represented by the weight, which must be as low as possible. Recently, some studies were presented by the Cluster of Excellence SE^{2}A – Sustainable and Energy-Efficient Aviation – and other research institutes about the application of such type of structures to the laminar flow control (LFC) with the purpose to reduce drag and the consequent fuel consumption delaying the laminar-turbulent transition [1]–[3]. Indeed, in the last years, the reduction of aircraft emissions has become more and more important to achieve sustainability since the aviation sector was responsible for 859 million tonnes of CO_2 in 2017 [4]. Moreover, from an economic standpoint, the expenses for fuel consumption represent 33.4% of the total cost of ownership (TCO) for an airline company [5]. The principle behind such concept consists in the suction of the boundary layer through the wing's outer skin. Indeed, previous studies demonstrated that a 45.5% of drag reduction can be obtained applying such a solution [6]. From an engineering standpoint, the solution can be implemented using porous wing panels supported by an internal porous core structure. A suction device is also needed to realise the air flow throughout the structure. The function of the porous filler can be performed by a TPMS structure, which is characterized by multiple cavities that allows the air to flow.

WIT Transactions on The Built Environment, Vol 215, © 2025 WIT Press
www.witpress.com, ISSN 1743-3509 (on-line)
doi:10.2495/SUSI250161

TPMS core structures also perform a structural role in addition to the aerodynamic one. The aircraft wing is subjected to different loads during flight. The leading edge plays a major role in aircraft safety. All parts oriented in flight direction can be subjected to bird strike loading, making this load case a design driving aspect. Therefore, the porous core needs to absorb a sufficient amount of the bird's kinetic energy to avoid damage to the wing's front spar. The front spar is one of the main load bearing parts of the aerodynamic loads. The structural integrity of the front spar and the wing box ensures a safe continuation of flight after a bird strike incident.

TPMS structures demonstrated excellent energy absorption capabilities [7], [8], which make them suitable for the aforementioned application. The implementation of LFC, and in particular the hybrid laminar flow control (HLFC), in which the boundary layer suction occurs also through the leading edge, requires the possibility to design and evaluate the structural strength and the energy absorption capability of the TPMS structures. Though a large number of experimental studies were carried out on the behaviour of TPMS structures [9]–[11], only few studies investigated the modelling strategies of such structures [12], [13].

In this work, different finite element (FE) modelling strategies are presented with the purpose to model the behaviour of TPMS structures, with focus on the gyroid type, in static and dynamic crushing conditions. The samples were produced by additive manufacturing in polymeric material. First, dogbone specimens were produced and experimentally tested under tensile loading to calibrate the material behaviour exploiting digital image correlation (DIC). Then, cubic specimens, representative of the gyroid structure, were manufactured and tested in static and dynamic compressive loading conditions. Different FE modelling approaches were compared based on the material's laws derived from the tests on the dogbone specimens.

This work is organized as follows: Section 2 describes the manufacturing process and the experimental methods adopted used in this study. Section 3 introduces the finite element modelling strategies. In Section 4, the numerical–experimental correlation is presented. Finally, Section 5 summarizes the main conclusions and outlines directions for future research.

2 EXPERIMENTS

This section outlines the methods adopted in this work to conduct the experimental part of the research activity. Specifically, in Subsection 2.1 the methodology to carry out the experiments and the manufacturing of the coupons is described, and in Subsection 2.2 the FE method to model the experiments is presented.

2.1 Manufacturing of the specimens

Two types of specimens were manufactured, the first type consisted of dogbone specimens, hereafter named specimen type A, which allowed calibrating the material model. The geometry is derived from the German norm DIN EN ISO 527-2 [14]. The second type of specimen consists of TPMS specimens, hereafter named specimens type B. The TPMS specimens are of a gyroid structure and cubic dimensions.

Specimens A were produced by additive manufacturing, using stereolithography technology. A Formlabs (Somervile) Form 3+ printer was used. A newly available photopolymer by Formlabs called Polyurethane Rigid 650 [15], which is characterized by great energy absorption properties according to the manufacturer was selected. This material is more challenging to manufacture compared to other SLA manufactured materials due to its tendency to interact with water. Therefore, the manufacturing process was conducted in a dry cabinet with a relative humidity below 5%. The material properties given by the manufacturer are presented in Table 1. The specimens were manufactured with a layer thickness of 50 μm. The printing orientation was varied in order to investigate any orientation dependency. Orientations of 0°, 45°, and 90° in respect to the printing platform were considered. The different printing orientations are depicted in Fig. 1. For each printing direction, six specimens were manufactured.

Table 1: Material properties of Formlabs Polyurethane Rigid 650 [15].

Property	Unit	Value
Young's modulus	GPa	0.67 ± 0.06
Ultimate tensile strength	MPa	34 ± 3.4
Elongation at break	%	170 ± 17
Density	G cm^{-3}	1.16
Notched Izod	J m^{-1}	375

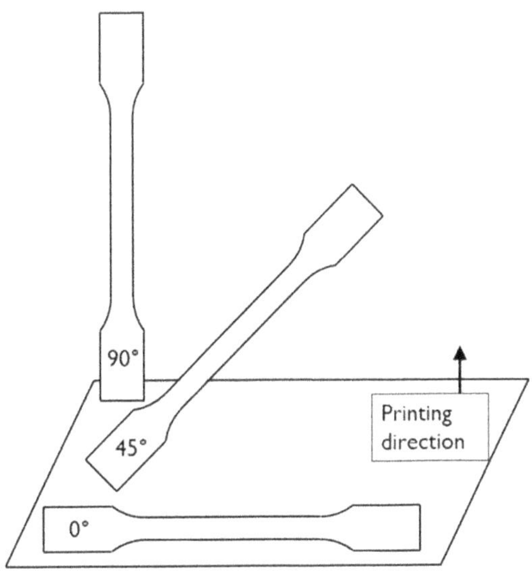

Figure 1: Different investigated printing orientations in respect to print bed.

After the printing, the specimens were washed in a propane-1,2-diyl diacetate (PGDA) bath and cured in an oven at a temperature of 46°C and a relative humidity of 70%.

Specimens type B were produced by the same additive manufacturing process as previously described, with a Formlabs Form 3+ and the same material adopted to produce the specimens type A.

A square glass-fibre-reinforced plastic plate with a thickness of 1.5 mm and an edge length of 53 mm was adhesively joined to the top and bottom of specimen B to ensure uniform loading into the specimen during the testing. The structural epoxy adhesive EC-9323 B/A by 3M Company (St. Paul, MN, USA) [16] was used. The adhesive was applied to the plates, which were then joined to the specimen. With weight placed on top of the specimen to ensure bonding, the structure was placed in an oven and cured at 70°C for a duration of 2 hours.

2.2 Static tests

The quasi-static tests were performed on a Zwick Roell GmbH & Co. KG (Ulm, Germany) universal test machine. The tensile tests using specimen A were conducted at a crosshead velocity of 10 mm/min. The strain was recorded using the DIC tool ARAMIS from Zeiss (Oberkochen, Germany), and the displacement of the crosshead was additionally tracked.

The quasi-static tests of specimen B were carried out on a universal test machine of Zwick-Roell with a load cell with a maximum load of 50 kN at a crosshead velocity of 2 mm/min. The displacement was recorded using the DIC tool ARAMIS from Zeiss. The test setup can be seen in Fig. 2(a).

2.3 Dynamic crushing tests

The dynamic testing of specimen type B was performed using a drop tower. The drop tower compromised a double-sided guided drop sled. The total mass of the drop sled including the load cell and impactor was 16 kg. The maximum drop height is 6 m. A load cell with a maximum load of 160 kN was used. The experiment was recorded using a high-speed camera with up to 40 k fps. The force and load were recorded at a rate of 500 k/s. The via displacement measured velocity before impact was 6.1 m/s. The drop sled and impactor are presented in Fig. 2(b).

(a) (b)

Figure 2: (a) Test setup of the quasi-static compression on the gyroid specimens; and (b) The dynamic crushing tests on the gyroid specimens.

3 FINITE ELEMENT MODELLING

In this section, the FE modelling methodology is described. First, in Subsection 3.1, the static tests carried out on specimens type A are modelled. Then, in Subsection 3.2, the FE modelling of the quasi-static and crushing tests is presented.

3.1 Material model calibration

The experimental material characterisation is modelled with FE using Simulia/Abaqus Explicit. The dogbone specimen is modelled with C3D8R type elements having characteristic size equal to 1 mm. The specimen was subjected to an imposed displacement at one end, while all the translational and rotational degrees of freedom of the other end were constrained. The model can be observed in Fig. 3. Different hyperelastic material models were compared. Uniaxial test data was provided as input in the material card.

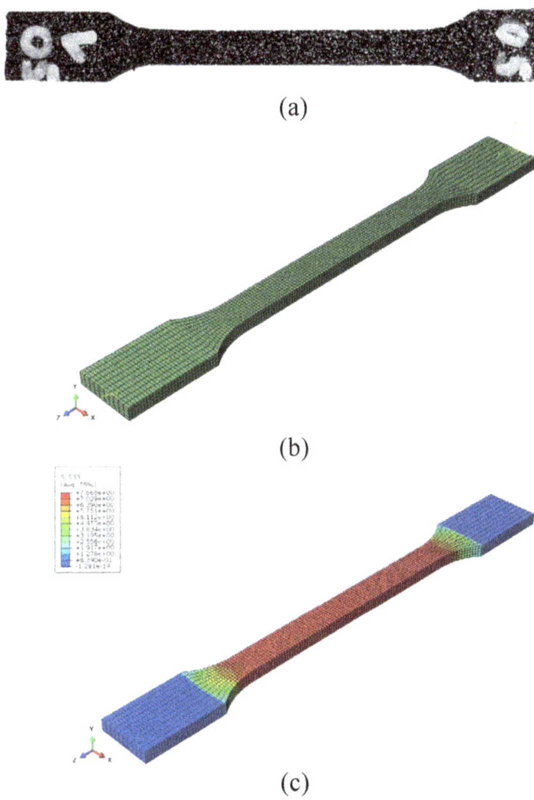

Figure 3: (a) Specimen A in finite element software. The mesh with C3D8 elements (b) and the resulting stress contour (c) are depicted.

3.2 Finite element modelling of triply periodic minimal surface structures

In general, TPMS structures are structures that divide a given volume into two volumes of equal size. This is achieved by an implicit function $\Phi(x,y,z)$ consisting of a combination of sine and cosine function, making the structure periodically duplicable in all three dimensions. An isosurface is generated in all given points where the function of Φ is true. The following equation describes the here investigated gyroid surface [17]:

$$\Phi(x,y,z)=\cos(\lambda_x x)\sin(\lambda_y y)+\cos(\lambda_y y)\sin(\lambda_z z)+\cos(\lambda_z z)\sin(\lambda_x x)=c, \qquad (1)$$

with

$$\lambda_i = 2\pi \frac{n_i}{L_{i,tot}}, \text{ for } i = x, y, z.$$

The appearance of the structure can be adjusted by its total edge length $L_{i,tot}$ and the size of a unit cell n_i. Those two parameters define the periodicity parameter λ_i. The constant c can be varied to shift the dividing isosurface. Additionally, a grading of the structure can be implemented this way. In the most common case, the constant c is zero.

Two distinct methodologies can be used to generate a volumetric structure from this isosurface. On one hand, material can be extruded in one direction perpendicular to the surface, creating a so-called solid network structure. On the other hand, material can be extruded in both directions perpendicular to the surface, generating a so-called sheet network structure. Sheet network structures tend to be isotropic and show greater potential for energy absorption. Therefore, they are the focus of this study. The behaviour and mechanical properties of gyroid structures are dependent on the material, manufacturing process, unit cell size and relative density. The relative density ρ^* is defined as the ratio between the density of the gyroid structure ρ_G and the density of the used material ρ_S.

$$\rho^* = \frac{\rho_G}{\rho_S}. \tag{2}$$

The correlation of the relative density ρ^*, the unit cell size L_i, and the wall thickness d can be approximated by the following equation:

$$d(L_i, \rho^*) = \frac{\rho^* \times L_i^{0,966}}{2.568}. \tag{3}$$

The chosen gyroid configuration is as follows. An edge length of 50 mm is chosen. A unit cell size of 10 mm and a relative density of 17% are selected. Using the above presented equation, a wall thickness of 0.612 mm is defined.

The finite element modelling of the TPMS structure is divided into smaller parts in order to enable a shorter computing time. The mesh generation is conducted in the software Hypermesh by Altair (Troy, MI, USA). The general procedure is as follows. Using the software Matlab R2022b, an .stl file of the isosurface of a gyroid structure is generated. The .stl file is then imported into Hypermesh. Here, the mesh is created and an Abaqus .inp file is created, which can be imported into Abaqus CAE. The mesh is generated using only one unit cell. This unit cell is then divided along all three planes through its centre, creating eight different parts. This enables a more structured mesh. These eight parts are then simultaneously meshed using S4R elements. The wall thickness that was derived above is implemented as the shell thickness. The number of nodes on opposite boundary lines must be equal. Opposing boundary nodes must be mirrored. This unit cell model is used for a mesh convergence study. The base element edge length is varied from 0.1 mm to 1.4 mm. The resulting meshes can be seen in Fig. 4.

The cubic gyroid structure is assembled as well as two rigid plates, which are positioned at the top and bottom of the gyroid structure. The connecting nodes and the given surface of the plate are connected using tie constraints. This is being done to model the adhesive bond between the gyroid structure and the plate in the experimental testing. The deformation is imposed using a smooth step and a displacement of a reference point on the upper plate. The equivalent reference point on the lower plate is pinned.

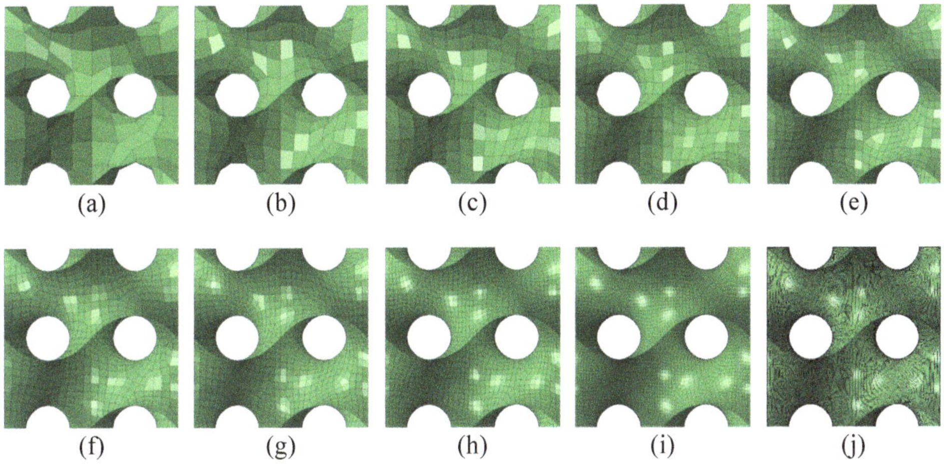

Figure 4: Different mesh sizes of one gyroid unit cell created of an isosurface (shell elements). The unit cell edge length is 10 mm. The mesh is refined from a coarse mesh with element edge lengths of 1.4 mm to a fine mesh. (a) 1.4 mm; (b) 1.2 mm; (c) 1.0 mm; (d) 0.8 mm; (e) 0.6 mm; (f) 0.5 mm; (g) 0.4 mm; (h) 0.3 mm; (i) 0.2 mm; and (j) 0.1 mm.

The large gyroid structure is constructed of five unit cells in each dimension. This is realized in the numerical modelling as follows. One in hypermesh meshed unit cell is imported into the model. This unit cell is then duplicated using the linear pattern function. All coincident nodes are merged, leaving a meshed gyroid structure. This structure is shown in Fig. 5(b). The imposed shell thickness of 0.612 mm is rendered in Fig. 5(c).

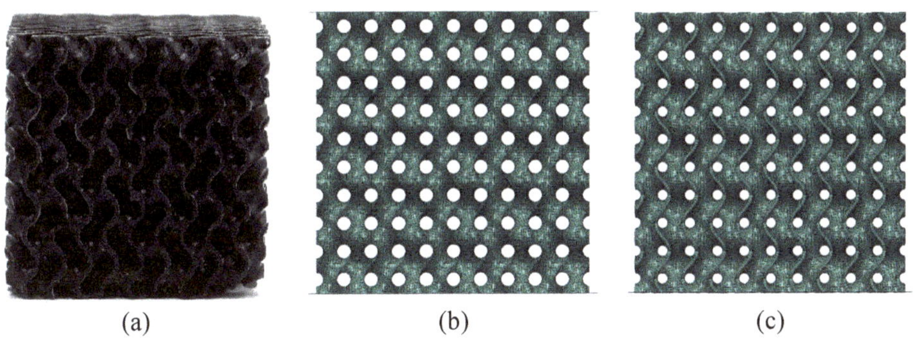

Figure 5: Additively manufactured gyroid structure of PU Rigid 650 material with a relative density of 17% and a unit cell size of 10 mm (a) as well as the numerical modelling of the same structure with shell elements (b). For better visualisation, the imposed shell thickness is rendered (c).

4 RESULTS

4.1 Calibration of the material model with standard dogbone specimens

The results of the experimental testing of specimen A are presented in Fig. 6 in regard to the printing orientation. A mean tensile modulus of 610 MPa was determined. After an initial linear increase in stress, the stress increase is lowered. The stress stays nearly constant over a large increase in strain. Towards the failure of the specimen, the stress increases again. The generated data is used to derive input data for the hyperelastic material model used in the FE modelling (Fig. 6).

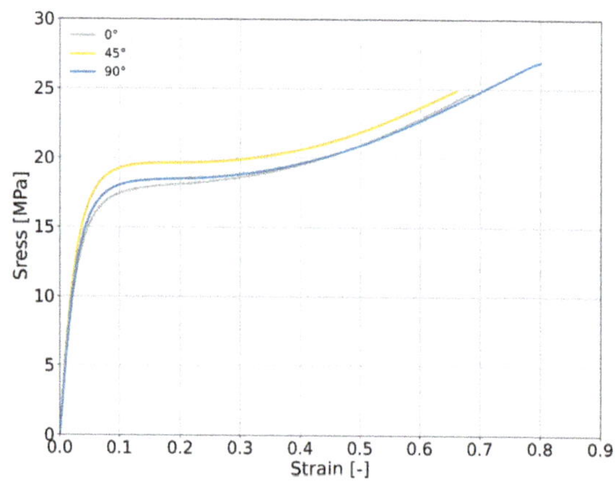

Figure 6: Results of tensile tests. All results of specimens manufactured in the same orientation are compromised into one graph for better comparison.

Overall, it can be observed that the models exhibiting a better numerical–experimental correlation are the ones based on the Ogden and Marlow model (Fig. 7). The result obtained with the Marlow model is however trivial, since such model is not parametric and does not rely on calibration coefficients. On the contrary, the model satisfies the condition for each given point in the calibration, and for any condition in between it interpolates between the two nearest points.

The Ogden model is a type of hyperelastic model describing the nonlinear stress–strain behaviour of materials, which is particularly useful to model polymeric materials, and in particular rubbers. A strain energy potential is defined on the basis of the principal stretches λ, as in eqn (4).

$$W=\sum_{p=1}^{N}\frac{\mu_p}{\alpha_p}\times\left(\lambda_1^{\alpha_p}+\lambda_2^{\alpha_p}+\lambda_3^{\alpha_p}-3\right), \tag{4}$$

where μ, α and N represent material constants. The principal stress components can be obtained deriving the strain energy potential W with respect to the principal stretch components, as in eqn (5):

$$\sigma_j=-p+\lambda_j\times\frac{dW}{d\lambda_j}, \tag{5}$$

where p represents a Lagrange multiplier enforcing the incompressibility constraint.

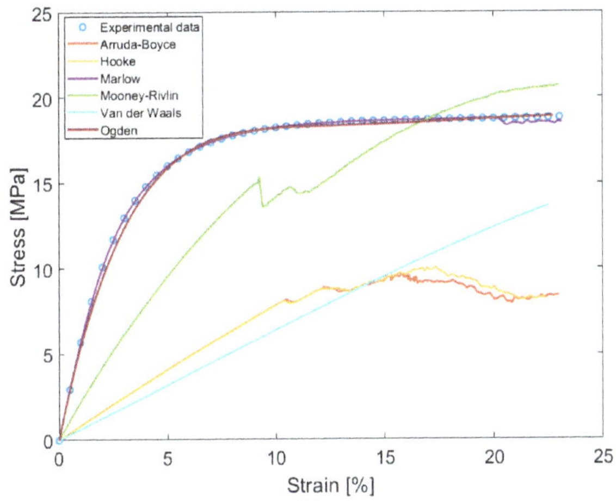

(a)

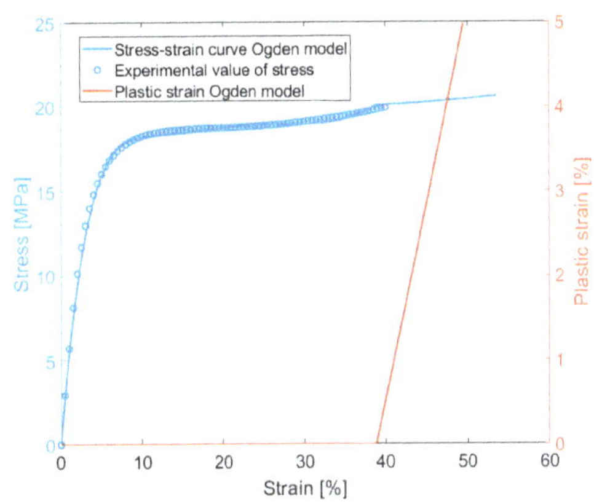

(b)

Figure 7: (a) Stress–strain curves of different hyperelastic material models with the given test input data; and (b) Detailed Ogden model with plastic strain function in regard to total strain.

4.2 Numerical–experimental comparison of static tests

The results of the mesh convergence study, which was conducted under quasi-static compression are presented in Fig. 8. It can be seen that the force–displacement curve converges with smaller element size. The initial peak force is the greatest with an element size of 1.4 mm and decreases with smaller element size. Starting at an element size of 0.5 mm and smaller, the force peak stays nearly constant. The behaviour after the initial failure varies with the element size and is influenced by element deletion of the failed elements. This result indicates that a general element size of 0.5 mm for a 10 mm gyroid unit cell or an element size of 5% of the unit cell length is sufficient to replicate the result achieved with a much finer mesh.

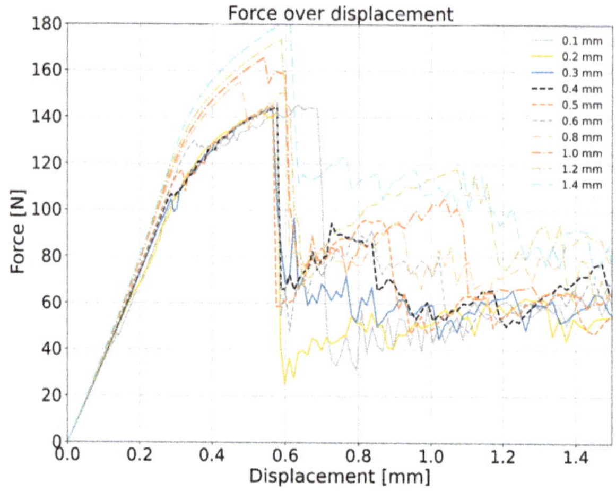

Figure 8: Results of mesh convergence study. The element size of the mesh is indicated in the legend. A sufficient convergence can be observed for an element size of 0.5 mm and smaller.

The result of the simulation of the whole gyroid structure compared to the experimental test is shown in Fig. 9. It can be seen that the numerical data closely follows the obtained force–displacement curve from the experiment. It slightly overestimates the curve. This can be caused by the fact that the material was modelled as isotropic but there were small differences between the specimens of different printing orientations. Additionally, there may have been some manufacturing induced inhomogeneities. When removing the structure from the plate, the specimen may have been damaged.

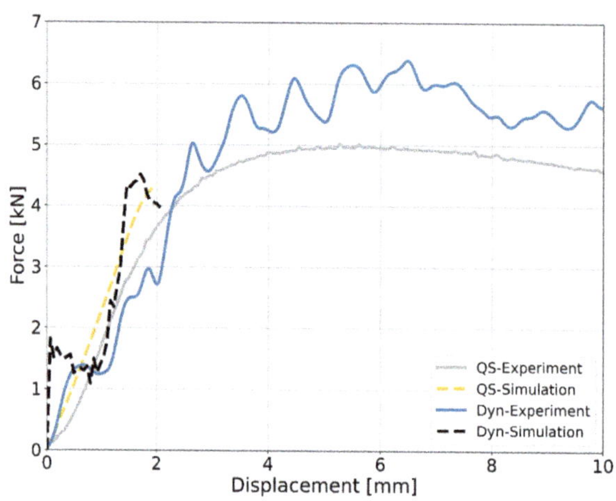

Figure 9: Comparison of experimental and numerical results of compression and crushing tests of a gyroid structure.

4.3 Numerical–experimental comparison of crushing tests

The results of the dynamic testing and numerical investigation of the specimen type B are presented in Fig. 9. The comparison of the numerical and experimental investigations of the quasi-static and dynamic testing is similar. The force is slightly overestimated. However, a general behaviour of the structure could be obtained for a small displacement. Although the oscillating behaviour of the dynamically tested structure, the dynamic curve is above the quasi-static one, most likely due to its strain rate sensitivity. Though, the simulations were aborted prematurely due to distorted elements on the boundaries, the initial slope is sufficiently captured from both the static and dynamic analysis.

5 CONCLUSIONS

In this experimental and numerical study, a recently available photopolymer for energy absorption applications has been investigated. In connection with a TPMS gyroid structure, this material can potentially be used for lightweight energy absorbing structures, e.g. a wing leading edge for bird strike loading. Therefore, a hyperelastic Ogden material model with plasticity was found to correctly reproduce the material behaviour. This material model was then implemented in a cubic gyroid structure of five unit cells of 10 mm length, resulting in a cubic structure with an edge length of 50 mm. This method showed to be promising for the applicability on real world structures. Future works will be aimed at validating this approach in order to demonstrate the applicability to represent typical aeronautical elements.

ACKNOWLEDGEMENT

We would like to acknowledge the funding by the Deutsche Forschungsgemeinschaft (DFG, German Research Foundation) under Germany's Excellence Strategy – EXC 2163/1 – Sustainable and Energy Efficient Aviation – Project-ID 390881007.

REFERENCES

[1] Lobitz, L. et al., Aircraft wing design for extended hybrid laminar flow control. *Aerospace*, **10**(11), p. 938, 2023.

[2] Traub, H., Wolff, J., Jose, S., Lobitz, L., Schollerer, M. & Hühne, C., Concept and design of extended hybrid laminar flow control suction panels, 2021. https://doi.org/10.21203/rs.3.rs-924184/v1.

[3] Beck, N., Landa, T., Seitz, A., Boermans, L., Liu, Y. & Radespiel, R., Drag reduction by laminar flow control. *Energies*, **11**(1), p. 252, 2018.

[4] Terrenoire, E., Hauglustaine, D.A., Gasser, T. & Penanhoat, O., The contribution of carbon dioxide emissions from the aviation sector to future climate change. *Environmental Research Letters*, **14**(8), 084019, 2019.

[5] Giannakeas, I.N., Khodaei, Z.S. & Aliabadi, M.F., Structural health monitoring cost estimation of a piezosensorized aircraft fuselage. *Sensors*, **22**(5), p. 1771, 2022.

[6] Wiedemann, M. et al., The suction panel-xHLFC and structural solution for energy efficient aviation. *AIAA Scitech 2022 Forum*, p. 0007, 2022.

[7] Zhang, L., Feih, S., Daynes, S., Chang, S., Wang, M.Y., Wei, J. & Lu, W.F., Energy absorption characteristics of metallic triply periodic minimal surface sheet structures under compressive loading. *Additive Manufacturing*, **23**, pp. 505–515, 2018.

[8] Sun, Q., Sun, J., Guo, K. & Wang, L., Compressive mechanical properties and energy absorption characteristics of SLM fabricated Ti6Al4V triply periodic minimal surface cellular structures. *Mechanics of Materials*, **166**, 104241, 2022.

[9] Yu, S., Sun, J. & Bai, J., Investigation of functionally graded TPMS structures fabricated by additive manufacturing. *Materials and Design*, **182**, 108021, 2019.

[10] Novak, N., Al-Ketan, O., Krstulović-Opara, L., Rowshan, R., Al-Rub, R.K.A., Vesenjak, M. & Ren, Z., Quasi-static and dynamic compressive behaviour of sheet TPMS cellular structures. *Composite Structures*, **266**, 113801, 2021.

[11] Wang, H., Tan, D., Liu, Z., Yin, H. & Wen, G., On crashworthiness of novel porous structure based on composite TPMS structures. *Engineering Structures*, **252**, 113640, 2022.

[12] Qiu, N., Wan, Y., Shen, Y. & Fang, J., Experimental and numerical studies on mechanical properties of TPMS structures. *International Journal of Mechanical Sciences*, **261**, 108657, 2024.

[13] Ma, X., Zhang, D.Z., Yu, X., Ren, Z., Mao, S. & Zheng, X., A novel design method for TPMS lattice structures with complex contour based on moving elements method. *The International Journal of Advanced Manufacturing Technology*, **123**(1), pp. 21–33, 2022.

[14] DIN Deutsches Institut für Normung e.V., Plastics: Determination of tensile properties. Deutsche Norm – DIN EN ISO 527-2. Beuth Verlag GmbH: Berlin, Germany, 2012.

[15] Formlabs, PU Rigid 650: For impact resistant and semi-stiff polyurethane parts. Formlabs: Somerville, MA, USA, 2022.

[16] 3M Company, 3MTM Scotch-WeldTM Structural Epoxy Adhesive EC-9323 B/A. Technical Data Sheet. 3M Company: Saint Paul, MN, USA, 2022.

[17] Schoen, A.H., Infinite periodic minimal surfaces without self-intersections. Technical Note NASA TN D-5541. National Aeronautics and Space Administration: Washington, DC, USA, 1970.

SECTION 5
DYNAMIC BEHAVIOUR OF MATERIALS AND STRUCTURES

MODELLING OF GLASS EXPOSED TO BLAST LOADING: STOCHASTIC FRACTURE INITIATION AND NODE SPLITTING

JONAS RUDSHAUG[1,2*], ODD STURE HOPPERSTAD[1] & TORE BØRVIK[1,2]
[1]Structural Impact Laboratory (SIMLab), Department of Structural Engineering, NTNU Norwegian University of Science and Technology, Norway
[2]Norwegian Defense Estates Agency, National Center for Protection of Buildings, Norway

ABSTRACT
In terms of mechanical behaviour, glass plates are known for their stochastic fracture initiation followed by rapid crack propagation. As structural components, glass plates exhibit linear elastic behaviour until fracture initiation. Due to microscopic surface cracks, often referred to as surface flaws, the fracture initiation behaviour varies from specimen to specimen. After fracture initiation, cracks propagate from the initiation point at speeds of approximately 1,500 m/s. These phenomena are challenging to model numerically, requiring stochastic models for fracture initiation and robust fracture models for crack propagation. In this study, we combine the glass strength prediction model for predicting fracture initiation with node splitting to propagate the resulting cracks. To evaluate the performance of this modelling approach, we compare numerical predictions with experimental data from blast-loaded monolithic and laminated glass plates.
Keywords: glass, GSPM, node splitting, blast loading.

1 INTRODUCTION

Glass is a material known for its excellent transparent properties, widely used in building facades and vehicles. In the event of an explosion close to a building or vehicle, the glass plates typically form the first protective shell of either the building or the vehicle. To serve as an effective protective shell, the glass plate must absorb as much energy as possible. Unfortunately, glass is a brittle material with stochastic fracture behaviour. Thus, the protective capabilities of the glass plate in terms of energy absorption are minor and uncertain.

Protective windows are typically constructed from a type of composite called laminated safety glass, which typically consists of two layers of glass plates and a polyvinyl butyral (PVB) interlayer. This composition offers significant advantages in energy absorption, reduced injuries during explosions and preventing burglary. While a single glass layer is prone to total failure upon cracking, laminated glass offers a fail-safe structure because the PVB interlayer maintains the integrity of the glass layers after initial failure. In the event of a window fracture, the PVB interlayer holds the glass fragments together, transforming the window from a laminated composite to a glass-reinforced polymer.

Understanding the structural behaviour of laminated glass is crucial, particularly in scenarios involving blast loading. The structural behaviour transitions through several phases: starting with the linear, elastic behaviour of the glass layers, progressing through the sequential failure of the glass layers, and culminating with the deformation and potential failure of the PVB interlayer. An important step in increasing this understanding is to model the glass plate fracture process in a sufficiently realistic manner in terms of both fracture initiation and crack propagation. Several strength prediction models with focus on the fracture initiation of glass have been introduced to reduce the glass strength uncertainties.

* *ORCID: https://orcid.org/0000-0001-5884-7624*

WIT Transactions on The Built Environment, Vol 215, © 2025 WIT Press
www.witpress.com, ISSN 1743-3509 (on-line)
doi:10.2495/SUSI250171

Rudshaug et al. [1] proposed the glass strength prediction model (GSPM), a strength prediction model based on the generation of microscopical surface flaws from the production process of float glass. The GSPM is integrated into the finite element solver LS-DYNA in the material model *MAT_GLASS_SPM in version R16 [2].

Conventional crack propagation modelling typically involves either element erosion or stiffness degradation to represent the crack path, which smears out the crack width over the element size, slowing down the crack speed. In this study, we combine the GSPM for predicting fracture initiation with the node splitting features in the Impetus Solver to propagate the resulting cracks [3]. To evaluate the performance of this modelling approach, we compare numerical predictions with experimental data from blast-loaded monolithic and laminated glass plates [4]–[8].

2 GLASS STRENGTH PREDICTION MODEL

The GSPM is a tool designed to predict fracture initiation in glass plates, integrating linear elastic fracture mechanics (LEFM) and sub-critical crack growth (SCG) through a Monte Carlo simulation framework [1]. The in-situ version of the model is integrated with finite element (FE) solvers. This integration enables real-time calculation of failure distribution during simulations, leveraging the stochastic nature of glass fracture to allow users to select specific flaw maps for failure initiation. The model provides insights into the variability of fracture initiation, offering valuable information on the repeatability of the scenarios under study.

A key feature of the GSPM is its capability to generate representative artificial flaw maps on glass surfaces. These maps include a variety of flaws characterized by their shape, depth, orientation, and location, which are critical for predicting fracture strength. The model tracks the sub-critical crack growth of each flaw based on environmental conditions and loading history, focusing specifically on mode I loading.

During the FE simulation, the GSPM generates a fracture initiation distribution from the representative sample of flaw maps, ranked from weakest to strongest. This allows users to select a flaw map to initiate failure, thereby determining the strength of the glass specimen. The model predicts scatter in fracture initiation location and timing, which provides an indication of the repeatability of the case studied. A small scatter suggests consistent strength for the given load case and geometry, enhancing the model's applicability in various design scenarios.

2.1 Implementation into the Impetus Solver

The GSPM is available as a standard material model in LS-DYNA R16, with the keyword *MAT_GLASS_SPM [1], [2]. In the LS-DYNA implementation, the GSPM predicts the fracture initiation location and timing, and activates the crack propagation features available in the glass material model *MAT_GLASS. The implementation is based on glass plates modelled with shell elements, where the outer integration points in the plate surface normal directions are located on the plate surfaces. The crack propagation features of *MAT_GLASS involve either element erosion or stiffness degradation to represent the crack path, smearing out the crack width over the element size.

An alternative way of representing the crack path is to use a node splitting technique, where the crack is represented by splitting the nodes at the element borders. The Impetus Solver is an explicit finite element solver that offers node splitting features [3]. However, it does not support deformable shell elements. For this reason, we must make some alterations in the way the glass plate surfaces are identified. The Impetus Solver offers cubic 3D elements, as illustrated in Fig. 1, where the outer integration points are close to the element surfaces. To determine which integration points that are associated with the two surfaces of a glass plate, we compare the normal vector of the three element surfaces, n_1, n_2 and n_3, related to each corner integration point to the normal vectors of the glass surfaces, n_{surf} (see Fig. 2). Further, we use the midpoints of the element surfaces associated with the glass surfaces as potential flaw locations for the generated flaw maps. The surface stress values, σ_1 and σ_2, used in the GSPM are estimated based on the mean stress values of the outermost integration point layers, $\sigma_{1,i}$ and $\sigma_{2,i}$, associated with the glass plate surfaces (see Fig. 3). We assume a linear stress distribution over the thickness and extrapolate the stress as presented in eqns (1)–(4):

$$\sigma_m = \frac{\sigma_{1,i}+\sigma_{2,i}}{2}, \tag{1}$$

$$\sigma_b = \frac{\sigma_{1,i}-\sigma_{2,i}}{2} \cdot \frac{l}{l_i}, \tag{2}$$

$$\sigma_1 = \sigma_m + \sigma_b, \tag{3}$$

$$\sigma_2 = \sigma_m - \sigma_b. \tag{4}$$

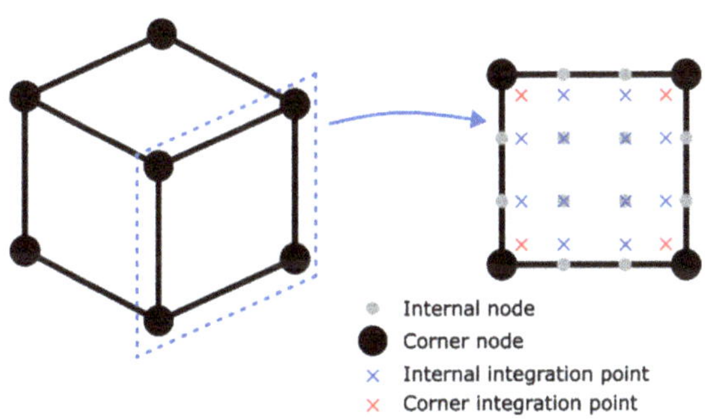

Figure 1: Illustration of a cubic 3D element showing nodes and integration points.

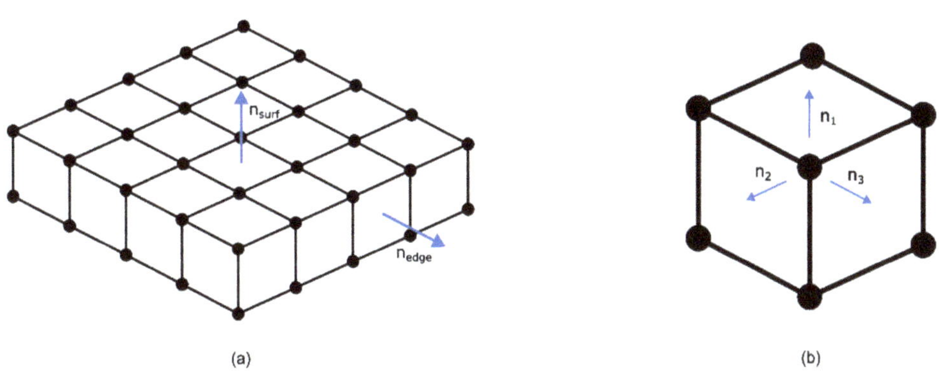

Figure 2: Illustration of the glass surface normal, n_{surf}, and edge normal, n_{edge} (a), and the element surface normal, n_1, n_2, and n_3, related to a corner integration point (b).

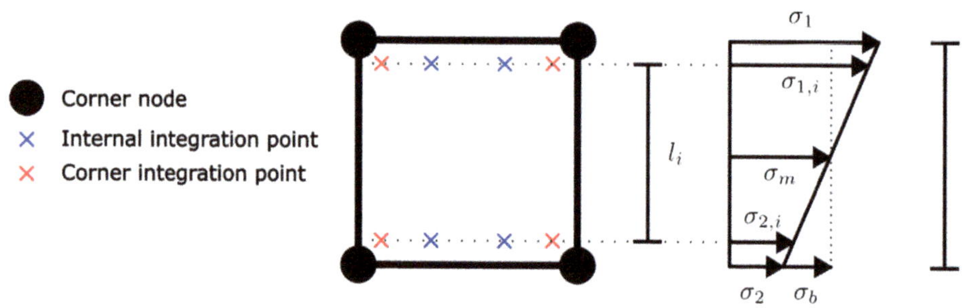

Figure 3: Illustration of the surface stress estimation used in the implementation. σ_1 and σ_2 represent the mean surface stress at surface 1 and 2 and l the plate thickness. $\sigma_{1,i}$ and $\sigma_{2,i}$ represent the mean stress at the integration point layer associated with the surface. l_i is the distance between the integration point layers. σ_m and σ_b denote the membrane and bending stress, respectively.

When the GSPM predicts fracture initiation, we initiate crack propagation from the specified element at the predicted timing. To model the fracture initiation, we split the nodes of the surface closest to the integration point where fracture initiation is predicted. Further, to initiate a sufficient number of crack paths in addition to the first fracture, we include the option to initiate fracture with a stress-based criterion relative to the predicted critical stress from the GSPM, $\sigma_{cr,GSPM}$. The stress-based criterion is controlled with the scaling parameter $\gamma > 0$, where the critical stress is set to $\gamma \cdot \sigma_{cr,GSPM}$ after the first fracture. After fracture initiation, the crack propagation process is modelled using the brittle damage model available in the Impetus Solver, *PROP_DAMAGE_BRITTLE, with the parameters presented in Table 1. In this study, the critical stress, σ_s, in *PROP_DAMAGE_BRITTLE is set to a high value to ensure that fracture initiates only from the GSPM.

Table 1: Input parameters for *PROP_DAMAGE_BRITTLE. t_s and α_s are taken from [4].

σ_s	K_c	t_s	α_s
MPa	MPa mm$^{1/2}$	s	–
200	23.7	2.0e-7	0.5

2.1.1 Validation

To validate the implementation of the GSPM into the user material interface of the Impetus Solver before coupling it to the brittle damage model, we use the experimental study on monolithic glass plates subjected to quasi-static punch loading performed by Osnes et al. [5]. The experimental study consisted of three test series with 30 monolithic glass plates in each. A wooden impactor provided the loading with loading rates of 3 mm/min, 100 mm/min and 300 mm/min. The in-plane plate dimensions were 400 mm × 400 mm and the thickness was 1.8 mm. An overview of the experimental setup and impactor dimensions are shown in Fig. 4. We model the setup in a similar way as presented by Osnes et al. [4] with cubic elements and use the GSPM parameters calibrated by Rudshaug et al. [1] for supplier 2. Both the wooden impactor and the rubber strips are modelled as linear elastic materials without any fracture models, while the glass plate is modelled as a linear elastic material with the GSPM. The failure percentile parameter, P_f, of the GSPM is set larger than one to not initiate fracture. Fig. 5 presents the discretized numerical model and the resulting strength predictions. To avoid a spurious penalty on the crack propagation paths in the diagonal direction of the elements and increase the number of possible crack paths in the glass plate, we use pentahedron elements as illustrated in comparison to the hexahedral element discretization in Fig. 6 for all glass plates in this study. The strength predictions for the three loading rates are consistent with what was found by Rudshaug et al. [1] and the experimental results of Osnes et al. [5].

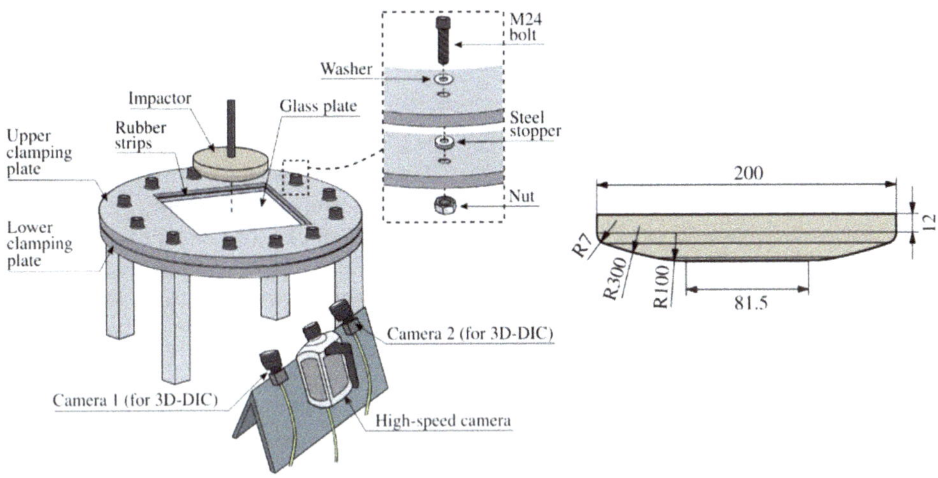

(a) (b)

Figure 4: (a) Overview of the experimental setup; and (b) Dimensions of the wooden impactor. *(Source: Adopted from [5].)*

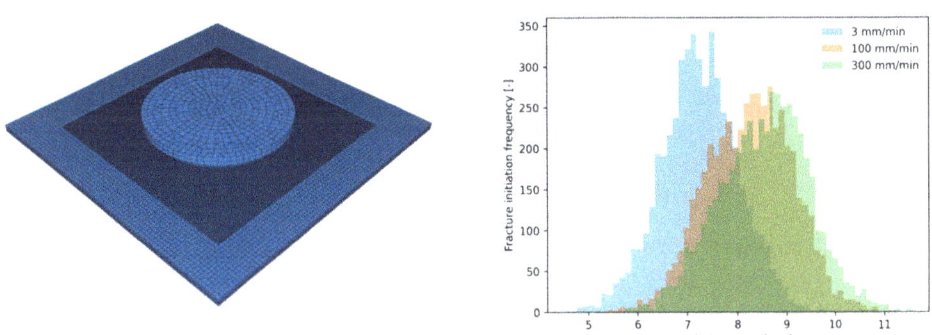

(a) (b)

Figure 5: (a) The numerical model of the quasi-static punch tests showing the element discretization; and (b) The resulting strength predictions from the GSPM.

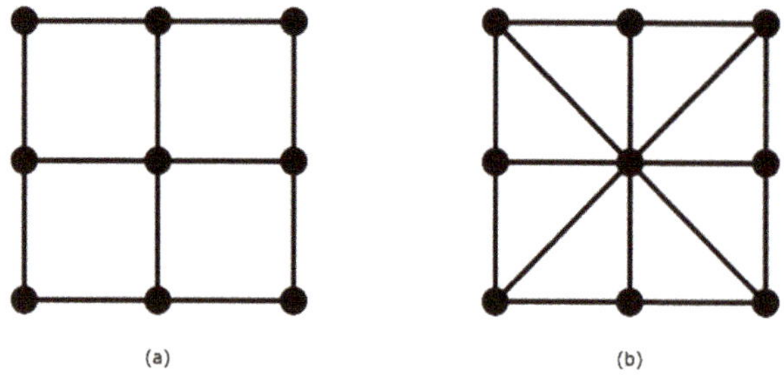

Figure 6: In-plane illustration of a hexahedral (a) compared to a pentahedral (b) discretization.

3 MONOLITHIC GLASS PLATES EXPOSED TO BLAST LOADING

Osnes et al. [6] and Brekken and Ingier [8] performed twelve experiments on monolithic glass plates of in-plane dimensions 400 mm × 400 mm and thickness 3.8 mm exposed to blast loading with three peak overpressures, ~50 kPa, ~60 kPa and ~70 kPa. In this study, we focus on the glass plates that were exposed to the ~70 kPa peak pressure blast loading. An overview of the experimental setup is illustrated in Fig. 7. We model the setup in a similar way as presented by Osnes et al. [4] with cubic pentahedron elements and use the GSPM parameters calibrated by Rudshaug et al. [1] for supplier 1.

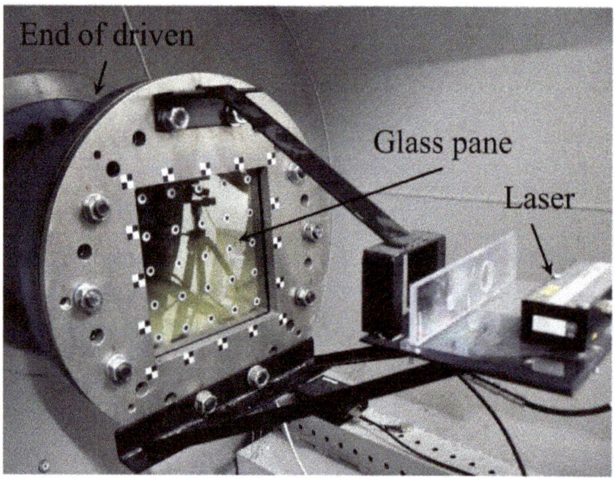

Figure 7: Overview of the experimental setup for the blast loaded glass plates. *(Source: Adopted from [8].)*

The failure percentile parameter, P_f, of the GSPM is set to 0.05 to initiate fracture for the virtual glass plate representing the strength at the 5% point of the weak tail of the predicted strength distribution. The critical stress after the first fracture is set to 50% of the predicted critical stress, i.e., $\gamma = 0.5$. Fig. 8 presents the predicted crack pattern compared to an experimental crack pattern. We note that the predicted crack pattern displays multiple features also seen experimentally with radial cracks propagating from the centre and circumferential cracks near the boundaries.

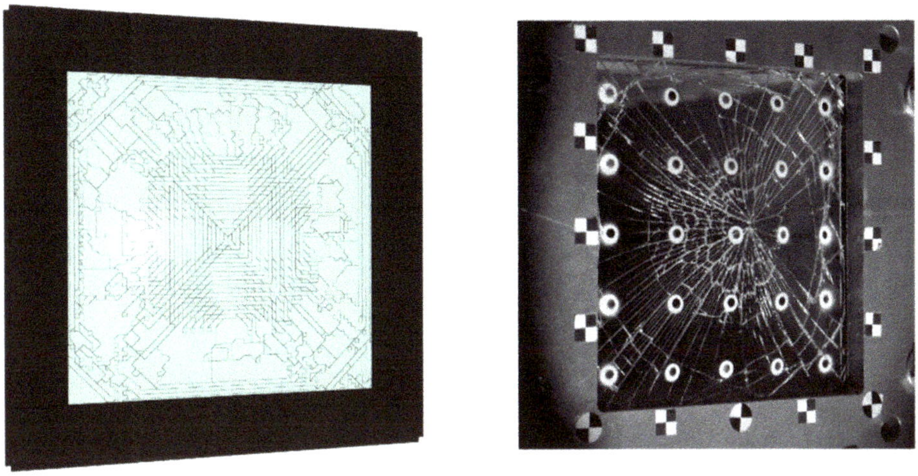

Figure 8: The predicted crack pattern with $P_f = 0.05$ (a) compared to an experimental crack pattern adopted from [7] (b).

4 LAMINATED GLASS PLATES EXPOSED TO BLAST LOADING

Osnes et al. [4] performed fifteen experiments on single-layered laminated glass plates consisting of two 3.8 mm thick glass plates and one 1.52 mm thick PVB interlayer. The in-plane plate dimensions were 400 mm × 400 mm. The single-layered laminated glass plates were exposed to blast loading in the same way as the previously discussed monolithic glass plates, with five peak overpressures, ~170 kPa, ~200 kPa, ~215 kPa, ~238 kPa and ~255 kPa. In this study, we focus on the four laminated glass plates that were exposed to the ~238 kPa peak pressure blast loading. For these laminated glass plates, fracture initiated in both glass layers around the centre of the plate. We model the setup in a similar way as presented by Osnes et al. [4] with cubic pentahedron elements and use the GSPM parameters calibrated by Rudshaug et al. [1] for supplier 1.

The failure percentile parameter, P_f, of the GSPM is set to 0.05 for the outer glass layer and 0.1 for the inner layer. The critical stress after first fracture is set to 50% of the predicted critical stress, i.e., $\gamma = 0.5$. Fig. 9 presents the predicted crack pattern compared to the experimental crack pattern for the outer layer. Also here, we note that the characteristics of the predicted crack pattern resemble what is seen experimentally, with higher crack density at the centre and corners including many radial cracks.

5 SUMMARY

We implemented the GSPM into the user material framework of the Impetus Solver, which offers higher order elements and node splitting features. To activate the crack propagation features with the node splitting technique, GSPM is coupled to the brittle damage model available in the Impetus Solver, *PROP_DAMAGE_BRITTLE. The implementation was validated using experimental data from three test series including monolithic glass plates subjected to quasi-static punch loading and monolithic and laminated glass plates exposed to blast loading. Overall, the predictions from the implemented framework compared well to the experimental data.

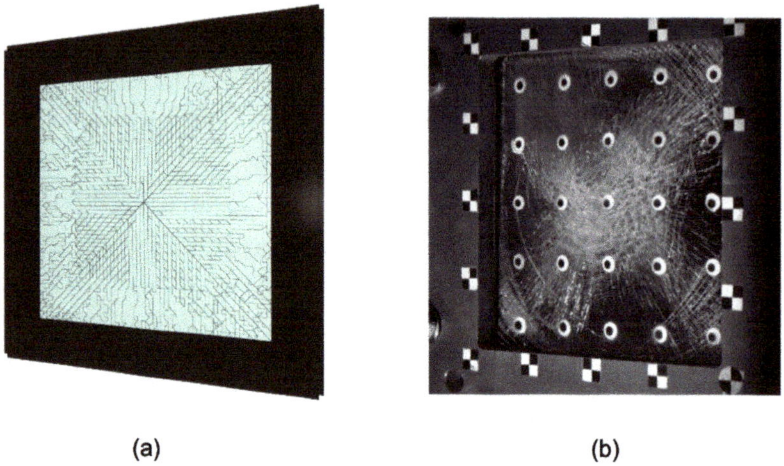

Figure 9: The predicted crack pattern with $P_f = 0.05$ for the outer glass layer and $P_f = 0.1$ for the inner glass layer (a) compared to an experimental crack pattern adopted from [4] (b).

ACKNOWLEDGEMENTS

The authors would like to acknowledge Mr. Henrik Lam and Dr. Lars Olovsson for their support in the implementation of the glass strength prediction model into the Impetus Solver.

REFERENCES

[1] Rudshaug, J., Aasen, K.O., Hopperstad, O.S. & Børvik, T., A physically based strength prediction model for glass. *International Journal of Solids and Structures*, **285**, 112548, 2023. https:/doi.org/10.1016/j.ijsolstr.2023.112548.
[2] Livermore Software Technology (LST), An ANSYS Company, LS-DYNA Manual Volume II R16, 2025.
[3] Impetus Afea, Impetus Afea Solver. https://www.impetus.no/. Accessed on: 22 Feb. 2025.
[4] Osnes, K., Holmen, J.K., Hopperstad, O.S. & Børvik, T., Fracture and fragmentation of blast-loaded laminated glass: An experimental and numerical study. *International Journal of Impact Engineering*, **132**, 103334, 2019. https:/doi.org/10.1016/j.ijimpeng.2019.103334.
[5] Osnes, K., Hopperstad, O.S. & Børvik, T., Rate dependent fracture of monolithic and laminated glass: Experiments and simulations. *Engineering Structures*, **212**, 110516, 2020. https:/doi.org/10.1016/j.engstruct.2020.110516.
[6] Osnes, K., Børvik, T. & Hopperstad, O.S., Testing and modelling of annealed float glass under quasi-static and dynamic loading. *Engineering Fracture Mechanics*, **201**, pp. 107–129, 2018. https:/doi.org/10.1016/j.engfracmech.2018.05.031.
[7] Bratsberg, O.S. & Kolsaker, M.R., Modelling of window glasses exposed to blast loading. Master's thesis, Norwegian University of Science and Technology, 2017.
[8] Brekken, K.A. & Ingier, P.T., Modelling of window glasses exposed to blast loading. Master's thesis, Norwegian University of Science and Technology, 2016.

DEFORMATION MODES OF GRADIENT HEXACHIRAL AUXETICS UNDER IMPACT LOADINGS

XIN LI[1] & CHUANQING CHEN[2]
[1]Nanjing University of Science and Technology, Nanjing, China
[2]Nanjing University, Nanjing, China

ABSTRACT

Chiral structures, a special type of negative Poisson's ratio structures, possess excellent energy absorbing capabilities and hold great promise in the field of impact protection. Compared with homogeneous structures, the gradient design of chiral structures can further optimize and enhance their energy absorbing capabilities. In this work, we experimentally and numerically investigated the deformation modes of gradient hexachiral auxetic metamaterials under quasi-static, dynamic and blast loadings. Five gradient design strategies were employed in the test, including uniform (UG), positive (PG), negative (NG), centre positive (CPG), and centre negative gradients (CNG). The effects of gradient distribution on deformation/failure modes under different loading conditions were analysed. Plane impact experimental results revealed that the gradient design can mitigate the shear deformation of hexachiral auxetics, which efficiently remain its negative Poisson's ratio effect under dynamic loading. Compare with the quasi-static loading, earlier failure of auxetics was observed under dynamic loads due to the strain rate and inertial effect, which diluted their negative Poisson's ratio effect and energy absorption capability. Among various gradient designs, the auxetics with negative gradient configurations exhibit the best energy absorption capability under dynamic loadings. Under blast loads, the various gradient types of auxetic cores exhibited distinct compression deformation behaviours. The hexachiral cores in sandwich panels with larger radius nodes on the blast surface are more likely to undergo compression deformation. The NG and CNG cores, featuring nodes resistant to flattening deformation, are more effective in generating NPR effect, resulting in a larger impact resistance. In addition, by using finite element method, the deformation mods of gradient hexachiral auxetics are further investigated under higher impact velocities and blast loadings to reveal loading rate effects on their deformation modes. This work offers profound insights into enhanced performance mechanisms and provides avenues for optimizing structural design of auxetic metamaterials.

Keywords: hexachiral, gradient, impact and blast loadings, deformation mode.

1 INTRODUCTION

Chiral structures, a unique class of auxetic mechanical metamaterials, are characterized by their negative Poisson's ratio (NPR) behaviour, which manifests as transverse expansion under longitudinal compression or shrinkage under tension [1]. This distinctive deformation pattern arises from their microstructural architecture rather than the intrinsic properties of the constituent materials. As a result, chiral structures exhibit a range of advantageous mechanical properties, including enhanced fracture resistance, superior shear resistance, improved indentation resistance, and exceptional energy absorption capabilities [2], [3]. These attributes make chiral structures highly promising for applications in impact protection, blast mitigation, and energy dissipation during extreme loading events [4], [5].

Hexachiral structures, first proposed by Prall and Lakes [6], exhibit an in-plane Poisson's ratio of −1, which has been both theoretically and experimentally validated. This unique behaviour, characterized by full-wave-shaped ligament deformation under loading, contributes to their superior energy absorption capabilities compared to other chiral configurations, such as anti-chiral structures, which exhibit half-wave-shaped deformation [7].

WIT Transactions on The Built Environment, Vol 215, © 2025 WIT Press
www.witpress.com, ISSN 1743-3509 (on-line)
doi:10.2495/SUSI250181

Recent advancements in chiral structure design have further diversified their applications [8]. In addition to traditional two-dimensional (2D) chiral structures, three-dimensional (3D) chiral configurations [9]–[12] have been developed, offering enhanced mechanical properties and dynamic performance. Moreover, the introduction of gradient design strategies has opened new avenues for optimizing the mechanical behaviour of chiral structures [13]–[15]. Gradient designs, which involve the integration of sub-structures with varying characteristics, such as cell wall thickness, material composition, or geometric parameters, have been shown to improve stiffness, strength, and energy absorption capabilities [16], [17]. These functionally graded auxetic structures exhibit superior performance under dynamic loading conditions, making them ideal for applications requiring impact resistance and energy dissipation.

Despite the growing interest in chiral structures, particularly in the context of gradient designs, there remains a significant research gap in understanding their deformation mechanisms and energy absorption capabilities under different loading conditions. Most existing studies have focused on uniform structures or have been limited to numerical simulations, often neglecting the effects of material damage and strain rate sensitivity. Furthermore, while gradient designs have shown promise in enhancing mechanical properties, their performance under extreme loading conditions, such as high-velocity and blast impacts, has not been thoroughly investigated.

In this work, based on our previous studies [14], [15], we address these gaps by conducting a comprehensive experimental and numerical investigation into the mechanical responses of gradient hexachiral auxetic metamaterials under quasi-static, impact, and blast loadings. Five distinct gradient design strategies – uniform (UG), positive (PG), negative (NG), centre positive (CPG), and centre negative (CNG) gradients – were employed to fabricate hexachiral structures using 3D printing technology. The deformation modes and Poisson's ratio effects of these structures were evaluated under plane compression conditions with various impact velocities. The deformation process and final deformation modes of these structures, when used as cores in sandwich panels under blast loads, were also studied. Additionally, based on the validated finite element (FE) models, we performed numerical analyses for higher impact velocities and more intense explosive loads.

2 GRADIENT AUXETIC CONFIGURATIONS

The hexachiral [6] configuration unit consists of six ligaments, each with a length denoted as L, and these are connected tangentially to a circular node with an outer radius of r, as illustrated in Fig. 1. The wall thicknesses of the node and the ligament are defined as t_n and t_l, respectively. The overall height of the structure is H. In this work, based on the uniform gradient (UG) configuration, an in-depth elucidation of the design methodologies for gradient hexachiral auxetics will be presented, encompassing positive gradient (PG), negative gradient (NG), centre positive gradient (CPG), and centre negative gradient (CNG). The gradient design principle is based on the ratio of node radii between adjacent layers, which are delineated by dashed lines and labelled as $n+1$, n and $n-1$ from top to bottom. The radii of these layers are denoted as r_{n+1}, r_n and r_{n-1}, with R representing the distance between node centres in adjacent layers. To quantify the gradient, a gradient factor ω is introduced [14], [15], defined as the ratio of node radii between upper and lower adjacent layers:

$$\omega = r_n / r_{n-1}. \quad (1)$$

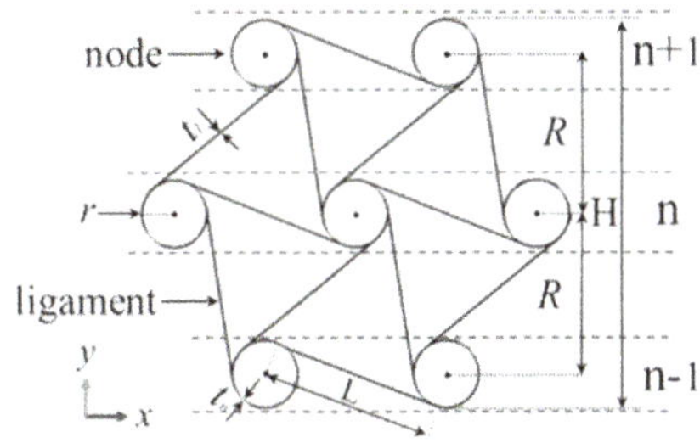

Figure 1: Geometric parameters of hexachiral auxetic cell structure [14].

The UG design is characterized by ω equal to 1, indicating uniform node radii throughout the structure. The PG design features ω greater than 1, with node radii decreasing from top to bottom. In contrast, the NG design has ω less than 1, with increasing radii values. The CPG and CNG designs are variations of the PG and NG configurations, respectively, with node radii exhibiting rotational symmetry around the structure's central point, leading to reciprocal gradient factors between the upper and lower sections. For CPG and CNG, the gradient factor above the central axis is considered the representative value.

3 EXPERIMENTAL PROCEDURE

3.1 Quasi-static and dynamic loadings

In this section, quasi-static and dynamic compression tests are conducted under plane compression velocities of 10 mm/min and 5 m/s for analysing the deformation modes of gradient hexachiral auxetics. To enhance the capture of detailed deformation characteristics during compression, the auxetics were designed with 5 node layers, as illustrated in Fig. 2. Arrows indicate the progression from larger to smaller node radii. The UG configuration served as the baseline, with specific parameters set as follows: r = 4.8 mm, $t_l = t_n$ = 1.22 mm, L/r = 3.65 and H = 73 mm. For the PG, NG, CPG, and CNG configurations, their respective ω values were set to be 1.1, 0.91, 1.1, 0.91, respectively. The cell wall thicknesses for PG, NG, CPG, and CNG were determined to be 1.22, 1.22, 1.20, and 1.26 mm, respectively. Detailed specimen information and experimental methods can be found in Chen et al. [14].

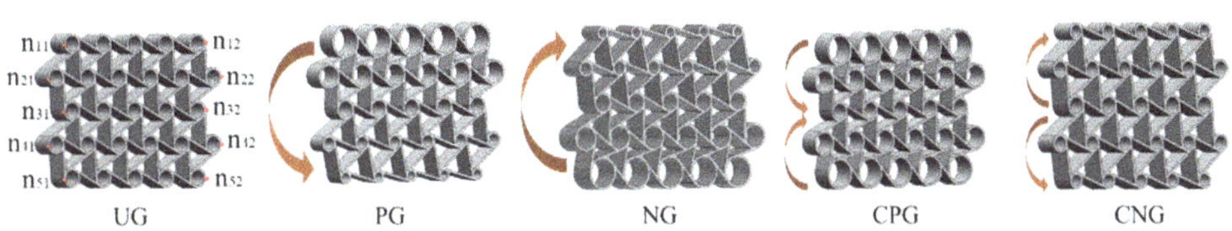

Figure 2: Different gradient types of hexachiral structures for plane compression test.

Fig. 3 depicts the deformation profiles of gradient hexachiral auxetics at vertical strains (ε_y) ranging from 0.1 to 0.6 under quasi-static load. At ε_y= 0.1, the deformation is localized mainly to the central node layer, where node rotation and ligament bending cause contraction. This inward shrinkage is the primary contributor to the NPR effect observed at this stage.

Deformation amplitudes differ across configurations, with UG and CNG showing the most significant changes, followed by PG and NG, and CPG exhibiting the least deformation. Larger node radii, which increase curvature, reduce bending stiffness, promoting node flattening in PG, NG, and CPG configurations. This results in shorter effective compression distances at the same strain, reducing rotational deformation in these designs. At the ε_y of 0.2 and 0.3, a more pronounced global NPR effect is observed, primarily due to the horizontal contraction caused by the rotational deformation of nodes in layers beyond the central one. The voids between the ligaments and nodes are crucial in facilitating the NPR effect in hexachiral auxetics. At ε_y of 0.4, these voids have largely disappeared, with compression occurring within the nodes themselves. Detailed observation indicates that all specimens' horizontal deformation at a strain of 0.5 exceeds that at 0.4, indicating a reduction in the NPR effect's intensity. At around a compressive strain of 0.6, the voids within the circular nodes are nearly completely compressed, leading to a fully densified structure.

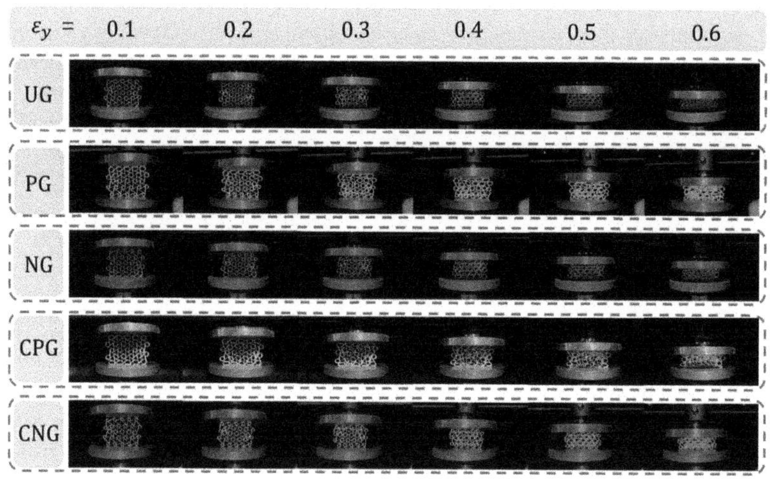

Figure 3: Deformation modes of diverse gradient hexachiral auxetics under quasi-static loading [14].

In addition, the Poisson's ratio υ of auxetics is defined as a negative ratio of the average horizontal strains of each layer to the vertical strain. At the vertical strain range 0.1–0.5, it can be written as

$$\upsilon = \sum_{i=1}^{5} {\varepsilon_x}^i / (5\varepsilon_y), \tag{2}$$

where ${\varepsilon_x}^i$ is the horizontal strain of ith node layer, which can be calculated by the ratio of the changing horizontal width of each node layer at varying compression levels to its initial width. The width of each node layer is ascertained by measuring the distance between the centre points of the side nodes at the layer's edges, marked by dots of UG in Fig. 2.

The quasi-static Poisson's ratio of different gradient hexachiral auxetics are listed in Table 1. A more negative value signifies a more pronounced NPR effect. Overall, the CNG configuration exhibits the most pronounced NPR effect, while the CPG configuration demonstrates the least. And the NPR effect in PG and NG is quite similar, slightly inferior to that of UG.

Table 1: Poisson's ratio values of gradient hexachiral auxetics under quasi-static load.

Gradient	Strain					
	0.1	0.2	0.3	0.4	0.5	0.6
UG	−0.40	−0.54	−0.53	−0.37	−0.22	−0.02
PG	−0.32	−0.49	−0.46	−0.32	−0.20	−0.05
NG	−0.35	−0.47	−0.39	−0.29	−0.16	−0.06
CPG	−0.08	−0.15	−0.12	−0.07	−0.04	0
CNG	−0.68	−0.58	−0.52	−0.43	−0.29	−0.04

Fig. 4 presents the deformation profiles of various gradient hexachiral auxetics at ε_y ranging from 0.1 to 0.6, subjected to an impact velocity of 5 m/s. At ε_y of 0.1, the deformation modes are similar to those under quasi-static loading (Fig. 3), with node rotation in the central layer and flattening of large-radius nodes. However, dynamic loading introduces inertial effects and stress wave propagation, leading to localized stress concentration and additional stress on the structure. As a result, structural damage occurs at lower strains (ε_y = 0.2), causing significant changes in the deformation modes compared to quasi-static conditions. Particularly in the UG configuration, shear deformation along the diagonal is observed, caused by the quasi-parallelogram distribution of nodes during auxetic deformation, allowing shear force transmission. Other configurations do not show shear deformation, despite experiencing local damage. This is likely due to their distinct node radius distributions, which form a quasi-trapezoidal structure, enhancing stability and reducing shear deformation. Localized damage in hexachiral auxetics weakens the NPR

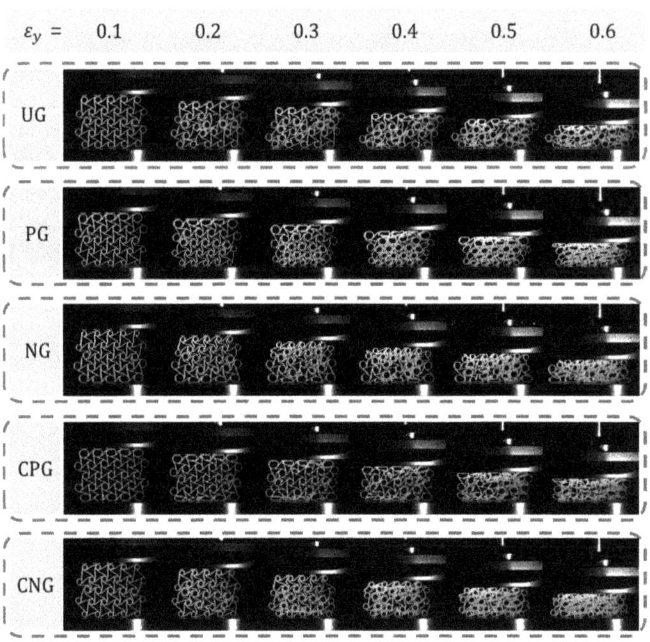

Figure 4: Deformation modes of diverse gradient hexachiral auxetics under dynamic loading [14].

effect, particularly in the UG configuration, where shear deformation splits the structure into two parts, deforming in opposite directions.

Table 2 presents the Poisson's ratio of different gradient hexachiral auxetics under dynamic compression. The NPR effect of the CNG configuration is comparable to that of the PG configuration, followed by the NG configuration, while the UG and CPG configurations exhibit the weakest NPR effect. In addition, through detailed numerical comparisons with quasi-static condition, the overall Poisson's ratio curves for UG, NG and CNG experience increases, signifying a weakening of the NPR effect under v = 5 m/s in these three configurations. It can be attributed to occurrence of the localized structure damage caused by inertial effects and wave propagation.

Table 2: Poisson's ratio values of gradient hexachiral auxetics under 5 m/s.

Gradient	Strain					
	0.1	0.2	0.3	0.4	0.5	0.6
UG	−0.55	−0.17	−0.05	0.07	0.11	0.24
PG	−0.32	−0.46	−0.48	−0.34	−0.20	0.24
NG	−0.34	−0.40	−0.32	−0.17	−0.05	0.10
CPG	−0.13	−0.16	−0.11	−0.06	0.02	0.03
CNG	−0.63	−0.48	−0.42	−0.28	−0.16	0.04

3.2 Blast loading

The compression modes of hexachiral auxetic cores under explosion load were investigated, with Fig. 5 illustrating the geometry and dimensions of the sandwich panel specimen and their gradient hexachiral cores. The face sheets, attached to both sides of the core, had a length (L) and width (W) of 300 mm and a typical thickness of 0.4 mm. The cores featured hexachiral structures of various gradient types with three node layers. These cores had a fixed height (H) of 21 mm, which is a common size used in practical engineering. The UG configuration was chosen as a reference standard, with the following parameters: the outer radius (r) of the hexachiral nodes was 2.4 mm, and both the ligaments (t_l) and nodes (t_n) had a thickness of 0.838 mm. For the PG, NG, CPG, and CNG configurations, ω values were set to 1.25, 0.8, 1.25, and 0.8, respectively. Correspondingly, the cell wall thicknesses (t) were 0.840, 0.840, 0.822, and 0.854 mm, respectively. Detailed specimen information and experimental methods can be found in Chen et al. [15].

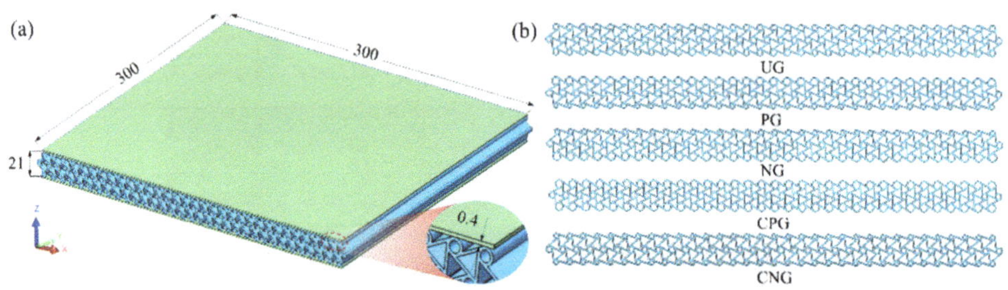

Figure 5: (a) The 3D structure of auxetic core sandwich panel [15]; and (b) Different gradient types of hexachiral cores for explosion test.

Fig. 6 shows the central cross-sections of five gradient hexachiral core specimens subjected to blast loads equivalent to 42 g of TNT. The specimens were exposed to impulses of 31.29, 36.90, 30.45, 28.02, and 37.80 Ns measured in blast experiment. Deformation was concentrated in three main areas: the central region (highlighted by red dotted rectangles) and the clamped regions on both sides (marked by red solid lines). The clamped areas deformed primarily due to the constraining effect of the steel frame, limiting the specimen's movement. In contrast, the central region experienced the most significant deformation, as it was directly affected by the blast load. The NPR effect was assessed by observing node rotation and ligament bending around them. Except for the PG configuration, all cores experienced deformation confined to the first ligament layer, restricting the NPR effect. The UG-3 and CPG-3 specimens exhibited similar compressive deformation, with node flattening near the front face sheet and bending in the first ligament layer. Little plastic deformation was observed in the deeper layers, and despite a smaller impulse, CPG-3 showed more node flattening than UG-3. NG-3, with an impulse similar to UG-3, demonstrated superior compression resistance, with minimal node deformation and slight bending in the first ligament layer, much less than UG-3. PG-3 experienced the most deformation, including extensive flattening and bending in both the first and second layers of nodes and ligaments, along with multiple joint fractures, due to a higher impulse. In contrast, CNG-3, subjected to a similar impulse as PG-3, showed less deformation, with bending primarily in the first ligament layer and fewer joint fractures. The findings indicate that node size significantly influences blast compression deformation. Configurations with larger nodes adjacent to the front face sheet, such as PG and CPG, were more prone to compression deformation. Conversely, configurations with smaller nodes, such as NG and CNG, were less susceptible to compression deformation. This is because larger nodes have a greater curvature radius, leading to lower bending stiffness, making larger-radius nodes more prone to deformation under external forces.

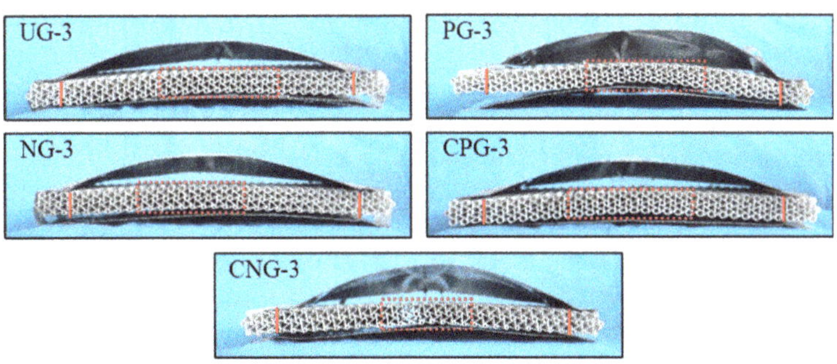

Figure 6: Compression deformation of specimens with different gradient auxetic cores [15].

4 SIMULATIONS

To gain deeper insights into the deformation mechanisms of gradient hexachiral auxetic structures and their compressive behaviour under higher impact velocity and blast loading, FE simulations were performed. Prior studies established validated FE models for plane compression using Abaqus [14] and for blast loading using LS-DYNA [15]. Building on these validated frameworks, simulations were conducted under planar compression at 20 m/s and 50 m/s, as well as blast loads equivalent to 70 g and 100 g TNT, to analyse the deformation modes of gradient hexachiral auxetics under extreme dynamic conditions. One noteworthy aspect of the simulations was the approach to material failure: it was considered in the planar compression simulations but excluded from the blast loading simulations. This decision was rooted in experimental observations showing that significant structural damage occurred during planar compression tests, while blast loading led to damage mainly seen as tearing between printed layers, with minimal in-plane structural damage. Two key reasons were identified for this phenomenon: (1) The wall thickness to node radius (or ligament length) of the specimens tested under blast loading was relatively large, leading to high stiffness, which made them less prone to in-plane structural damage. (2) Applied blast loading is a localized type of load, which more easily causes shear or tensile tearing between layers when subjected to a 3D printed plate. Here, we primarily focus on in-plane compression modes.

4.1 FE results

4.1.1 Impact loading

Numerical simulations of gradient hexachiral structures subjected to high-velocity planar compression revealed distinct deformation patterns. Figs 7 and 8 depict the structural response at compressive strains ranging from 0.1 to 0.5 under impact velocities of 20 m/s and 50 m/s, respectively. At 20 m/s, damage initiation occurred at significantly smaller strains (< 0.1) compared to lower-impact scenarios as shown in Figs 3 and 4, accompanied by a pronounced deformation gradient across the structure – a phenomenon attributed to the micro-inertial effect. This effect highlights how dynamic loading accelerates localized deformation and stress redistribution. The geometry of nodal regions played a critical role in governing structural integrity. Structures with larger node radii near the impactor (e.g., UG, PG, CPG) exhibited severe damage, whereas designs with smaller radii (e.g., NG, CNG) demonstrated reduced nodal failure. This underscores the sensitivity of hexachiral auxetics to node geometry, as their energy dissipation primarily relies on node rotation and ligament bending. Node damage may disrupt the quasi-static constitutive relationships, altering failure modes and energy absorption mechanisms, which ultimately impacts macroscopic mechanical performance. A key observation was the influence of gradient design of hexachiral structures on deformation uniformity. The NG structure, characterized by a stiffness gradient decreasing from top to bottom, achieved more homogeneous deformation. The stiffer upper region absorbed the majority of the impact energy, while the compliant lower region accommodated deformation, mitigating stress concentration and delaying structural collapse. In contrast, the CNG structure, despite sharing identical stiffness near impactor region with NG, suffered severe localized damage due to its uniformly high stiffness in the lower region. This rigidity restricted structure deformation and energy dissipation, leading to stress accumulation and premature failure near the impactor.

Under a high-velocity impact of 50 m/s, all gradient hexachiral structures exhibit a distinct layered 'I'-shaped deformation mode that propagates sequentially from the impact end to the fixed end (Fig. 8). This deformation behaviour arises from the synergistic interaction between inertial effects and stress wave propagation. At the impact end, localized inertial resistance causes rapid stress accumulation, forming a crushed zone (evident at a strain of 0.1), while the delayed arrival of stress waves temporarily preserves the undeformed state in distal regions, dynamically separating the structure into a 'crushed zone' and a 'delayed deformation zone'. It can be anticipated that further increasing the impact velocity would narrow the gap between the stress wave propagation speed and the impact velocity, dividing the structure into crushed and delayed deformation zones at the shock front. Under such

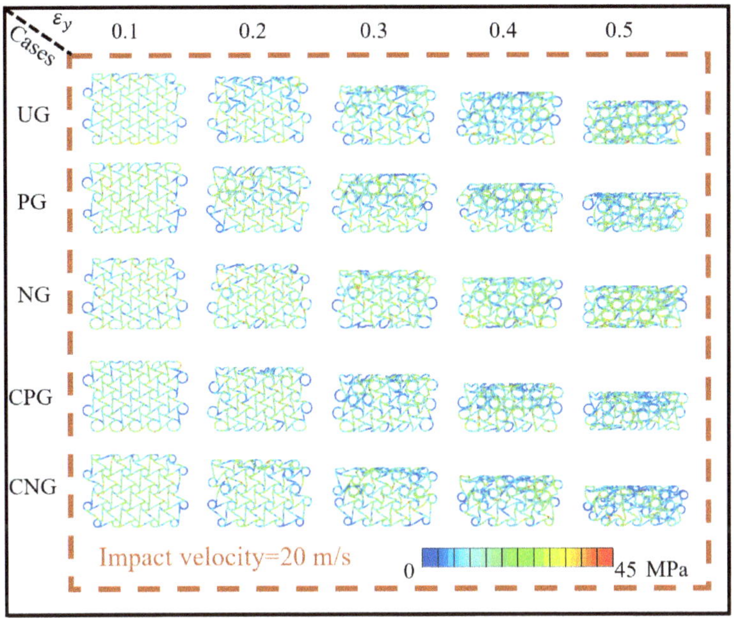

Figure 7: Compression deformation of specimens with different gradient auxetic cores under impact velocity of 20 m/s.

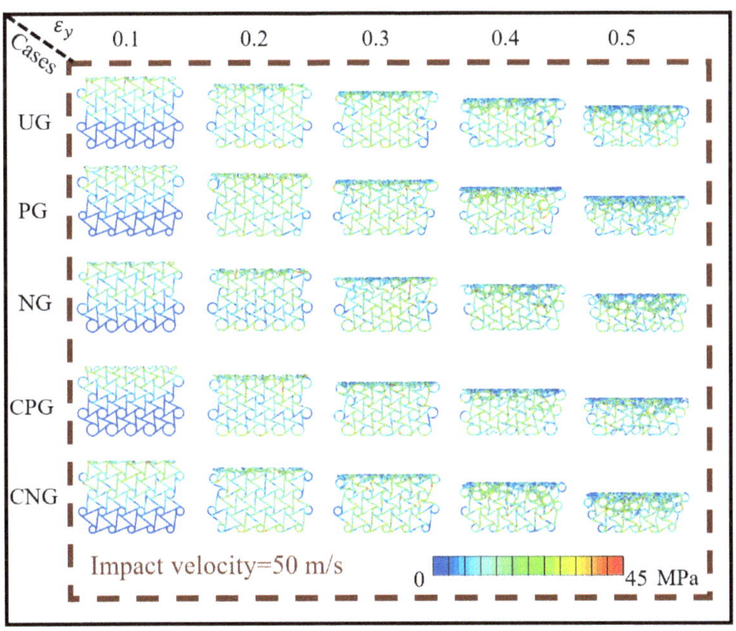

Figure 8: Compression deformation of specimens with different gradient auxetic cores under impact velocity of 50 m/s.

conditions, the crushed zone experiences significantly higher stress concentrations, with a discontinuous stress and deformation behaviour across the whole structure.

4.1.2 Blast loading

This section investigates the dynamic response of hexachiral core sandwich panels under 70 g and 100 g TNT blast loads using validated FE models. The deformation processes and final modes are illustrated in Figs 9 and 10. The deformation processes of PG specimen under 100 g TNT blast load are illustrated in Fig. 9. The blast shockwave, propagating as compressive waves, initially impacts the front face sheet, inducing transient deformation (Fig. 9(a)). Subsequently, interfacial separation occurs between the front face sheet and the core due to their divergent deformation rates (Fig. 9(b)). During this phase, the core and the back-face sheet remain coupled, moving synchronously to the maximum displacement position without significant separation (Fig. 9(c)). Afterward, the core and the back panel rebound together. In the rebound stage, the rebound amplitudes of these two components are smaller compared to the maximum compression displacement and the compressed core retaining permanent plastic deformation. This is in contrast to the larger rebound observed under smaller blast loads, as reported in previous literature, mainly due to the more significant plastic deformation under larger blast loads. Notably, complete separation between the core and back face sheet emerges during rebound due to rebound velocity mismatch (Fig. 9(d)). Ultimately, the structure stabilizes in a residual deformation state through damping oscillations.

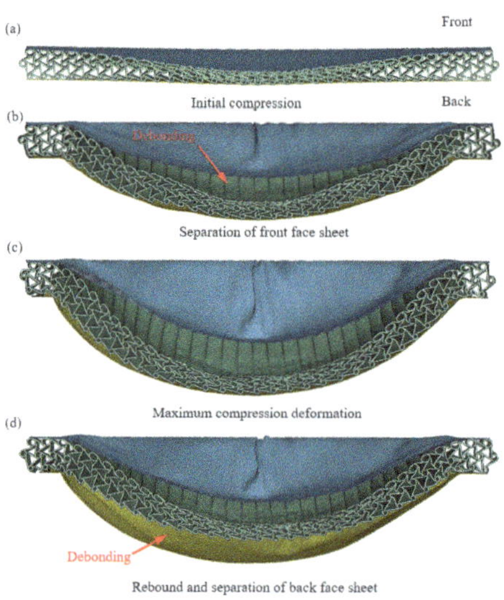

Figure 9: Cross section profiles of deformation progress under blast load (100 g TNT) of PG specimen.

Fig. 10 provides the final deformation modes of different gradient hexachiral cores under 70 g (represented by '1') and 100 g (represented by '2') TNT blast loads. The frontal (δ_1) and back (δ_2) displacements of core are measured, and then the core compression amounts $\Delta\delta$ can be calculated as $\delta_1 - \delta_2$. Under 70 g TNT blast load, The CPG core exhibits the largest compression ($\Delta\delta$ = 9.29 mm), followed by UG (8.70 mm), PG (8.10 mm), NG

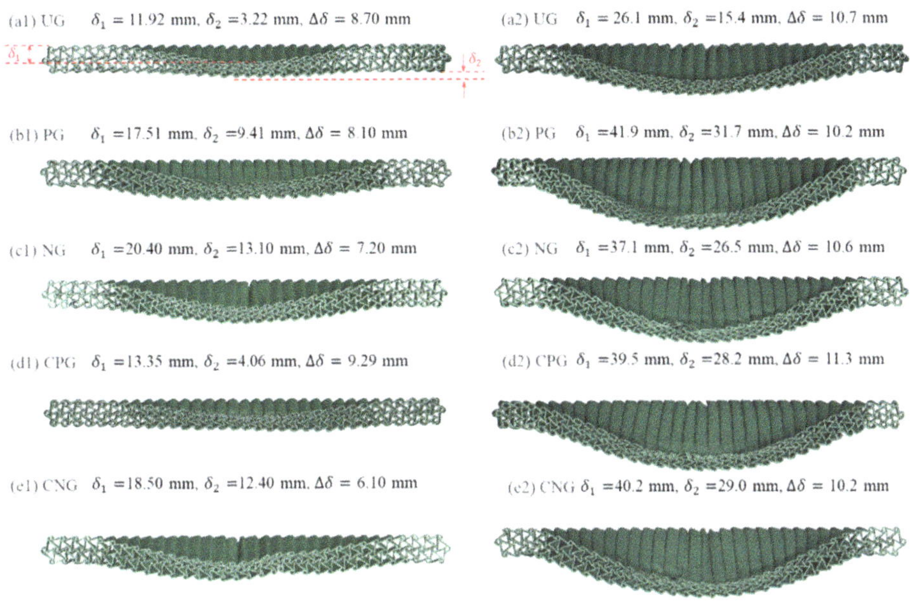

Figure 10: Compress deformation of cores with different gradient strategies: (1) 70 g [15]; and (2) 100g TNT blast loads.

(7.20 mm), and CNG (6.10 mm). While UG, PG, and CPG cores undergo greater compression, their deformation is dominated by first-layer node flattening, which undermines the rigid-node assumption critical for achieving the NPR effect in chiral auxetics. This flattening restricts NPR behaviour by limiting node rotation and ligament bending – key mechanisms for auxetic. Conversely, NG and CNG cores, with smaller node compression, preserve nodal rigidity, enabling pronounced NPR deformation through coordinated node rotation and ligament bending. The enhanced NPR effect in NG and CNG cores correlates with higher structural stiffness.

Under the 100 g TNT blast load, the compression and compressed regions of the cores increased compared to the 70 g TNT load. However, the compression amounts ($\Delta\delta$) across the different gradient hexachiral cores did not exhibit significant differentiation. The CPG core having the highest compression ($\Delta\delta$ = 11.3 mm), while the compression values for the other structures fell within the range of 10.2 to 10.7 mm. This is primarily due to the larger blast load causing plastic compression behaviour across the entire node layers of the core. Notably, the gradient cores exhibited significantly larger plastic deformation regions compared to the UG structure, which has a uniform node distribution. This difference is reflected in the δ_1 values: UG has δ_1 = 26.1 mm, while the largest value, observed in PG, is δ_1 = 41.9 mm, showing a 41.3% difference. A larger plastic deformation area may imply greater energy dissipation, which ultimately enhances the protective performance of the structure.

5 CONCLUSION

Gradient strategies play significant roles in the deformation modes of hexachiral auxetics under different loading conditions. Dynamic loading introduces earlier and more structural damage than quasi-static conditions, reducing NPR effects. Gradient design can mitigate the shear deformation under dynamic loadings. In both quasi-static and dynamic loading conditions, the CNG configuration exhibits a substantial NPR effect, while the CPG configuration performs the least. Under blast loads, configurations with larger nodes adjacent to the front face sheet are more prone to compression deformation. Numerical results show that at a 20 m/s impact velocity, the NG gradient design promotes more uniform deformation across the structure, while other designs exhibit a gradient deformation pattern. Under higher impact loads (50 m/s), all structures exhibit a sequential 'I'-shaped collapse mode driven by inertial effects and stress wave propagation. Similarly, under a larger 100 g TNT blast load, the cores experience more severe compression, with PG, NG, CPG, and CNG showing larger plastic deformation regions.

ACKNOWLEDGEMENT

This work was supported by the National Natural Science Foundation of China (Grant No. 12172164).

REFERENCES

[1] Evans, K.E., Nkansah, M.A., Hutchinson, I.J. & Rogers, S.C., Molecular network design. *Nature*, **353**, p. 124, 1991. https://doi.org/10.1038/353124a0.

[2] Chen, C., Airoldi, A., Caporale, A.M., Sala, G. & Yin, X., Impact response of composite energy absorbers based on foam-filled metallic and polymeric auxetic frames. *Composite Structures*, **331**, 117916, 2024. https://doi.org/10.1016/j.compstruct.2024.117916.

[3] Airoldi, A., Novak, N., Sgobba, F., Gilardelli, A. & Borovinšek, M., Foam-filled energy absorbers with auxetic behaviour for localized impacts. *Materials Science and Engineering: A*, **788**, 139500, 2020. https://doi.org/10.1016/j.msea.2020.139500.

[4] Linforth, S.J., Auxetic armour system for protection against soil blast loading. The University of Melbourne, 2020.

[5] Imbalzano, G., Tran, P., Ngo, T.D. & Lee, P.V., Three-dimensional modelling of auxetic sandwich panels for localised impact resistance. *Journal of Sandwich Structures and Materials*, **19**, pp. 291–316, 2017. https://doi.org/10.1177/1099636215618539.

[6] Prall, D. & Lakes, R.S., Properties of a chiral honeycomb with a Poisson's ratio of −1. *International Journal of Mechanical Sciences*, **39**, pp. 305–314, 1997. https://doi.org/10.1016/S0020-7403(96)00025-2.

[7] Alderson, A. et al., Elastic constants of 3-, 4- and 6-connected chiral and anti-chiral honeycombs subject to uniaxial in-plane loading. *Composites Science and Technology*, **70**, pp. 1042–1048, 2010. https://doi.org/10.1016/j.compscitech.2009.07.009.

[8] Wu, W., Hu, W., Qian, G., Liao, H., Xu, X. & Berto, F., Mechanical design and multifunctional applications of chiral mechanical metamaterials: A review. *Materials and Design*, **180**, 107950, 2019. https://doi.org/10.1016/j.matdes.2019.107950.

[9] Caporale, A.M., Airoldi, A. & Novak, N., A novel body centered cubic 3D auxetic chiral geometry. *Smart Mater. Struct.*, **34**, 015050, 2024. https://doi.org/10.1088/1361-665X/ad9dc9.

[10] Jiang, Y. & Li, Y., Novel 3D-printed hybrid auxetic mechanical metamaterial with chirality-induced sequential cell opening mechanisms. *Advanced Engineering Materials*, **20**, 1700744, 2018. https://doi.org/10.1002/adem.201700744.

[11] Fu, M., Liu, F. & Hu, L., A novel category of 3D chiral material with negative Poisson's ratio. *Composites Science and Technology*, **160**, pp. 111–118, 2018. https://doi.org/10.1016/j.compscitech.2018.03.017.

[12] Frenzel, T., Kadic, M. & Wegener, M., Three-dimensional mechanical metamaterials with a twist. *Science*, **358**, pp. 1072–1074, 2017. https://doi.org/10.1126/science.aao4640.
[13] Sadikbasha, S. & Pandurangan, V., Crashworthiness of auxetic hexachiral core under in-plane loading. *Materials Today: Proceedings*, 2023. https://doi.org/10.1016/j.matpr.2023.02.267.
[14] Chen, C. et al., Quasi-static and dynamic responses of gradient hexachiral auxetics: Experimental and numerical analysis. *Materials Today Communications*, **41**, 110670, 2024. https://doi.org/10.1016/j.mtcomm.2024.110670.
[15] Chen, C., He, Y., Xu, R., Gao, C., Li, X. & Lu, M., Dynamic behaviors of sandwich panels with 3D-printed gradient auxetic cores subjected to blast load. *International Journal of Impact Engineering*, **188**, 104943, 2024. https://doi.org/10.1016/j.ijimpeng.2024.104943.
[16] Qi, D., Lu, Q., He, C., Li, Y., Wu, W. & Xiao, D., Impact energy absorption of functionally graded chiral honeycomb structures. *Extreme Mechanics Letters*, **32**, 100568, 2019. https://doi.org/10.1016/j.eml.2019.100568.
[17] Qi, C., Jiang, F. & Yang, S., Advanced honeycomb designs for improving mechanical properties: A review. *Composites Part B: Engineering*, **227**, 109393, 2021. https://doi.org/10.1016/j.compositesb.2021.109393.

ON THE USE OF ARTIFICIAL NEURAL NETWORKS FOR INVERSE ANALYSIS OF SINGLE DEGREE OF FREEDOM RESPONSE OF BLAST LOADED STRUCTURES

SAMUEL E. RIGBY[1,2], DANNY SMYL[3], TYLER RHODES[2], ANQI CHEN[2] & GAUTHIER STIEGLER[2]
[1]School of Mechanical, Aerospace and Civil Engineering, University of Sheffield, Sheffield, UK
[2]Arup Resilience, Security and Risk, London, UK
[3]School of Civil and Environmental Engineering, Georgia Institute of Technology, Georgia, USA

ABSTRACT

The Single Degree of Freedom (SDoF) method is widely used to evaluate the response of structures or structural elements under blast loading. Here, the structure in question is transformed into a single-point equivalent for which a single equation of motion can be solved. While the SDoF method is orders-of-magnitude faster than alternatives such as the finite element method, both are focussed only on solving the forward problem (inputs → outputs). In practise, however, a required performance limit is known (peak displacement, support rotation, etc.) and an adequate structure should be provided so as to not exceed that limit. This necessitates some form of iteration as the SDoF equation of motion cannot be simply inverted. Alternatively, machine learning techniques such as artificial neural networks (ANNs) may be used. ANNs are agnostic to input data type and therefore can just as easily learn patterns between input and output data as they can between output and input data. This paper presents a novel application of ANNs to rapidly solve the inverse problem (outputs → inputs) for SDoF structures subjected to blast loads. 180,000 SDoF analyses were run for 36 British Steel Universal Column sections (5,000 runs for each). The subsequent data was used to train an ANN classification network to suggest a section size deemed to meet a target value of support rotation, for a given set of basic inputs (span, peak force, impulse). The ANN is able to perform to a high degree of accuracy (> 90% correct classification of the test data set) and performs well in unseen 'design cases', suggesting that machine learning could be a highly valuable tool to aid in solving inverse problems relating to blast loading.

Keywords: artificial neural network, blast, classification, inverse modelling, SDoF.

1 INTRODUCTION

Protection of critical national infrastructure is predicated on the ability of engineers to accurately model the effects of blast on structures. Whether from large-scale industrial explosions such as Tianjin (2015) and Beirut (2020), acts of terror, or during ongoing conflicts such as those in Ukraine and the Middle East, explosions impart loads on structures that are several orders of magnitude larger than those which civil infrastructure is typically designed to resist. Furthermore, blast loading parameters are highly sensitive to the exact composition, size, shape, and location of the explosive, and are significantly influenced by non-linear physical processes which occur when a blast wave reflects off and diffracts around obstacles (e.g. in an urban environment), or when two wavefronts coalesce [1].

Whilst tools such as computational fluid dynamics (CFD) and the finite element method (FEM) provide solutions to the underlying conservation equations (e.g. they are 'physics-based') with a potentially very high degree of spatiotemporal resolution, they cannot easily incorporate the underlying uncertainties in explosion events. That is, whilst they may provide a near-exact answer to a given set of well-defined inputs, they provide only *one* answer (for, say, a few hours of computational time), and thus their use in probabilistic, risk-based design is currently limited. CFD/FEM therefore finds a natural place more in the detailed design stage, rather than the scheme design stage, and there remains a very clear need for a tool which provides 'good enough and quick enough' results.

WIT Transactions on The Built Environment, Vol 215, © 2025 WIT Press
www.witpress.com, ISSN 1743-3509 (on-line)
doi:10.2495/SUSI250191

This is where we find the natural habitat of the Single Degree of Freedom (SDoF) method; at the stage where engineers need to rapidly assess a large number of potential designs and quickly ascertain which would work in principle, which would be prohibitively expensive (in terms of cost, or carbon) to implement, which simply would not work, etc. The SDoF method works by transforming a 'real life' system with distributed mass, stiffness, loading, and resultant displacements (and stresses, strains, etc.) into an 'equivalent' single point [2], with one value of mass, stiffness, applied load, and damping coefficient (although this is typically ignored in blast applications due to the fact that its influence on first peak displacement is negligible [3]). This single point normally represents the point of maximum displacement of the 'real life' structure (e.g. midspan of a beam or column), and it is allowed to translate through one degree of freedom only, hence the name.

Transforming a structure into an equivalent single point has the effect of reducing a large system of simultaneous equations:

$$\boldsymbol{M}\ddot{\boldsymbol{z}}(t) + \boldsymbol{K}\boldsymbol{z}(t) = \boldsymbol{F}(t), \tag{1}$$

into a single equation:

$$K_M m\ddot{z}(t) + K_L kz(t) = K_L F(t), \tag{2}$$

where $\boldsymbol{M}$ and $\boldsymbol{K}$ are mass and stiffness matrices, and $\ddot{\boldsymbol{z}}$, $\boldsymbol{z}$ and $\boldsymbol{F}$ are vectors of time-varying acceleration, displacement and applied force at each node, whereas m, k, $\ddot{z}$ and z are scalar values of the above. Mass and load transformation factors, K_M and K_L, are employed in the SDoF method to ensure that kinetic energy, internal strain energy, and work done are conserved when moving from 'real life' to 'equivalent' systems [2]. The SDoF method offers orders-of-magnitude reductions in run-time compared to FEM and has demonstrated high levels of agreement with experimental data (see Rigby et al. [4] as an example).

2 THE DESIGN PROCESS AS AN INVERSE METHOD

The methods described previously (CFD, FEM, SDoF), although offering substantial differences in terms of fidelity of solution and computational expense, all still solve the problem in a 'forward model' sense. That is, for a given set of inputs (load, span, support conditions, etc.) and design variables (material type, depth, width, section modulus, etc.), the response of that system can be calculated; for the SDoF method this is peak displacement or support rotation.

In terms of protective *design*, however, the aim is to limit response to some acceptable value, and whilst the inputs are normally constrained (e.g. the element *has* to span a certain distance), the design variables are chosen such that the response limit is not exceeded. These response limits are given for support rotation and ductility ratio (peak deflection divided by elastic deflection limit), for either Category 1 (for the protection of personnel and equipment from the effects of primary and secondary fragments and falling portions of the structure under the action of blast loading) or Category 2 (for the protection of structural elements themselves from collapse under the action of blast loading) protection levels (see Table 1).

This means that the design process can effectively be viewed as an inverse problem (outputs → inputs), with *analysis* being the corresponding forward problem (inputs → outputs), a conceptual point introduced in Gallet et al. [6] and expanded upon in Gallet et al. [7]. The aim of this paper is to explore this idea in the context of blast response of structural elements. In particular, we devise a machine learning model to rapidly solve the

Table 1: Rotation and deformation limits for common structural components for Category 1 and Category 2 protection levels. *(Source: Adapted from Cormie et al. [5].)*

Element	Protection category	Support rotation	Ductility ratio
Structural steel beams/plates	1	2°	10
	2	12°	20
Reinforced concrete beams/ slabs	1	2°	–
	2	4°	–

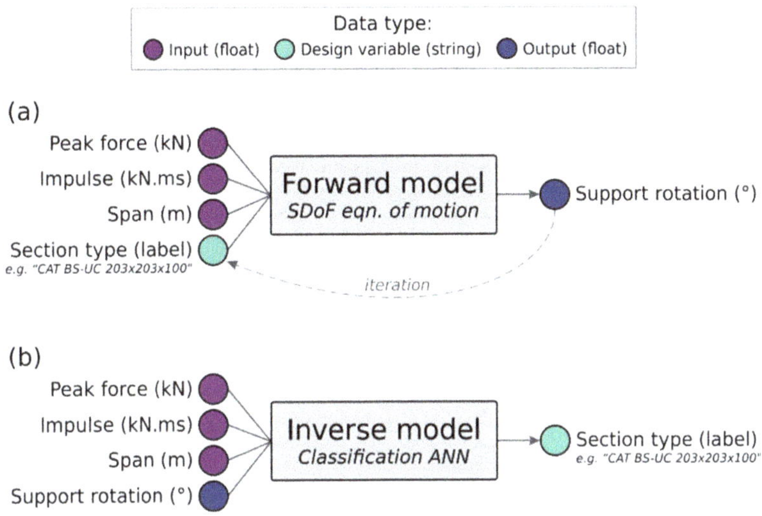

Figure 1: (a) Conceptual overview of the standard 'forward model' approach; and (b) the 'inverse model' approach adopted in this study.

inverse/design problem, as in Fig. 1, meaning that the design process can be undertaken in a single step, rather than through iteration of the forward model, which would otherwise be the approach adopted.

When considering steel columns specifically, as can be seen in Fig. 1, the forward model requires an initial guess of section type, e.g. a choice of one of the sections available from the BS-UC (British Steel Universal Column) catalogue [8] (see the table in the Appendix for the full list). This section is then run through the forward model (which, although typically inexpensive when solving the SDoF equation of motion, may still require several thousand timesteps to be computed), the results extracted, compared against the response limit, and a new section chosen if necessary. Clearly, this may require several iterations before the most suitable section is chosen, and the required computational steps could number in the many thousands. Alternatively, an inverse model will 'jump' straight to section type for a given response limit in a *single* computation.

3 OVERVIEW OF DATASET

A total of 180,000 individual SDoF analyses were performed, corresponding to 5,000 runs for each of the 36 BS-UC column types. Analyses were performed using *Ergo Compute*, Arup's in-house SDoF solver. The columns were simply supported, and the applied loading was a linearly-decaying reverse-ramp load, uniformly distributed along the column's length, acting on the column's major axis. The steel was modelled with a yield strength of 355 MPa, elastic modulus of 200 GPa, and a *dynamic increase factor* of 1.17 to incorporate strain rate sensitivity of the steel [5]. The columns had a bilinear elastic-perfectly-plastic resistance function, i.e. they performed elastically (linearly increasing resistance) until the plastic moment capacity was reached (plastic modulus, Z_{pl}, multiplied by yield stress, σ_y), and then deformed plastically with a constant resistance thereafter. No damping was specified, and analyses were terminated when velocity of the single degree of freedom reached zero (and therefore displacement had reached a maximum). The forcing function was input as a peak pressure and positive phase duration, which can simply be converted to peak force (*Ergo* allows the user to specific a *loaded length*, with force then given by the multiple of pressure, span, and loaded length. For simplicity, in this study the loaded length was set to 1m, and the peak force was controlled by peak pressure and span only) and impulse (integral of force with respect to time. Since the loading decays linearly, this is simply half the peak force multiplied by positive phase duration) in post-processing. An example output is shown in Fig. 2.

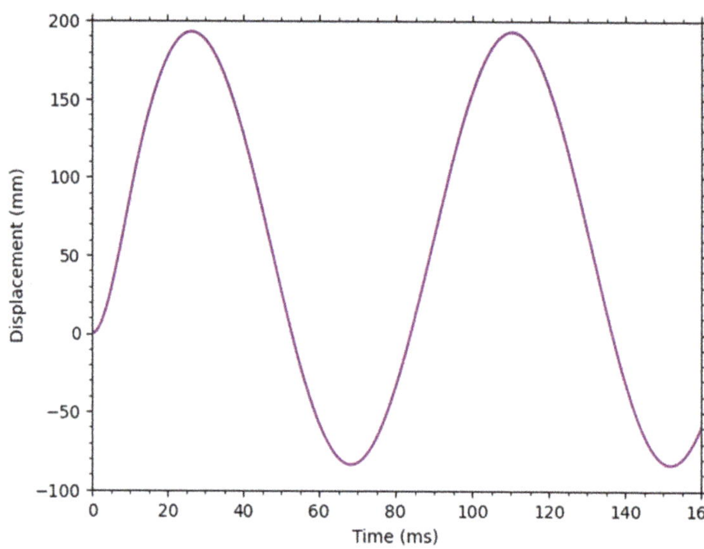

Figure 2: Example *Ergo* output for an 8 m spanning 203 × 203 × 113 column with 1 MN peak force and 10 ms duration. Peak displacement = 193.1 mm, peak support rotation = 2.76°. Note: analysis was run beyond first zero velocity (~30 ms) for clarity of presentation.

Monte Carlo sampling was used to initialise each simulation. Span was randomly selected from the interval 2–10 m (uniform probability). Peak pressures were randomly selected from the interval 9–5000 kPa, and positive phase durations from the interval 0.1–40 ms (both with a uniform probability in *log*-space), loosely corresponding to Hopkinson-Cranz scaled distances of 1–20 kg/m$^{1/3}$ and charge masses of 0.1–1,000 kg [9]. Basic logic was used to reject any results where peak rotation exceeded 45°, and several 'passes' were undertaken to ensure the outputs were reasonably well distributed between 0–45°, with a slight preference towards rotations of 2°, 4°, and 12°, as these are common limits used in design (e.g. as in

Table 1). This naturally resulted in the tendency for higher values of force and impulse to be associated with stiffer columns (larger plastic modulus).

In summary, each independent *Ergo* analysis ultimately returned five pieces of information: peak force (kN), impulse (kN.ms), span (m), section classification (in the form 'CAT BS-UC XXXxYYYxZZZ'; one of 36 available BS-UC sections), and maximum support rotation (°). This comprises the entire 'labelled' dataset which was used to train the machine learning inverse model described in the next section. Examples of the labelled data for the columns with largest and smallest section modulus are shown in Fig. 3.

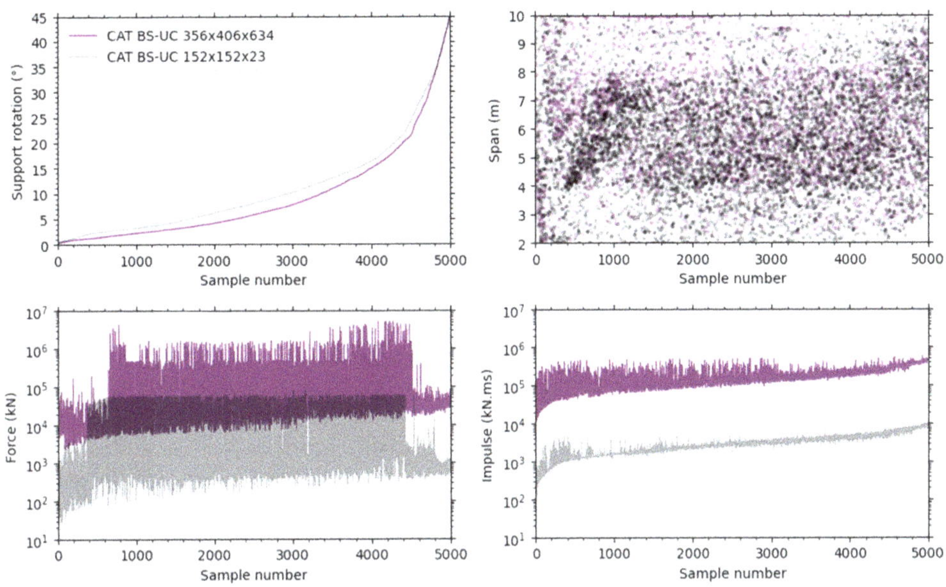

Figure 3: Labelled data for columns with largest (356 × 406 × 634) and smallest (152 × 152 × 23) plastic modulus, sorted by increasing support rotation.

4 CLASSIFICATION ARTIFICIAL NEURAL NETWORK

Classification artificial neural networks (ANNs) are a type of supervised machine learning model used to classify labelled data into a number of discrete categories and have recently been used to predict failure mode of reinforced concrete walls under near-field explosions [10].

In general, ANNs are designed to 'learn' the connections between input data and output data through successive 'training'. A network is formed of three main structures: input, hidden, and output layers (see Fig. 4). Each of these layers is formed of connected nodes, or 'neurons', and an input signal (in the form of numerical values fed into the input layer) is then propagated through the network until, ultimately, the ANN produces a prediction in the output layer. Each neuron will send a signal along a pathway to every (if the network is what is known as 'fully connected') node in the next layer, and whether this node then sends the signal onwards, and how strong the resulting signal is, is a function the combined strengths of the incoming signals. This strength is adjusted using tuneable parameters – the 'weights' and 'biases' – along the pathways between neurons.

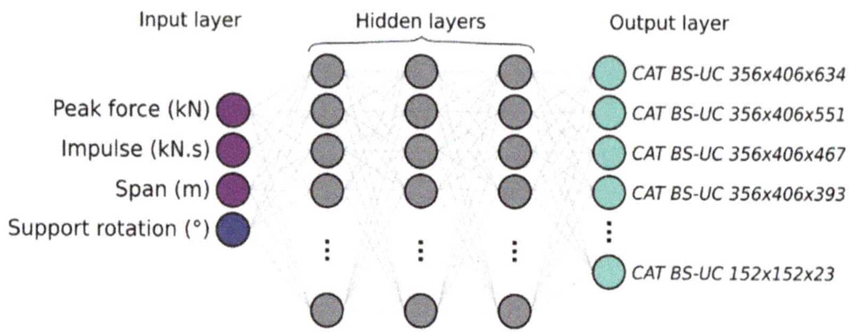

Figure 4: Architecture of the classification neural network adopted in this study.

Mathematically, the output signal of a given node, y, is given as [11]:

$$y = f\left(\sum_{i=1}^{n} w_i x_i + b\right) \tag{3}$$

where x_i is the signal received from the ith node on the previous layer, w_i is the weight, b is the bias, n is the number of nodes in the previous layer, and $f(\cdots)$ indicates that the composite signal is passed through an 'activation' function before going on to the next layer. The activation function (with ReLU, the 'rectified linear unit', used in this study) introduces non-linearity to the network (allowing complex relations between input and output to be learned), as well as suppressing low magnitude, potentially irrelevant connections and also ensuring computational efficiency.

First, the labelled data is separated into training, validation, and test data. The training dataset is available to the model to learn from throughout, whereas the validation and test datasets are kept aside and do not form part of the model's training. The model is sequentially asked to predict the validation dataset during training, so its performance can be monitored, whereas the test dataset is used only after training to assess how well the final model can predict *unseen* data, i.e. 'generalise'.

The network is then trained through successive passes of labelled data, where for a given set of inputs the network is asked to produce an output (or 'prediction'), and its performance compared against the known output (or 'target'). Training is typically done in 'batches'; the number of predictions the network is asked to make before weights and biases are adjusted. Once the network has attempted to predict the entire training dataset, one 'epoch' is said to have passed, and training can continue for a specified number of epochs or until some performance criteria has been met (hence the need for the validation dataset).

In regression networks, the output is typically numerical, with the output predictions readily assessed against targets using measures such as mean absolute error or root mean square error. In classification tasks, however, the network is asked to assign a *confidence value* to each of the neurons in the output layer. The 'true' target, therefore, would be a series of zeros for each incorrect label, and a single 1.00 for the correct label. The classification network therefore outputs a series of numbers which sum to 1.00 (achieved using the *softmax* function on the output layer).

In this study, column classes were listed in order of decreasing plastic section modulus (see Appendix) in a slight reworking of the order in which they appear in the BS-UC catalogue. This was deemed necessary to ensure that similar predictions (in terms of input) propagated through the network in a similar manner and ultimately arrived at predictions in close proximity to one another. In that sense, the index of each class has some physical significance, i.e. it is the rank order of plastic section modulus.

Imagine the converse, where two datapoints share all but one input, say span, yet their classes occupy the top and bottom nodes of the output layer (with the inputs associated with the shorter span requiring a slightly less stiff section for a given support rotation). Here, we are asking one set of inputs to return [1, 0, 0, … 0] and the other only marginally different set to return [0, 0, 0, … 1], requiring each signal to be sent to drastically different parts of the network; a difficult task that is sure to hamper the training process. As another example, if we trained a network to classify types of pet from a photograph of an animal, we would hope that connections representing 'Springer Spaniel' and 'Cocker Spaniel' occupy similar areas of the network, whereas those representing 'Goldfish' and 'Horse' should occupy wildly different areas.

ANN training is achieved by minimising the 'loss function', effectively a measure of how close the predictions are to the targets, with a loss of zero suggesting a network that makes perfect predictions every time. Here, the categorical cross entropy loss function is used:

$$L = -\frac{1}{N}\sum_{j=1}^{N}\sum_{c=1}^{C} y_{j,c}\ln(\hat{y}_{j,c}) \tag{4}$$

where N and C are the number of samples (batch size) and the number of classes (36, in this study), $y_{j,c}$ is a binary indicator that represents whether class c is the correct class for sample j, and $\hat{y}_{j,c}$ is the predicted probability of class c, sample j. The term 'accuracy' refers to the number of correct classifications (the neuron in the output layer with the highest numerical value is deemed the 'predicted' class, and the numerical value of this output (from 0 to 1) is called 'confidence'. The aim is for the network to predict with 100% accuracy *and* 100% confidence) divided by the total number of samples.

The machine learning model was generated in Python using TensorFlow and the Keras Sequential Model. Input features were scaled to have zero mean and unit variance to prevent excessively large inputs from dominating the network's learning. The loss function was minimised using the Adam optimizer with a learning rate of 0.001 and a batch size of 64. The model was trained for 200 epochs as this was deemed sufficient to ensure proper training whilst avoiding overfitting (manifesting as a divergence in the loss/accuracy curves of the training and validation datasets). Hyperparameter tuning determined that three hidden layers of 256 neurons in each layer achieved the best balance between accuracy and computational time, whilst also avoiding overfitting. Dropout, which is commonly adopted to reduce a network's tendency to overfit, was not required. L1 and L2 regularisation, which encourage sparsity in the network and penalise very large weights respectively, were found to have a detrimental effect on model accuracy.

The train/test split was 80/20, with a further 10% of the training data set aside for validation, giving 129,600/14,400/36,000 datapoints in the training/validation/test sets. Fig. 5 shows the loss and accuracy of the network with increasing epochs, with modest improvement in performance beyond epoch 100, but importantly with overfitting having been avoided. The final accuracy of the test dataset was reported as 90.41% (loss: 0.1899 (training), 0.2093 (validation), 0.2047 (test); accuracy: 91.53% (training), 90.55% (validation), 90.41% (test)).

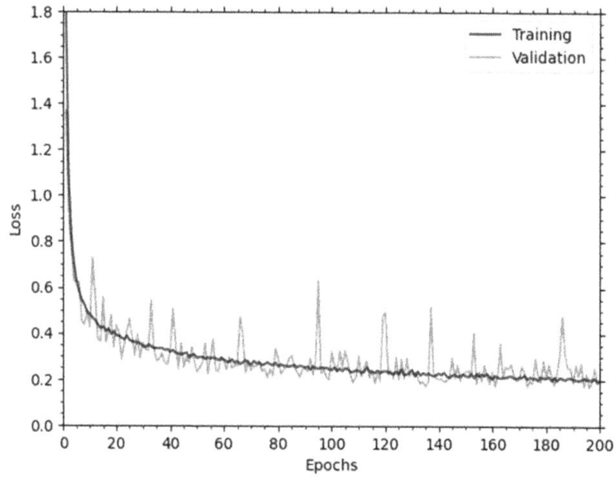

(a)

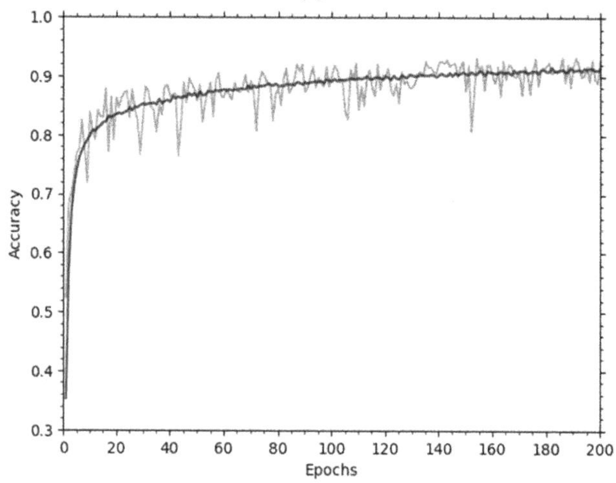

(b)

Figure 5: Training progress of the classification network. (a) Loss; and (b) Accuracy.

5 PERFORMANCE ASSESSMENT

The trained network was evaluated against the test dataset (approximately 1,000 datapoints per class), the results of which are compiled in the 'confusion matrix' shown in Fig. 6. Here, true classes are shown on the y-axis, and predicted classes are shown on the x-axis. A model with 100% predictive accuracy would occupy only the diagonal spaces.

Generally, the model is able to correctly classify the test data around 90% of the time, however this is not consistent among the classes. Of the 36 section types, 13 are predicted with >95% accuracy, and 21 are predicted with >91% accuracy. There are five classes that score an accuracy of <80%, however on closer inspection of Fig. 6 it appears as though these are formed of two main clusters, each containing three sections with similar plastic modulus.

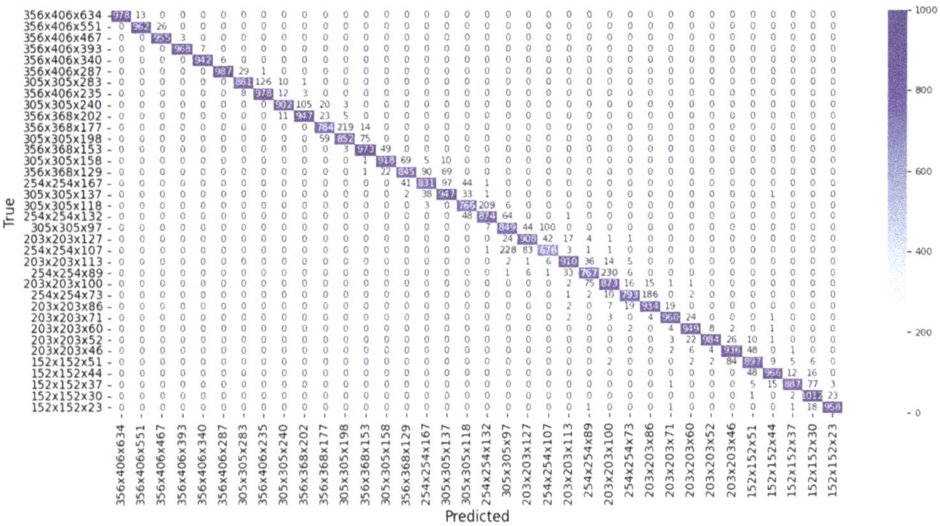

Figure 6: Confusion matrix of the trained model against the test dataset.

For the two lowest-accuracy sections (254 × 254 × 107 and 254 × 254 × 89), confidence in the true class was calculated separately for cases where the model predicted correctly and incorrectly (see Table 2). Even when the model misclassifies, it still has around a 30% confidence in the true classification, meaning that in these cases it would almost certainly have picked the correct class as its second choice (second highest confidence value). Therefore, whilst the accuracy of the model's *first* guess may be around 90% overall, the accuracy of its *top two* guesses would be approaching 100%

Table 2: Low-accuracy section classifications and corresponding misclassifications.

Section	Plastic modulus, Z_{pl} (cm^4)	Mass per metre (kg/m)	Accuracy (%)	Confidence in true class (%)		Most frequent misclassification
				When correct	When incorrect	
254 × 254 × 107	1480	107.1	68.08	78.62	33.11	305 × 305 × 97
305 × 305 × 97	1590	96.9	84.90	80.60	36.83	203 × 203 × 127
203 × 203 × 127	1520	127.5	91.07	91.75	29.06	254 × 254 × 107
254 × 254 × 89	1220	88.9	73.47	88.29	26.21	203 × 203 × 100
203 × 203 × 100	1150	99.6	88.81	88.81	29.81	203 × 203 × 113
203 × 203 × 113	1330	113.5	93.43	97.20	23.23	254 × 254 × 89

Each time the model misclassified, the most frequent incorrect class was also stored, again see Table 2. The two lowest accuracy classes are each joined by two sections with remarkably similar properties, and the model understandably has some difficulty in distinguishing between them. Currently, the ANN is not given these properties (plastic modulus and mass) in advance, but ongoing work is exploring whether further improvements in accuracy can be achieved if the model is allowed to incorporate this information into the training process.

6 THE MODEL IN USE

Whilst the model has shown good accuracy against the test dataset, its intended use will be subtly different. Rather than predicting a section that is known to *exactly* match a set of inputs, the model will be given span, loading, and a *target* rotation, and asked to provide a section which, to its best knowledge, matches that rotation most closely. For example, running the model for the inputs [8, 1000, 5000, 2], in the order of span (m), peak force (kN), impulse (kN.ms), and support rotation (°) returns: CAT BS-UC 254 × 254 × 73. Running this section back through the *Ergo* forward model returns a peak support rotation of 1.85°, within 7.5% of the target value. These are the same inputs as in Fig. 2, but with a larger section.

In order to assess how close the rotation of the predicted section is to the target rotation, in a more general sense, a grid search was performed with peak force ranging from 5e3–5e5kN (20 values spaced logarithmically), impulse ranging from 5e3–1e5kN.ms (25 values

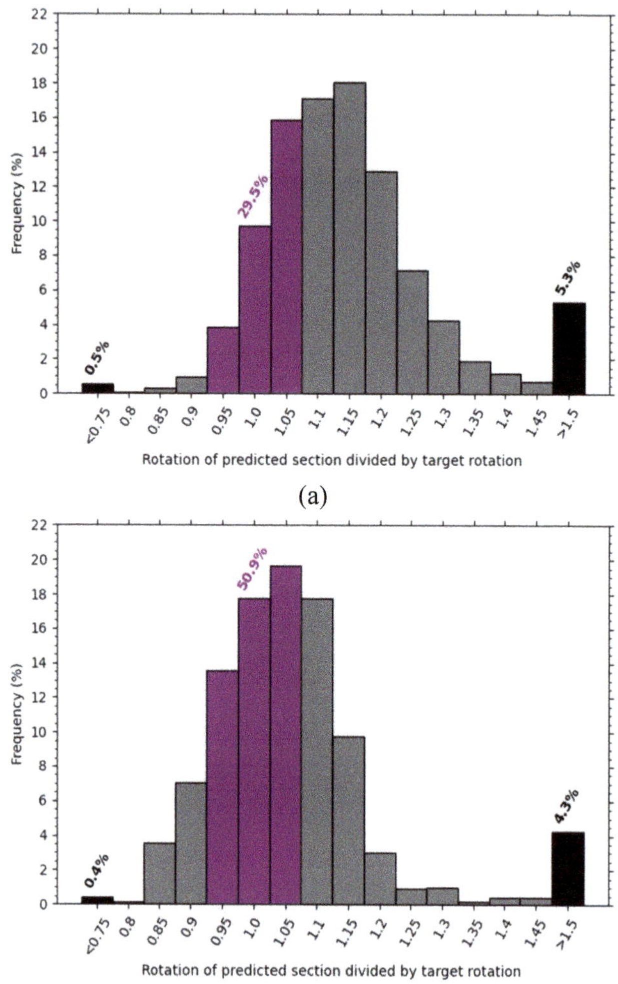

Figure 7: Histograms of peak rotation of predicted section divided by target peak rotation. (a) First choice only; and (b) Closest match from the model's top two choices.

spaced logarithmically), span ranging from 4–8 m (increments of 1 m), and two target rotations: 2° and 12°, as per Table 1, resulting in 5,000 total unseen design cases. For each case, the top two sections from the predictor (in terms of % confidence) were run through the *Ergo* forward model. The results from this grid search are presented in Fig. 7, where (a) shows the output from only the top classification, whereas (b) shows the closest match from either of the top two classifications. It can be seen that in 29.5% of the design cases, the top choice section results in a rotation that is within ±7.5% of the target, and in 50.9% of the design cases the top two choices return a rotation within ±7.5% of the target. These are both indicated by the purple regions of each histogram.

Excluding the underflow and overflow bins, the mean deviation is +13.0% when considering only the top choice, and +4.2% when considering the top two choices. These are associated with mean confidence values of 95.95% (top choice only) and 99.79% (top two choices). In total, only 5.8% of the design cases sit within the under/overflow bins (representing >25% underprediction or >50% overprediction respectively) when considering the top choice, which reduces to 4.7% when considering the top two choices.

7 SUMMARY AND OUTLOOK

This paper presents the development of a classification artificial neural network to perform rapid inverse analysis of blast loaded steel columns. 180,000 individual SDoF analyses are leveraged to train the network, which takes inputs of span, force, impulse, and target support rotation, and outputs one of 36 BS-UC section classifications. The trained model is able to correctly classify the test dataset 90.41% of the time. The model was further tested on 5,000 unseen design cases, where it was able to pick a section – which resulted in an output rotation within 7.5% of the target – in 29.5% of the cases if a single output was taken from the inverse model, which increased to 50.9% of the cases if the top two outputs were taken from the model and the closest match selected.

As a proof-of-concept, this work can be considered highly successful, and it is hoped that future studies will benefit by reframing the design process as an inverse method and leveraging machine learning tools to aid in this. This will particularly benefit scenarios where the forward model is much more computationally expensive than the SDoF method used herein (e.g. CFD/FEM), and therefore any computational steps saved by reducing or eliminating the need for iteration, will have considerable savings.

The model itself could be further improved by: (a) incorporating prior knowledge of relevant section properties such as plastic modulus and mass per unit length; (b) training the model on more section types; and (c) developing a non-symmetrical loss function to attempt to correct the inherent conservativism of the model, which showed average deviations in the design case of >1.00. Work is ongoing in this regard.

ACKNOWLEDGEMENT

The first author gratefully acknowledges financial support from the Royal Academy of Engineering under the *Industrial Fellowships Round 17* programme.

APPENDIX

Full list of BS-UC sections sorted by decreasing plastic modulus [8].

Section	Plastic modulus, Z_{pl} (cm^4)	Mass per metre (kg/m)	Section	Plastic modulus, Z_{pl} (cm^4)	Mass per metre (kg/m)
356 × 406 × 634	14,200	633.9	254 × 254 × 132	1,870	132
356 × 406 × 551	12,100	551	305 × 305 × 97	1,590	96.9
356 × 406 × 467	10,000	467	203 × 203 × 127	1,520	127.5
356 × 406 × 393	8,220	393	254 × 254 × 107	1,480	107.1
356 × 406 × 340	7,000	339.9	203 × 203 × 113	1,330	113.5
356 × 406 × 287	5,810	287.1	254 × 254 × 89	1,220	88.9
305 × 305 × 283	5,110	282.9	203 × 203 × 100	1,150	99.6
356 × 406 × 235	4,690	235.1	254 × 254 × 73	992	73.1
305 × 305 × 240	4,250	240	203 × 203 × 86	977	86.1
356 × 368 × 202	3,970	201.9	203 × 203 × 71	799	71
356 × 368 × 177	3,460	177	203 × 203 × 60	656	60
305 × 305 × 198	3,440	198.1	203 × 203 × 52	567	52
356 × 368 × 153	2,960	152.9	203 × 203 × 46	497	46.1
305 × 305 × 158	2,680	158.1	152 × 152 × 51	438	51.2
356 × 368 × 129	2,480	129	152 × 152 × 44	372	44
254 × 254 × 167	2,420	167.1	152 × 152 × 37	309	37
305 × 305 × 137	2,300	136.9	152 × 152 × 30	248	30
305 × 305 × 118	1,960	117.9	152 × 152 × 23	182	23

REFERENCES

[1] Larcher, M. & Casadeia, F., Explosions in complex geometries – A comparison of several approaches. *International Journal of Protective Structures*, **1**(2), pp. 169–196, 2010.

[2] Biggs, J., *Introduction to Structural Dynamics*, McGraw-Hill: New York, 1964.

[3] Krauthammer, T. & Altenberg, A., Negative phase blast effects on glass panels. *International Journal of Impact Engineering*, **24**(1), pp. 1–17, 2000.

[4] Rigby, S.E., Tyas, A., Bennett, T., Warren, J.A. & Fay, S., Clearing effects on plates subjected to blast loads. *Engineering and Computational Mechanics*, **166**(EM3), pp. 140–148, 2013.

[5] Cormie, D., Mays, G. & Smith, P., Blast *Effects on Buildings*, 3rd ed. Thomas Telford: London, 2019.

[6] Gallet, A., Rigby, S., Tallman, T.N., Kong, X., Hajirasouliha, I., Liew, A., Liu, D., Chen, L., Hauptmann, A. & Smyl, D., Structural engineering from an inverse problems perspective. *Proceedings of the Royal Society A*, **478**, 20210526, 2021.

[7] Gallet, A., Liew, A., Hajirasouliha, I. & Smyl, D., Machine learning for structural design models of continuous beam systems via influence zones. *Inverse Problems*, **40**(5), 055011.

[8] British Steel, Sections: Universal columns (UC) datasheet. https://britishsteel.co.uk/media/nm1kuplv/british-steel-universal-columns-datasheet-100723.pdf. Accessed on: 21 Jan. 2025.

[9] Kingery, C.N. & Bulmash, G., Airblast parameters from TNT spherical air burst and hemispherical surface burst. Technical Report ARBL-TR-02555. Aberdeen Proving Ground, MD, USA: U.S Army BRL, 1984

[10] Holgado, D., Mourão, R., Montalva, A. & Florek, J., On the assessment of reinforced concrete (RC) walls under contact/near-contact explosive charges: A deep neural network approach. *Buildings*, **14**(9), 2683, 2024.

[11] Dennis, A., Pannell, J.J., Smyl, D.J. & Rigby, S.E., Prediction of blast loading in an internal environment using artificial neural networks. *International Journal of Protective Structures*, **12**(3), pp. 287–314, 2021.

SECTION 6
HAZARD MITIGATION AND SECURITY

HOW CAN WE ENHANCE CRITICAL INFRASTRUCTURE AND TRANSPORT SECURITY THROUGH AN INNOVATIVE METHODOLOGY OF OBSERVATION AND SHOCK WAVE TRACKING IN HARSH ENVIRONMENTS?

MICHEL-OLIVIER STURTZER[1] & SOPHIE TRÉLAT[2]
[1]Department of Explosive Protection, French–German Research Institute of Saint-Louis (ISL), Saint-Louis, France
[2]Department of Explosion and Weapons Effects, French Institute for Protection and Nuclear Security (DEND), Fontenay-aux-Roses, France

ABSTRACT

The paper deals with an innovative methodology aimed at enhancing security of critical infrastructure and transport against the effects of explosive phenomena, in particular in the field of overseas and homeland security, as well as in the nuclear industry. In this context, the French–German Research Institute of Saint-Louis (ISL) and the French Institute for Protection and Nuclear Safety (DEND, previously IRSN) cooperate since 2017 by sharing their expertise and experimental capabilities. The methodology developed in this article consists of performing applied research studies and reduced-scaled experiments on reference targets facing representative threats. They rely on (i) the use of real-time pressure measurements to assess the mechanical loads at the targets surface and (ii) the capture of high-speed images of the supersonic flow interacting with the structures. One objective of the studies is to analyse efficiency of protective means and materials in terms of blast effects mitigation so as to propose guidelines on geometrical and structural dimensioning. This will contribute to building practical and normative recommendations for protection system design against explosive threats.

Keywords: blast effects, pressure measurements, high-speed imaging, scaling effects.

1 INTRODUCTION

The French–German Research Institute of Saint-Louis (ISL) is a bi-national research institute established by the Federal Republic of Germany and the French Republic on the basis of a treaty signed in 1958. The core mission of ISL is 'Research, scientific studies and basic predevelopment in the field of defence and security'. Among other tasks, ISL focuses on the physical protection of personnel and assets against the effects of various explosive charges. To improve general survivability, fundamental studies on blast wave propagation and target interaction are continuously being conducted at ISL.

Nuclear security includes preventive measures to detect and counteract theft, tampering, illegal transportation of nuclear materials or any other malevolent act relating to nuclear materials, any radioactive substances or facilities harbouring them. In this context, the French institute for protection and nuclear security (DEND, Direction de l'Expertise Nucléaire de Défense et de sécurité in French) provides assistance and technical support for authorities in the field of sensitive activities, and assesses the safety of defence related facilities, the protection of civil nuclear installations and transports against any act of malevolence.

Both institutions consequently conduct research to study the effects of explosive phenomena (new mode of malevolence such as terrorism, unmanned aerial vehicles (UAVs), internal threats) on critical transport and infrastructure to ensure safety and security of either military assets or nuclear industry. The considered sources of explosions include weapons (metal fragments), industrial chemicals (ammo-nitrates) and emerging threats (homemade explosives). To achieve this goal, scaled-down experiments are being designed to analyse explosive blast propagation and interaction with reference targets such as parallelepipedic or

WIT Transactions on The Built Environment, Vol 215, © 2025 WIT Press
www.witpress.com, ISSN 1743-3509 (on-line)
doi:10.2495/SUSI250201

cylindrical shapes for instance, or to evaluate protection systems' blast resistance under real conditions. Diagnostic methods include pressure gauges and high-speed imaging.

Since 2017, DEND and ISL are cooperating on guidelines development relative to reduced-scale explosive tests. Both institutes are presently developing a high-speed imaging system demonstrating a large luminosity range capable to cope with intense light emission and pressure gradients, especially adapted to high-dynamic range and explosive phenomena. In 2022 a patent application was consequently filed. After focusing on the general experiments design methodology based on real information, the paper shows up some applications of the groundbreaking system: air blast propagation in the very close field of a cylindrical container or an access control system, subaquatic explosion, evaluation of actual and emerging threats, etc. Such applicative studies directly aim at contributing to maintain and further strengthen national nuclear security.

2 REDUCED-SCALE EXPERIMENTS: DESIGN METHODOLOGY

The BMAX (Bureau d'étude et Modélisation des effets des Armes et eXplosifs in French) office in charge of blast and weapons effects assessment at DEND develops models and tools to evaluate the effects of a malevolent explosion on nuclear facilities and transports. Different types of reference shapes are being investigated on a blast table: for instance, small explosive charges can be detonated in front of reduced-scale obstacles equipped with pressure sensors. Reference obstacles include vertical walls, cylinders, hemicylinders, protective barriers, etc., and can be representative of real infrastructures or containers thanks to the scaling principle of Hopkinson–Cranz [1], [2].

The PCE (Protection contre les effets des Charges Explosives in French) team responsible for protection against the explosive charges at ISL conducts studies in the field of threat characterization, threat defeat and target protection to provide physical protection of personnel and assets. Two operational environments are considered: foreign and homeland security. Reference explosive charges are being investigated on dedicated blast pads equipped with characteristic scaled down structures and vehicles.

The facilities on which the design methodology is based are described in the following sections.

2.1 DEND test table

The blast table has been principally designed to study shock waves reflection phenomena and interaction with different non-deformable structures [3]. It measures 1.60 × 2.40 m and features an array of mounting holes that facilitates the placement of modular 0.40 × 0.40 × 0.05 m wooden plates, and pressure transducers (Figs 1 and 2). Different types of pressure transducers (piezo-electric and piezo-resistive, Figs 3(b) and 3(c)) are mounted on an elastic support inserted into the holes (each separated by 133 mm) provided for this purpose. Each pressure transducer is calibrated prior to the tests. Explosive charges (Fig. 3(a)) are installed and ignited on a dedicated reinforced steel table plate to generate a hemispherical blast wave using a Davey-Bickford SA 4201A detonator [5].

Experimental campaigns are performed at the ArianeGroup's research centre located near Paris (Vert-le-Petit, France). ArianeGroup handles all the experiment pyrotechnics and also provides the data recording system (Nicolet Genesis data acquisition system, sampling rate of 500 kHz). During experiments, the modular table is mounted and placed at the centre of a closed bunker (Fig. 2), so as to avoid the perturbation from shock reflection on the bunker walls.

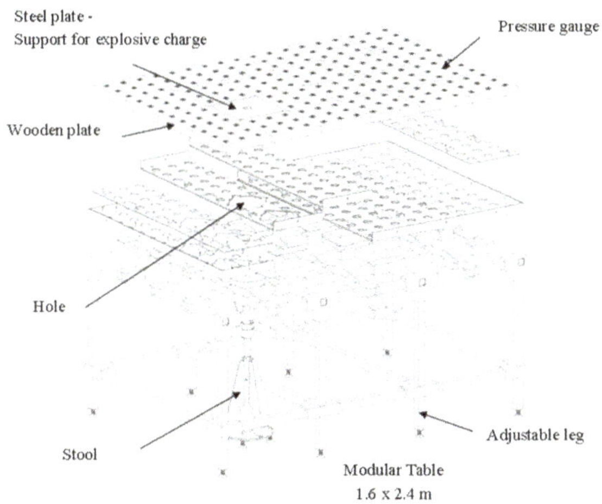

Figure 1: Schematic view of the DEND laboratory scale table [4].

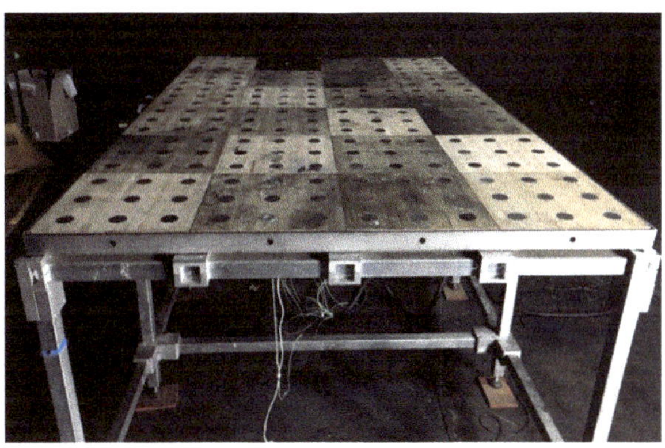

Figure 2: DEND blast table in its bunker [3].

(a) (b) (c)

Figure 3: (a) Plastic explosive charge; (b) and (c) Pressure transducers [3].

2.2 ISL blast pad

ISL developed a dedicated outdoor blast pad located at its own explosive range: explosive charges are detonated in a factor 2 up-scaled version of DEND test configuration (position actually rated up to 10 kg of TNT equivalent). Sensors' integration is presented in Fig. 4 [3].

Figure 4: ISL hemispherical blast pad [3].

Different types of pressure transducers (piezo-electric and piezo-resistive) are mounted in a polypropylene support inserted in one of the integrated rail ports, each separated by 266 mm. Each pressure transducer is calibrated prior to the tests. All data are recorded using a Transcom system running at 2 MHz. Explosive charges are installed and ignited on a dedicated reinforced steel ground plate to generate a hemispherical blast wave (Figs 4 and 5), using a RP83 detonator [5].

Figure 5: Hemispherical explosive charges (C-4, Semtex, Hexomax and Comp-B) [3].

In addition, ISL offers the possibility of investigating spherical blast waves consecutive to above ground explosive charge detonation using similar pressure transducers installed on 1.5 m high poles as shown in Fig. 6.

Figure 6: ISL spherical blast pad [6].

3 APPLICATIVE STUDIES

3.1 Industrial chemical: ammo-nitrate detonation

On 4 August 2020, a large quantity of ammonium nitrate storage located in a hangar of the Beirut harbour exploded, causing few hundred deaths, more than 5,000 injuries, and dramatic fallouts for the Lebanese economy [7]. Following a legal procedure, a few thousand tons of ammonitrate had been 'temporarily' stored in a warehouse in non-adequate conditions: mixed with fireworks, motor oil, wooden pallet, and chemical products, etc. [8]. The explosion was preceded by a fire in the same warehouse. After some time, a large orange fireball became visible engulfing the storage building (Fig. 7(a)), followed by the transmission of a large blast wave materialized by the colossal white Wilson cloud seen in Fig. 7(b).

The experimental methodology was used to replicate the effects of the Beirut explosion, particularly the fireball development and the propagation of the blast wave into the

Figure 7: Webcam views of Beirut explosion [9]. (a) Fireball; and (b) Wilson cloud.

surrounding complex environment. A publicly available Lebanese webcam video (Fig. 7) was analysed to extract the trajectory of the blast wave and a consequent approximate TNT equivalent of the ammonitrate explosion. Based on these data, a non-deformable parallelepipedic wooden mock-up (Fig. 8(a)) was designed to provide a simplified model of the vertical cylindrical concrete silo stack seen in Fig. 7(a): its dimensions are 721 mm in length, 223 mm in height and 126 mm in width.

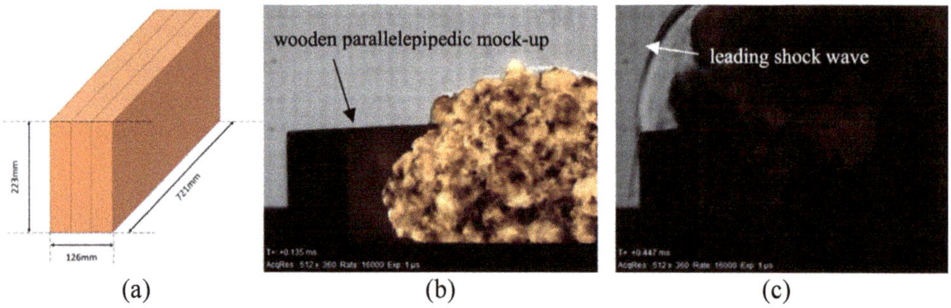

Figure 8: Beirut reduced-scale test. (a) Wooden mock-up; (b) Fireball development; and (c) Leading shock.

The scaled-down structure was exposed to the detonation of a few hundred grams of plastic explosive charge: the hemispherical fireball can be precisely observed as it impinges the front face of the stack of silos (Fig. 8(b)). In Fig. 8(c), the leading shock wave, invisible on the original video, is rendered visible once it has cleared from the post-combustion products. It is then enveloping the rear face of the concrete structure, undergoing diffraction phenomena from the building surfaces. High speed imaging highlights the importance of such observations for the assessment of urban structures vulnerability and protective characteristics, as well as the evaluation of prevention or mitigation measures.

3.2 Weapons effects

3.2.1 Contact explosive charge

The effects of a high explosive charge detonating (by accident or malevolence) directly on contact with a steel plate were studied. Prediction of the dynamic response and failure modes of a metallic protection represents a key knowledge notably in the field of critical infrastructure and vehicle protection and design. To that extent, both institutes recently joined their efforts in improving the knowledge in this field by conducting experimental and analytical work.

A series of tests was conducted using S235 steel plates commonly used for metallic structures (and similar to the outer steel jacket used for some storage container design [10], [11]). A cylindrical shaped plastic explosive weighing a few grams was placed at the surface of the metallic plate, as shown in Fig. 9.

Real-time deformation and metal fragments projections were analysed by the mean of high-speed images presented in Fig. 10. On the first image, we observe deformation and tearing of the steel plate rear surface leading to the formation of a large flying spall. A leading shockwave visible from the second image on corresponds to the shock transmitted to the air through the metallic plate by the detonation wave. A second visible shock is certainly

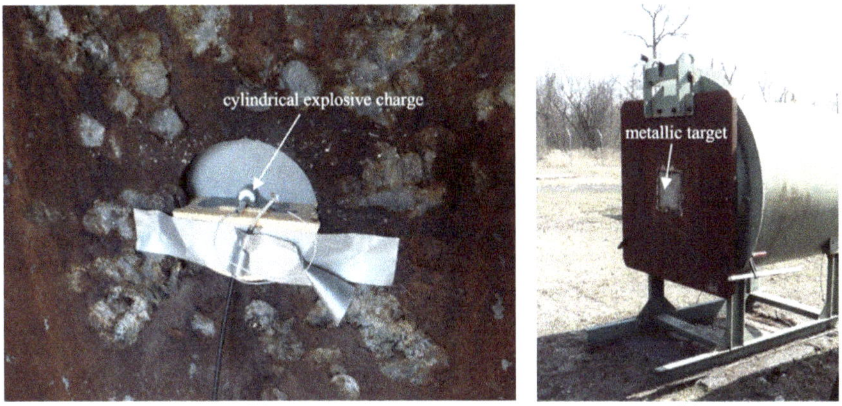

(a) (b)

Figure 9: Contact explosive blast pad. (a) Explosive charge; and (b) Experimental setup [5].

generated by the spall supersonic flight. In the wake of the leading shock, a cloud of ejecta particles is being progressively dispersed. Conducted tests allowed the determination of some elements linking the charge/plate characteristics to the topology of the projected materials and the formed crater. The position of the leading shock could also be tracked, as well as that of the fastest visible ejecta and the spall to determine their trajectories. By estimating their relative weight through direct measurements of collected fragments or visual analysis, it is then possible to contribute to the knowledge of backside effects behind a protection solution based on a metallic plate.

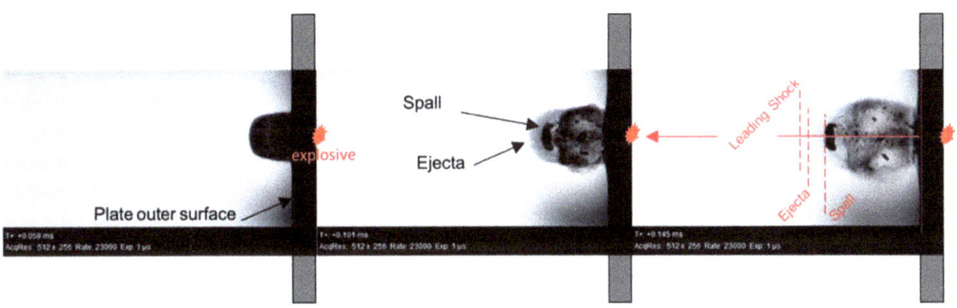

Figure 10: Series of three high-speed images of contact detonation [5].

3.2.2 Near-field explosive charge

DEND brings its expertise in the assessment of security levels for defence facilities, and protection of civil nuclear installations and transports against any act of malice. To achieve these goals, the institute manages active research projects in various nuclear safety and security fields in order to maintain top-level expertise and ensure on a long-term perspective the necessary knowledge and skills. ISL on the other hand has been investigating the effects of multiphase explosives for more than 15 years. The delayed combustion of reactive particles and the bare impulse of projected unreacted or partially reacted significantly modify the blast load characteristics in close range on vehicles panels or infrastructure walls.

The research work of both institutions contributes for example to developing or validating models, and identifying the most adapted tools or methods for assessing vulnerability of critical targets under explosive loading, in light of the fact it is a complex topic to deal with: walls of a shipping cask, as well as security doors or armoured vehicles panels by the way, are indeed made of one or several materials including steel, lead, resins, etc. [10]. These challenges in explosion area of expertise notably require determining the TNT equivalency of various explosive compositions or devices in peak overpressure, which represents a usual way of measuring the magnitude of an explosive event, accidental, experimental or terroristic. The determination of TNT equivalency in close range may be critical when the objective is to assess the effects on infrastructures. Fig. 11 shows an example of an ISL blast test for material resistance evaluation. A test plate is attached to the side of a test-bunker designed to lateral side blast effects.

Figure 11: Near-field detonation test [5].

A plastic explosive charge is detonated above ground in the vicinity of the test metallic sample. A large fireball is enveloping the target as the leading shockwave (rendered visible thanks to the contrast of the dark trees on the bright background) has already departed from the afterburning region. If the resulting effects are lower than predicted (i.e. an overestimation of the plastic explosive TNT equivalent), survival of the candidate protective system being evaluated may be misinterpreted.

3.3 Emerging threats: aluminized home-made explosive

The research institutes may also be required to issue their opinion on the threat characterization and sensitivity studies for critical assets or several fuel cycle facilities. In this context, DEND must be able to assess the assumptions adopted by a nuclear operator to model its capacity for action for the design-basis threats of sabotage as defined by the national security directive applicable to nuclear facilities [11]. These are rapidly changing fields with regard to threat evolution on communication networks, which implies a continuous effort on the part of the institutes to maintain their expertise. ISL on the other hand may be asked to assess the physical effects of certain types of explosive compositions in characteristic configurations such as its mobile-roof internal detonation bunker [12].

For instance, in order to characterize the effects generated by the same mass of C-4 and an aluminized explosive composition (Al-Ex), the propagation of a hemispheric blast wave in free environment was studied. One charge of each type with the same mass was detonated at the centre of ISL hemispheric blast pad (Fig. 12). Side-on pressure gauges placed at a few meters from the charge centre recorded the pressure evolution for both explosive compositions. High-speed images presented in Fig. 13 demonstrated the fundamental difference in the fireball behaviour of the two explosives: the reactive aluminium particles ensure a long-lasting light emitting combustion reaction leading to enhanced quasi-static pressure effects. This observation was confirmed by the overpressure and positive impulse evolutions with distance. Results thus proved to be critical to estimate this new type of explosive threat's degree of strength in free field in comparison to a reference plastic explosive such as C-4 (or even TNT) used for previous vulnerability studies.

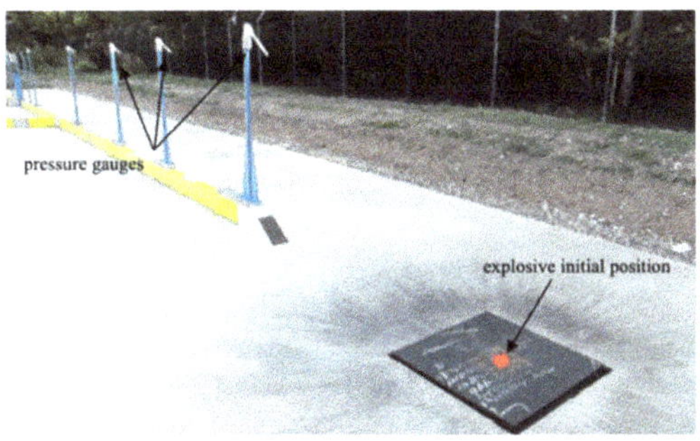

Figure 12: ISL hemispheric blast pad [5].

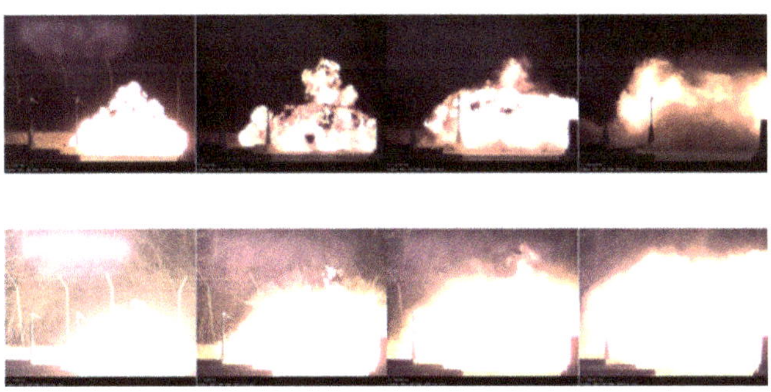

Figure 13: Two series of high-speed images: C-4 on top, Al-Ex on bottom [5].

3.4 Air blast propagation close to a cylindrical shape

Understanding the effects of a blast generated by an accidental or a terroristic explosion nearby a critical structure is a main concern for both DEND and ISL. Hemicylindrical shapes are commonly encountered on industrial sites, military facilities and also representative of certain infrastructures and transport containers in the nuclear sector. Full-scale reactive phenomena are however seldomly compatible with long-term studies due to cost and regulation issues. Blast wave interaction with a convex structure was consequently investigated through a twin reduced scale experimental approach. The effects of explosive charges were characterized at the surface of a hemi-cylindrical obstacle, shown in Fig. 14, inspired by nuclear and radioactive containers, or at larger scale, by the nuclear reactor building or a hangar to store vehicles or aircrafts. Based on the 50 g TNT equivalent DEND installation described in Section 2.1, ISL designed a factor 2 up-scaled setup by using 400 g TNT equivalent explosive charges. Blast wave propagation was analysed through the peak overpressures recorded along the hemi-cylindrical structure's central line with numerous sensors placed at its surface. An innovative parameter named the reduced shifted direct distance (RSDD) was introduced to take into account the geometry of the obstacle and the 3D free-field pressure decay. A physical model (Model S) [13] allows then the prediction of the peak overpressure evolution on whole circumference of the hemi-cylinder. Further results confirmed the extension of this conjecture to other charge masses and ratio between the distance to the charge and the hemicylinder radius. This rapid tool will be useful to perform structural analysis or to assess the resistance of transport and infrastructure.

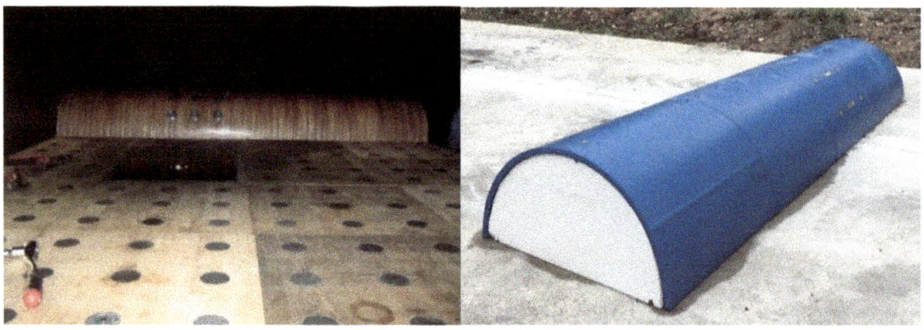

Figure 14: Hemispherical mock-ups at two different scales [13].

3.5 Subaquatic blast propagation

This section delves into understanding the impact of detonations caused by underwater explosives, with a specific focus on shallow water-filled nuclear pools and liquid-filled containers as parts of various vehicle fuel systems. These reservoirs are designed to maintain water tightness, often achieved through the presence of a thin stainless-steel liner. The need for research relies on investigating the potential loss of this functionality due to damage to the concrete structure and perforation of the liner. Such damage could result from local deformations caused by the overall response of the structure as well as the direct effects of the explosion. While the former aspect can be explored using techniques like, for instance, the method of images (MOI) [14], the latter occurs only when the explosive device is close enough to generate pressures capable of causing volumetric damage to materials.

This secondary effect complements the primary one and is exemplified through experimental simulations that analyse local conditions, combining pressure measurements and high-speed images recordings, thereby enhancing the understanding gained from global analyses. As an illustration, the series of images in Fig. 15 presents the initial propagation of the shock generated by the detonation of a small underwater high-explosive charge. The orange fireball of detonation products is visible on the first two images whereas the shock is transmitted to the surrounding water until it reaches the water tank walls leading to their rupture visible on the last image. Shock waves' reflection on the free surface can also be analysed thanks to the high quality of the capture image.

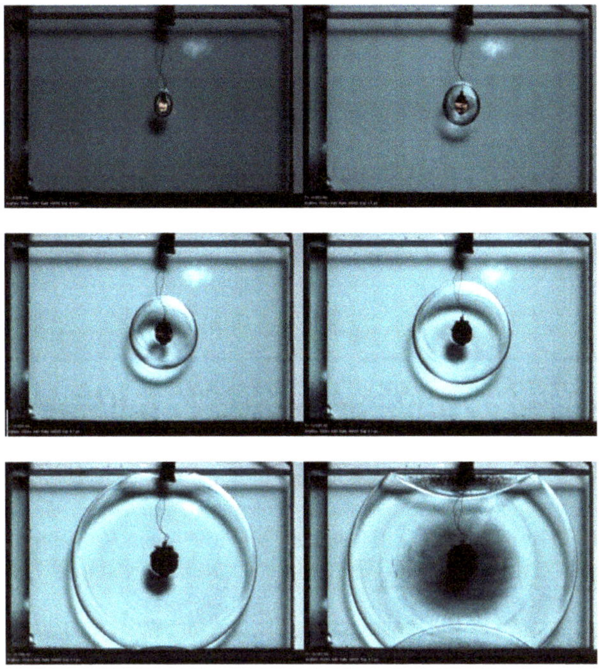

Figure 15: High-speed images for the free-field (i.e. without obstacle) subaquatic tests [5].

4 CONCLUSIONS

In conclusion, the paper sheds light on an innovative methodology aimed at enhancing the security of critical infrastructure and transportation systems against explosive threats, particularly in the military and nuclear industry domains. The methodology involves conducting applied research studies and reduced-scaled experiments on reference targets to assess blast effects and evaluate the efficiency of protective measures and materials. By utilizing real-time pressure measurements and high-speed imaging techniques, the study aims to provide guidelines for protection systems design and normative recommendations development to mitigate the impact of explosive phenomena. The collaboration between the French–German Research Institute of Saint-Louis (ISL) and the French Institute for Protection and Nuclear Security (DEND) has been instrumental in advancing this research, with a focus on developing a high-speed imaging system capable of capturing the dynamics of explosive interactions with critical structures. The paper also presents several applicative studies, including investigations into industrial chemicals, weapons effects, and emerging threats such as homemade explosives. These studies provide valuable insights into the behaviour of structures and materials under explosive loading conditions, facilitating the development of robust protection strategies and enhancing national nuclear security.

Overall, the research outlined in this paper contributes to advancing the understanding of explosive phenomena and provides practical tools and methodologies for enhancing the safety and security of critical infrastructure and transports in the face of evolving threats. Further research in this area will continue to strengthen authorities' ability to contribute mitigating the impact of explosive events on vital assets and personnel.

REFERENCES

[1] Hopkinson, B., British Ordnance board minutes, Report No. 13565. British Ordnance Office: London, 1915.

[2] Cranz, C., *Lehrbuch der Ballistik*, Springer: Berlin, 1926.

[3] Gavart, R., Trélat, S., Sturtzer, M.-O. & Chaumeix, N., A two-scale approach to widen a predictive blast propagation model around a hemicylindrical obstacle. *Shock Waves*, **33**(3), pp. 287–297, 2023.

[4] Cheval, K., Loiseau, O. & Vala, V., Laboratory scale tests for the assessment of solid explosive blast effects. Part I: Free-field test campaign. *Journal of Loss Prevention in the Process Industries*, **23**(5), pp. 613–621, 2010.

[5] Trélat, S. & Sturtzer, M.-O., IRSN and ISL, Explosive compositions and blast effects, Quality Manager Master in Security, Environment in Pyrotechnic Depollution and Remnants of War, Class Lecture, 2023.

[6] Sturtzer, C.-A., Etude des mécanismes de dispersion par choc et des régimes de combustion de nuages de particules d'aluminium. Doctoral thesis, Poitiers, 2014.

[7] Rigby, S.E. et al., Preliminary yield estimation of the 2020 Beirut explosion using video footage from social media. *Shock Waves*, **30**, pp. 671–675, 2020.

[8] Hubbarb, B., Abi-Habib, M., El-Naggar, M., McCann, A., Singhvi, A., Glanz, J. & White, J., How a massive bomb came together in Beirut's port. *The New York Times*, 2020.

[9] *Global News*, Beirut explosion: Video shows new angle of the massive blast. 2020. www.youtube.com/watch?v=LNDhIGR-83w.

[10] Delmaire-Sizes, F., Autrusson, B., Nicaud, A., Brochard, D., Gil, F., Guerin, J.M. & Chaffard, P.Y., Behaviour of transport casks under explosive loading. Presented at *13th International Symposium on the Packaging and Transportation of Radioactive Materials*, Chicago, USA, 2001.

[11] IRSN Annual Report, pp. 33, 37, 2019. https://en.irsn.fr/institutional/annual-reports.

[12] Sturtzer, M.-O., Eckenfels, D. & Legendre, J.-F., Enhanced blast evaluation in confined bunker. ISL Report no. 6092. Presented at *14th International Symposium on Interaction of the Effects of Munitions with Structures*, Seattle, USA, 2011.

[13] Trélat, S., Sturtzer, M.-O. & Eckenfels, D., Multi-scale experimental study of blast propagation around a hemi-cylindrical barrier. *WIT Transactions on The Built Environment*, vol. 198, WIT Press: Southampton and Boston, pp. 53–64, 2020.

[14] Delmaire-Sizes, F., Autrusson, B., Nicaud, A. & Brochard, D., Behaviour of a pool filled with water and loaded by an internal explosion. *Transactions*, 1366, 2001.

MYTH OR FACT: IS THERE A LINK BETWEEN WAVELENGTH AND BARRIER MITIGATION FACTORS?

SOPHIE TRÉLAT[1], MICHEL-OLIVIER STURTZER[2], LAURE HEUDIER[3] & JÉRÔME DAUBECH[3]
[1]Department of Explosion and Weapons Effects, French Institute for Protection and Nuclear Security, France
[2]Department of Explosive Protection, French German Research Institute of Saint-Louis, France
[3]Institut National de L'environnement Industriel et des Risques, Parc Technologique ALATA, France

ABSTRACT

In this work, we focus on the third item: the use of obstacles to mitigate the effects of an explosion. The aim of the study is to precisely characterize the interaction of an explosive blast wave with an obstacle used as a barrier. The obstacle is considered as non-deformable. In particular, the effect of the shape of the obstacle on the blast wave propagation and maximum overpressure is investigated through the notion of wavelength. To achieve this, various shapes of rigid reduced-scale obstacles were exposed to overpressure waves generated by different types of blast sources: parallelepipedic, cylindrical and hemicylindrical reference shapes and a straight wall. This innovative approach will contribute to proposing guidelines on geometrical and physical dimensioning of protection barriers so as to improve the efficiency of protective means in terms of blast effects mitigation.

Keywords: blast, explosion, mitigation, barrier, security and safety, innovative methodology.

1 INTRODUCTION

Nowadays, security of personnel and goods is of major interest. In order to improve the level of security of civilian, industrial or military infrastructures, it is necessary to predict blast effects consecutive to the effects of a strong explosion (accidental or malevolent) in a complex environment. In general, these effects can be suppressed or at least mitigated by means of:

- inhibition of blast source (chemically, physically, …),
- confinement of blast source (porous material, water, …),
- use of obstacle disrupting the blast propagation in the vicinity of a target,
- addition of a protective layer at the target surface,
- requirements in terms of stand-off distances.

2 BLAST INTERACTION WITH OBSTACLES

The following various shapes of rigid reduced-scale obstacles were exposed to overpressure waves generated by different types of sources (Fig. 1): parallelepipedic, cylindrical and hemicylindrical, straight wall.

Several previous studies investigated the use of rigid mock-ups facing explosive charges at small scale as a way to characterize their mechanical load (i.e. evolution of overpressure at their surface), as for example, Lapébie et al. [1]. An alternative technique consists in placing mock-ups directly inside a shock tube. This setup is particularly adapted to reproduce and visualize steady planar shock waves, representative of far-range blast. However, most shock tube setups are unable to reproduce a spherical divergent blast wave (tri-dimensional), representing a close-range explosive event.

In Trélat [2], the transmission coefficient (C_t in Fig. 2) determined for a specific point at the surface or downstream of an obstacle and for a physical characteristic (peak overpressure, arrival time, impulse, positive phase duration, etc.), is defined as the ratio between this

WIT Transactions on The Built Environment, Vol 215, © 2025 WIT Press
www.witpress.com, ISSN 1743-3509 (on-line)
doi:10.2495/SUSI250211

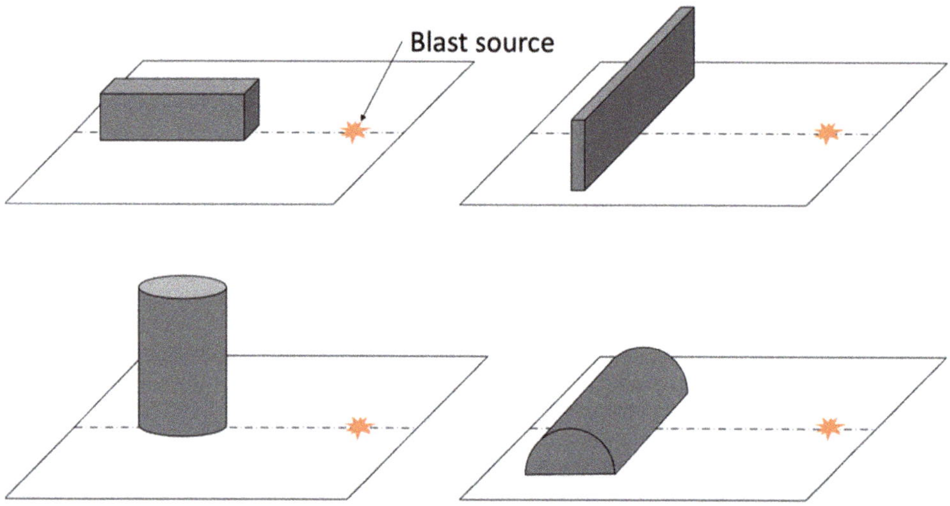

Figure 1: Examples of reference reduced-scale mock-ups exposed to a blast wave.

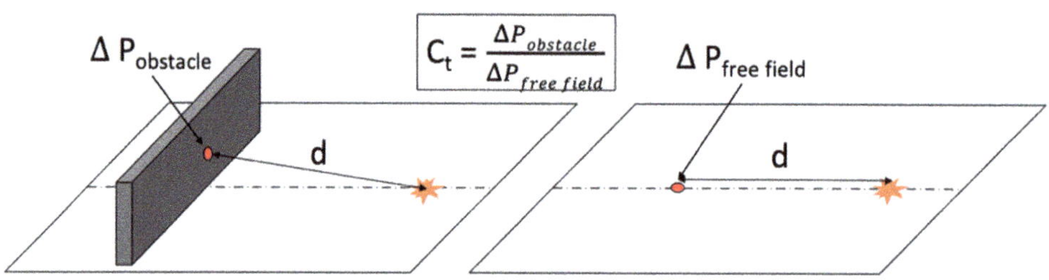

Figure 2: Transmission coefficient C_t at an obstacle surface point exposed to a blast wave, at a distance d from the source. Example with overpressure parameter (ΔP).

characteristic value and the value that would be obtained at the same distance in free field. Concerning peak overpressures, this ratio is commonly denominated as reflection coefficient for a surface point directly exposed to blast.

Overpressures on different obstacle surfaces collected from our work quantify the ability of the structure shape to alter (increase or decrease) the wave intensity by their deviation to the values in free field at equivalent distances.

3 BLAST WAVE LENGTH

Blast wave length and specific length are, respectively, defined by the product of the positive phase duration with the speed of sound in air and the local blast wave speed at a specific time of the wave propagation. These parameters (acting as scale factors for free field detonation) were useful to interpret local levels of amplification or attenuation around obstacles, once they were divided by a characteristic length depending on the obstacle geometry (radius, height, width, etc.) and the relative position to the point of interest. Such ratios are usually defined to evaluate the average blast load on an exposed obstacle surface.

Fig. 3 aggregates evolutions of transmission coefficient Ct measured in the wake of three barrier shapes (cylindrical in red, parallelepipedic in green, pyramidal in blue) extracted from different studies (dealing with detonation and deflagration propagation regimes) as a function of a dimensionless ratio LoS+/L (where LoS+ represents the specific wave length at the beginning of the expansion phase and L the direct distance 'deployed' from this point to the point of interest, see Fig. 4). The considered reactive sources and propagation regime couples are the following (and also summarized in Table 1):

- Trélat [3] detonated gaseous C_3H_6/O_2 volumes next to cylindrical shapes,
- IRSN and ISL conducted a joint work [4] on blast generated by Hexomax charges, propagating in front of hemicylinders placed on a blast table and blast pad surface,
- Trélat et al. [5] reported the blast loading characteristics on parallelepipedic obstacles resulting from the detonation of a stoichiometric C_3H_6/O_2 mixture,
- INERIS [6] designed an experimental campaign to investigate the effect of deflagrating gaseous charges ($H_2/O_2/N_2$ mixture) in the vicinity of concrete L-wall segments,
- Vyazmina et al. [7] proposed a numerical study to assess the protective characteristics of walls from the effects of a vapor cloud explosion,
- Pellegrinelli [8] exposed a straight merlon to the blast effects from gaseous H2/Air deflagrations,
- Triangular section barricades were studied by Allain with exposure to TNT detonating charges,
- Eveillard [10] detonated gaseous C_3H_6/O_2 charges in front of a similarly shaped merlon,
- INERIS [11], [12], respectively, exposed reduced and large-scale walls to the pressure effects of deflagrating $H_2/O_2/N_2$ mixtures,
- For [13], a large-scale wall was placed in front of a 50L shock wave generator containing a deflagrating C_2H_4/O_2 reactive composition.

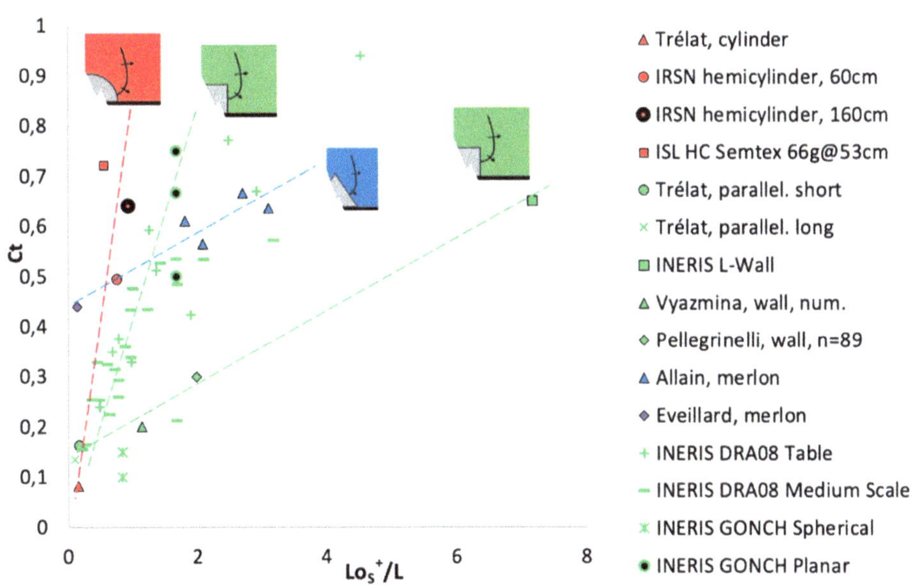

Figure 3: Evolution of transmission coefficient of C_t as a function of Los^+/L and the obstacle shape.

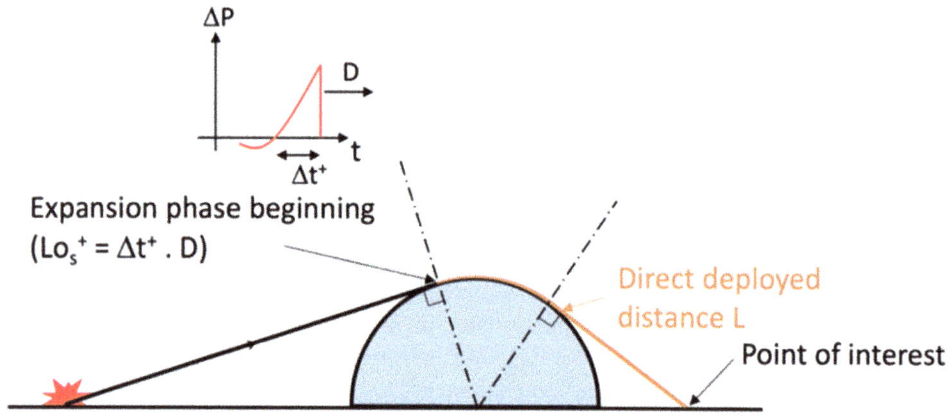

Figure 4: Illustration of beginning of expansion phase and direct deployed distance L (D: shock wave velocity in m/s, Δt+: positive phase duration in s).

This representation links transmission coefficients estimated downstream of the various shapes obstacles to a dimensionless parameter based on obstacle and source physical characteristics.

The influence of wave length on the attenuation ability of an obstacle exposed to a blast wave (generated either by detonation or deflagration combustion regimes) will require further studies in the future: a concrete parallelepipedic barrier will be placed in the vicinity of various exploding sources to investigate its attenuation factor on blast propagation, depending on the explosion intensity.

We have to keep in mind that the positive phase of deflagration waves is generally of longer duration and of different temporal evolution than for detonation phenomena. We can thus expect that effects on a large target may strongly differ with the various reactive regimes, for a similar peak overpressure level for instance. Geng and Thomas [14] already illustrated that for the same peak overpressure: attenuation behind parallelepipedic obstacle decreased with the wave length increase.

4 CONCLUSION

Effects of blast wave propagating around obstacles were investigated by studying the transmission coefficient for peak overpressure, i.e. the attenuation factor due the presence of the obstacle impacted by the blast. The influence of the wave and the obstacle physical characteristics were taken into account through the ratio between the specific wavelength and the deployed propagation distance.

Results showed that depending on the type and shape of barrier, correlations can be observed. In order to confirm these links, further investigations will require to vary the scale of the targets for similar types of explosive sources and also to explore other types of blast profiles generated, for example, by gaseous deflagration waves.

Table 1: Reactive sources and propagation regime couples.

Obstacle shape	**Reactive source and propagation regime**	**Obstacle dimensions**
Trélat vertical cylinder [3]	Gaseous detonation	Diameter 0.36 m Height 0.36 m
Trélat hemicylinder [4]	Hexomax detonation	Diameter 0.40 m Length 1.60 m
IRSN hemicylinder [4]	Hexomax detonation	Diameter 0.40 m Length 1.60 m
Trélat parallelepiped [5]	Gaseous detonation	Length 0.40 m Width 0.18 m Height 0.14 m
INERIS L-wall [6]	Gaseous deflagration	Length 6 m Thickness 0.15 m Height 2 m
Vyazmina wall [7]	Numerical simulation (gaseous detonation)	Length infinite Thickness 0.15 m Height 2.50 m
Pellegrinelli wall [8]	Gaseous deflagration	Length infinite Thickness 0.06 m Height 0.10 m
Allain merlon [9]	TNT detonation	Height 1.50 m Triangular section 45° sloping
Eveillard merlon [10]	Gaseous detonation	Height 0.20m Triangular section 45° sloping
INERIS DRA08 table wall [11]	Gaseous deflagration	Height 0.30–0.50–0.70 m Thickness 0.015 m
INERIS DRA08 medium scale wall [12]	Gaseous deflagration	Height 1.80–2.40–3.60 m Thickness 0.60 m
INERIS GONCH spherical wall [13]	Gaseous detonation	Height 1.80 m Thickness 0.60 m
INERIS GONCH planar wall [13]	Gaseous detonation	Height 1.80 m Thickness 0.60 m

REFERENCES

[1] Lapébie, E., Youinou, L., Osmont, A., Saurel, R. & Vides J., Fast-running models for blast and QSP in complex geometries. *17th International Symposium on the Interaction of the Effects of Munitions with Structures (ISIEMS)*. Bad Neuenahr, Germany, 2017.

[2] Trélat, S., Effets des ondes de souffle d'origine accidentelle ou malveillante sur des installations et transports nucléaires. Mémoire d'habilitation à diriger les recherches, Université de Technologie de Compiègne, 2021.

[3] Trélat, S., Impact de fortes explosions sur les bâtiments représentatifs d'une installation industrielle. Doctoral thesis, Université d'Orléans, 2006.

[4] Trélat, S., Sturtzer, M.-O., Eveillard, S., Eckenfels, D., Mory, J., Legendre, J.-F., Braina, F. & Soubiès, B., Strong explosion near a convex structure: A multi-scale experimental study. *25th MABS*, The Hague, The Netherlands, 2018.
[5] Trélat, S., Sochet, I., Autrusson, B., Loiseau, O. & Cheval, K., Strong explosion near a parallelepipedic structure. *Journal of Shock Waves*, **16**, pp. 349–357, 2007.
[6] Heudier, L., Lecocq, G., Gregoire, Y. & Proust, Ch., Investigation on the diffraction of a medium scale gaseous deflagration pressure wave behind a protective wall. *26th ICDERS*, Boston, USA, 2017.
[7] Vyazmina, E., Jallais, S., Beccantini, A. & Trélat, S., Protective walls against effects of vapor cloud fast deflagration: CFD recommendations for design. *Process Safety Progress*, **37**(1), 2018. https://doi.org/10.1002/prs.11930.
[8] Pellegrinelli, B., Barrière physique de protection face à une deflagration. Doctoral thesis, Université d'Orléans, 2014.
[9] Allain, L., Barricade influence on blast wave propagation. SNPE/Division Defense and Space, private communication, 1994.
[10] Eveillard, S., Propagation d'une onde de choc en présence d'une barrière de protection. Doctoral thesis, Université d'Orléans, 2013.
[11] Nauton, A., Explosion: Étude paramétrique de dispositifs de protection. Rapport de stage TN 09, Université Technologique de Compiègne, 2017.
[12] INERIS, DRA08_effet protecteur d'un merlon: Effet d'un mur sur les surpressions engendrées par une explosion. Ineris-178290-830722-0.5, 2019.
[13] INERIS, Etude de l'effet atténuateur d'un mur: Essais d'explosion en présence d'un mur – P181 DSC39. Ineris-203543-2724872-v1.0, 2022.
[14] Geng, J. & Thomas, J., Diffracted blast loads behind structures. *Pressure Vessels and Piping Conference (ASME-PVP 2020)*, 2020.

FUNCTIONALLY-GRADED SSIC/AA6082/HDPE COMPOSITE SHIELDS FOR MMOD HYPERVELOCITY IMPACT MITIGATION

KAYLEIGH MOORE & FILIPE TEIXEIRA-DIAS
Institute for Infrastructure and Environment (IIE), School of Engineering, University of Edinburgh, UK

ABSTRACT

Protecting low Earth orbit infrastructure from micrometeoroid and orbital debris impacts remains a critical challenge in modern space engineering. Traditional shielding systems often face challenges in balancing weight and impact resistance. This study explores an impedance-graded shielding approach, incorporating sintered silicon carbide (SSiC), aluminium alloy 6082 (AA6082) and high-density polyethylene (HDPE) to enhance hypervelocity impact mitigation. The combination of graded mechanical impedance and multiple material failure mechanisms within the system offers significant potential for minimising damage to space structures by enhancing projectile fragmentation and maximising energy dissipation. The smoothed particle hydrodynamics method was implemented to model a 4 mm diameter spherical AISI 304 projectile impacting multiple arrangements of AA6082, SSiC and HDPE at 5 km/s. The resulting debris cloud data was analysed to visually assess the performance of different configurations. The findings demonstrate that the graded structure significantly improves stress wave dispersion and reduces the kinetic energy of the impactor, providing insights for the design of next-generation spacecraft shielding systems.

Keywords: impedance-grading, hypervelocity impact, smoothed particle hydrodynamics, Whipple shield, debris cloud.

1 INTRODUCTION

The proliferation of artificial satellites and the increasing density of orbital debris have amplified the risk posed by micrometeoroid and orbital debris (MMOD) impacts, particularly in low Earth orbit (LEO). Traditional shielding solutions, such as the Whipple shield (Fig. 1), provide a degree of protection against hypervelocity impacts (HVI) by using a sacrificial front bumper plate to fragment incoming projectiles and generate a debris cloud, which then disperses its kinetic energy across a rear wall [1]. However, optimising these systems to maximise impact resistance while maintaining mass efficiency restraints remains an ongoing challenge.

Recent research has explored multi-material, impedance-graded shielding systems to address the limitations of traditional configurations. These systems employ layers of materials with varying mechanical impedance and distinct failure modes to enhance energy dissipation and promote projectile fragmentation under HVI conditions [3], [4]. The mismatch in mechanical impedance between adjacent material interfaces can result in an increase in peak shock pressure and pressure pulse duration when graded from high-to-low impedance due the wave interaction within each layer. This was demonstrated by Cherniaev and Telichev [3] using a ceramo-aluminium bumper with a SiC coating, and by Wen and Chen [4] using a layered shield composed of aluminium alloy, magnesium alloy, and steel. Zhang et al. [5] investigated a Ti6Al4V/AA2024-T4/PA66 shielding system under an impact velocity of 6.5 km/s. The graded metal–polymer configuration generated high shock pressures and elevated temperatures, contributing to increased projectile fragmentation. Additionally, the wave propagation path and duration were significantly influenced by the graded structure, allowing more time for temperature rise and material break-up. The extreme

www.witpress.com, ISSN 1743-3509 (on-line)
doi:10.2495/SUSI250221

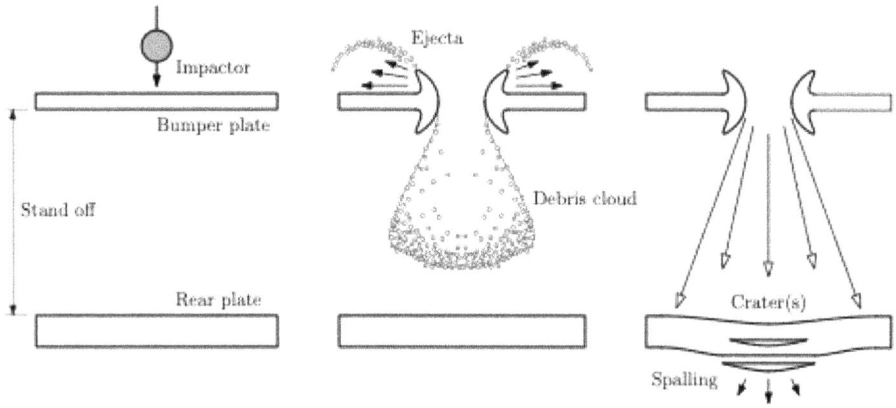

Figure 1: Whipple shield schematic. *(Source: Reprinted from Fowler and Teixeira-Dias [2], licensed under CC BY 4.0.)*

strain-rate response of high-density polyethylene (HDPE) under HVI has only recently been studied and modelled, by Rogers et al. [6]. While HDPE is likely unsuitable for standalone HVI protection, its low weight, cost-effectiveness, bulk melting capability, and capacity for large-scale plastic deformation suggest potential for its use within a layered shielding system.

This study investigates the potential of using sintered silicon carbide (SSiC), aluminium alloy 6082 (AA6082) and high-density polyethylene (HDPE) in combination to optimise projectile fragmentation and reduce residual kinetic energy, thereby enhancing overall protective performance. The proposed ceramic–metal–polymer configuration incorporates a high-impedance ceramic front layer to amplify shock pressure within the shield, a metallic layer to facilitate plastic deformation and a polymer backing layer to contribute to bulk melting and additional energy absorption through plastic deformation. While ceramic–metal systems have been widely studied and proven effective, the incorporation of HDPE in such multi-layered configurations remains largely unexplored.

2 MODELLING

A numerical model was developed, and the corresponding simulations were carried out in IMPETUS Afea using the surface smoothed particle hydrodynamics (SPH) solver for the target configurations listed in Table 1. The thickness of each layer was adjusted to maintain an approximately constant areal density across all configurations. An AISI 304 spherical projectile, 4 mm in diameter, was used in all simulations – except for the HDPE validation model discussed in Section 2.1 – impacting the target plates at a velocity of 5 km/s under normal incidence. The SSiC plate was modelled using the Johnson–Holmquist II (JH2) constitutive model to capture its dynamic brittle behaviour, alongside a polynomial equation of state (EoS). The AA6082 aluminium alloy was represented using the Johnson–Cook constitutive model, which incorporates work hardening, strain-rate sensitivity, and thermal softening, along with the Johnson–Cook damage model and the Grüneisen EoS.

HDPE was also modelled using the Johnson–Cook constitutive and damage models to account for its response under high strain rates. The material parameters for all constitutive models and equations of state are provided in Table 2. A particle density of approximately 10 particles/mm was used to ensure convergence, with negligible variation observed in the results.

Table 1: Target composition and thickness.

Layer configuration	Thickness (mm) $L_1/\ldots/L_n$
AA6082	3.62
HDPE	11.00
SSiC/AA6082	1.50/2.30
SSiC/HDPE	1.50/6.00
SSiC/AA6082/HDPE	1.50/1.27/2.00

Table 2: Material properties for SSiC, AA6082 and HDPE.

		SSiC [7], [8]		AA6082 [9]	HDPE [10]	AISI 304 [11]–[13]
	ρ (kg/m^3)	3150	ρ (kg/m^3)	2700	954	7950
	G (GPa)	179.825	*G* (GPa)	26.923	0.276	75.391
Constitutive model	Johnson–Holmquist		Johnson–Cook			
	A	0.960	*A* (MPa)	250.00	31.27	310
	B	0.350	*B* (MPa)	243.60	0.00	1000
	C	0.009	*N*	0.17	0.75	0.65
	M	1.000	*C*	0.00747	0.04	0.07
	N	0.650	*M*	1.31	0.85	1
	σ_H (GPa)	11.70				
	T (GPa)	0.750				
Damage	Johnson–Holmquist		Johnson–Cook			
	D_1	0.48	D_1	0.0164	0.05	0.53
	D_2	0.48	D_2	2.2450	1.48	0.5
			D_3	−2.798	−1.50	−6.8
			D_4	0.007	0.25	0.014
			D_5	3.65	0.00	0
Equation of state	Polynomial		Mie–Grüneisen			
	K_1 (GPa)	220	γ_0	1.97	0.86	1.93
	K_2 (GPa)	361	S_1	1.339	7.9	1.49
	K_3 (GPa)	0	S_2	0.000	2.296	0
	β	1.0	S_3	0.000	43.36	0

2.1 Validation

To validate the material models used, monolithic HDPE targets were modelled using parameters consistent with the experimental work of Rogers et al. [6]. A comparison between the experimental results and the current numerical work is shown in Fig. 2. A strong correlation is observed in both debris cloud formation and ejecta geometries at impact velocities of 3.75 km/s and 5 km/s, confirming the validity of the numerical approach.

3 RESULTS AND DISCUSSION

The absolute values for target perforation, debris cloud velocity, and geometric characteristics are presented in Table 3. The maximum head velocity – measured at the tip of the debris cloud – is greater for the monolithic targets compared to the multi-layered

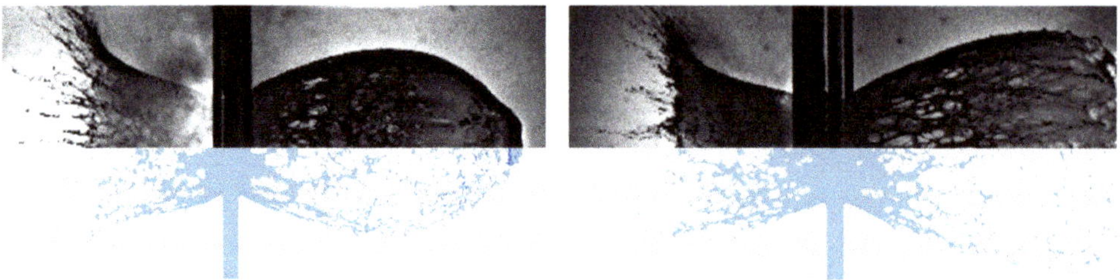

Figure 2: Experimental and numerical results of impact on HDPE target: at 3.75 km/s (left) and 5.00 km/s (right).

Table 3: Target composition and thickness.

Layer configuration	Perforation diameter (mm)	Max. head velocity (m/s)	Debris cloud dimensions at 30.4 μs		
			Angle, α (°)	Length, l (mm)	Height, h (mm)
AA6082	11.3	4,040.5	35	125.0	45
HDPE	26.7	3,811.4	9	118.6	61
SSiC/AA6082	8.6	3,589.7	41	103.3	64
SSiC/HDPE	30.6	3,554.3	15	100.0	56
SSiC/AA6082/HDPE	25.3	3,585.1	43	105.3	63

configurations. Similarly, the debris cloud length, measured from the rear surface of the target to the leading edge of the cloud, is also longer for both monolithic targets. These results indicate reduced effectiveness of the monolithic shields in dissipating impact energy and containing the debris cloud relative to the multi-layered systems.

The SSiC/HDPE target shows a slightly lower maximum head velocity compared to the SSiC/AA6082 and SSiC/AA6082/HDPE configurations; however, the difference among these three configurations is less than 1% (0.98% and 0.86%, respectively). This suggests that the front layer plays a more dominant role in influencing the residual velocity of the projectile than the rear layers. Notably, targets with HDPE as the rearmost layer display significantly larger perforation diameters than those with AA6082 in the same position. This indicates that HDPE may serve as a beneficial rear-layer material, likely due to its capacity for large-scale plastic deformation, which can contribute to increased material displacement and energy absorption.

An increase of 67% in debris cloud angle is observed when comparing the HDPE target to the SSiC/HDPE configuration, while a 17% increase is noted from the AA6082 target to the SSiC/AA6082 configuration. The addition of the high-impedance ceramic layer results in a greater spread of the debris cloud, particularly in the configurations incorporating HDPE.

The relative performance in terms of projectile residual kinetic energy and mass reduction is assessed by comparison with the monolithic AA6082 target, used as the baseline solution, as shown in Fig. 3. The monolithic HDPE target is excluded from this comparison, as it exhibited a residual kinetic energy 3.8 times greater than that of the AA6082 target, highlighting the unsuitability of monolithic HDPE as an effective shielding material under HVI conditions.

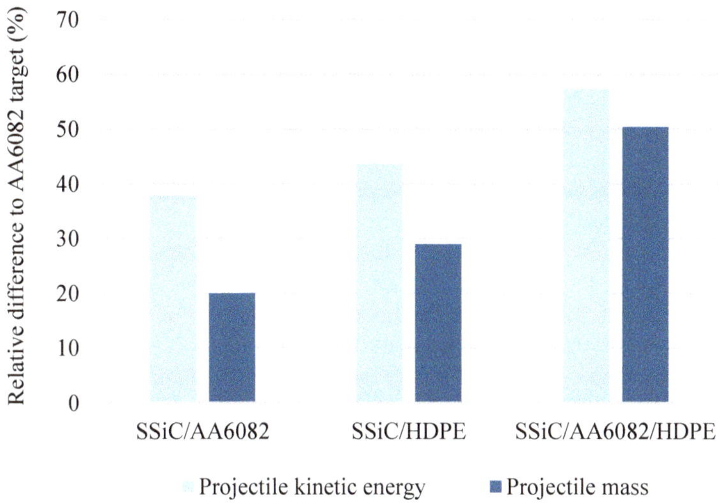

Figure 3: Relative difference of projectile kinetic energy and mass compared with AA6082 target.

The relative differences in projectile kinetic energy for the SSiC/AA6082, SSiC/HDPE, and SSiC/AA6082/HDPE target configurations are 37.8%, 43.7% and 57.3%, respectively. In terms of mass reduction, the SSiC/AA6082, SSiC/HDPE, and SSiC/AA6082/HDPE targets exhibit reductions of 20.0%, 29.0% and 50.4%, respectively, compared to the AA6082 baseline solution. Despite the poor performance of the monolithic HDPE target, the layered SSiC/HDPE configuration outperforms the SSiC/AA6082 configuration in both reducing projectile kinetic energy and mass loss, while maintaining the same target areal density. The SSiC/AA6082/HDPE shield further reduces both projectile kinetic energy and mass. These improvements can be attributed to the interaction of stress waves and the combined failure mechanisms within the layered structure, which enhance energy dissipation and material fragmentation.

4 CONCLUSION

Multiple target configurations comprising SSiC, AA6082, and HDPE components were modelled using SPH to evaluate their protective performance under HVI. Analysis of the resulting debris cloud data revealed that layered targets with an SSiC front layer provided enhanced protection compared to monolithic AA6082 and HDPE targets. The SSiC/AA6082/HDPE configuration demonstrated a 57.3% reduction in projectile residual kinetic energy and a 50.4% reduction in projectile mass relative to the AA6082 baseline solution. These findings highlight that, although monolithic HDPE shields are ineffective for HVI protection, the inclusion of HDPE within layered systems offers promising potential for the development of advanced passive protection solutions and requires further investigation.

ACKNOWLEDGEMENT

The authors acknowledge the support given to this project by the Engineering and Physical Sciences Research Council (EPSRC) (grant: EP/W524384/1).

REFERENCES

[1] Whipple, F., Meteoroid and space travel. *Astron. J.*, **1161**, p. 132, 1947.
[2] Fowler, K. & Teixeira-Dias, F., Optimisation of hybrid composite shields for hypervelocity micro-meteoroid and orbital debris (MMOD) impact. *Adv. Space Res.*, **73**(12), pp. 6194–6208, 2024.
[3] Cherniaev, A. & Telichev, I., Sacrificial bumpers with high-impedance ceramic coating for orbital debris shielding: A preliminary experimental and numerical study. *Int. J. Impact Eng.*, **119**, pp. 45–56, 2018.
[4] Wen, K. & Chen, X., Influence of the impedance gradient on the debris cloud produced by hypervelocity impact. *Int. J. Impact Eng.*, **159**, 104034, 2022.
[5] Zhang, P.L. et al., Study of the shielding performance of a Whipple shield enhanced by Ti-Al-nylon impedance-graded materials. *Int. J. Impact Eng.*, **124**, pp. 23–30, 2019.
[6] Rogers, J.A., Mead, P., Wilkerson, J., Lacy, T. & Williams, N., Hypervelocity impact response of monolithic UHMWPE and HDPE plates. *Int. J. Impact Eng.*, **161**, 104081, 2022.
[7] Liu, Y., Li, B., Wu, C. & Zheng, Y., Simulation-based evaluation of surface microcracks and fracture toughness in high-speed grinding of silicon carbide ceramics. *Int. J. Adv. Manuf. Technol.*, **86**, pp. 799–808, 2016.
[8] Saint-Gobain, Hexoloy® SA Silicon Carbide – Technical Data Sheet. www.ceramicsrefractories.saint-gobain.com/files/4731/view. Accessed on 8 Apr. 2025.
[9] Adeyinka, A.M. & Olaleke, M.O., Predictive model for thermal response during dry machining of Al 6082-T6 using FEM. *Int. J. Eng. Res.*, **9**(6), pp. 1261–1271, 2020.
[10] Mead, P., Rogers, J., Williams, N., Wilkerson, J. & Lacy, T., Simulating hypervelocity impacts to monolithic high-density polyethylene plates. 2023. https://doi.org/10.31224/3205.
[11] Zou, Z., He, L., Zhou, T., Zhang, W., Tian, P. & Zhou, X., Research on inverse identification of Johnson–Cook constitutive parameters for turning 304 stainless steel based on coupling simulation. *J. Mater. Res. Technol.*, **23**, pp. 2244–2262, 2023.
[12] Cai, S., Liu, J., Cheng, Y., Hao, W. & Zhang, P., Numerical investigation into the effect of stand-off distance on the blast performance of metallic corrugated-core sandwich panels. Presented at *ASME 2018 37th Int. Conf. Ocean, Offshore and Arctic Eng.*, American Society of Mechanical Engineers Digital Collection, 2018.
[13] Junxiang, Q., Guanghong, M., Jiuying, A. & Yu, H., Study on numerical simulation of TA1-304 stainless steel explosive welding. *China Weld.*, **30**(2), pp. 11–16, 2021.

Author index

www.ingramcontent.com/pod-product-compliance
Ingram Content Group UK Ltd.
Pitfield, Milton Keynes, MK11 3LW, UK
UKHW052004051026
15624UKWH00003B/564

9 781784 664954